Soil Specific Crop Management

Research and Development Issues

Proceedings

of

Soil Specific Crop Management

A Workshop on
Research and Development Issues

Editors
P. C. Robert, R. H. Rust, and W. E. Larson

April 14-16, 1992
Sheraton Airport Inn
Minneapolis, MN

Conducted by the
Department of Soil Science and Minnesota Extension Service
University of Minnesota

Published by:
American Society of Agronomy, Inc.
Crop Science Society of America, Inc.
Soil Science Society of America, Inc.
Madison, Wisconsin, USA

Copyright © 1993 by the American Society of Agronomy, Inc.
Crop Science Society of America, Inc.
Soil Science Society of America, Inc.

ALL RIGHTS RESERVED UNDER THE U.S. COPYRIGHT ACT OF 1976 P.L. (94-553)

Any and all uses beyond the limitations of the "fair use" provision of the law require written permission from the publisher(s) and/or the author(s); not applicable to contributions prepared by officers or employees of the U.S. Government as part of their official duties.

Reprinted in 1993

American Society of Agronomy, Inc.
Crop Science Society of America, Inc.
Soil Science Society of America, Inc.
677 South Segoe Road, Madison, WI 53711, USA

Library of Congress Cataloging-in-Publication Data

Proceedings of soil specific crop management : a workshop on research and development issues : April 14-16, 1992, Sheraton Airport Inn, Minneapolis, MN / editors, P.C. Robert, R.H. Rust, and W.E. Larson ; conducted by the Department of Soil Science and Minnesota Extension Service, University of Minnesota.
   p. cm.
  ISBN 0-89118-116-4
  1. Crops and soils--Congresses. 2. Soil management--Congresses. 3. Agronomy--Congresses. I. Robert, P. C. (Pierre C.) II. Rust, Richard H. (Richard Henry), 1921- . III. Larson, William E., 1921- .
IV. University of Minnesota. Dept. of Soil Science.
V. Minnesota Extension Service.
S596.7.P76 1993
631.4--dc20   93-12790
                                                                        CIP

Printed in the United States of America

CONTENTS

Preface ix
Acknowledgements xi

SECTION I. SOIL RESOURCES VARIABILITY

1. **Keynote Paper** Origin and nature of soil resource variability.
   J. Bouma and P. A. Finke 3

2. Mapping and managing spatial patterns in soil fertility and crop yield.
   D. J. Mulla 15

3. Terrain analysis for soil specific crop management.
   I. D. Moore, P. E. Gessler, G. A. Nielsen, and G. A. Peterson 27

4. Application of soil survey information to soil specific farming.
   M. J. Mausbach, D. J. Lytle, and L. D. Spivey 57

5. **Working Group Report**, C. S. Holzhey, Chair 69

SECTION II. MANAGING VARIABILITY

6. **Keynote Paper** Some practical field applications.
   R. E. Ascheman 79

7. Yield mapping and application of yield maps to computer-aided local resource management.
   E. Schnug, D. Murphy, E. Evans, S. Haneklaus, and J. Lamp 87

8. Tillage considerations in managing soil variability.
   W. B. Voorhees, R. R. Allmaras, and M. J. Lindstrom 95

9. Weed distribution in agricultural fields.
   D. A. Mortensen, G. A. Johnson, and L. J. Young 113

10. Value of managing within-field variability.
    F. Forcella 125

11. **Working Group Report**, R. R. Johnson, Chair 133

SECTION III. ENGINEERING TECHNOLOGY

12. **Keynote Paper** Sensing and measurement technologies for site specific management.
    S. C. Borgelt 141

13. Positioning technology (GPS).
    D. A. Tyler 159

14. Importance of spatial variability in agricultural decision support systems.
    G. W. Petersen, J. M. Russo, R. L. Day, C. T. Anthony, and J. Pollack 167

15. **Working Group Report**, J. Schueller, Chair 181

SECTION IV. PROFITABILITY

16. **Keynote Paper** Profitability of farming by soils.
    N. C. Wollenhaupt and D. D. Buchholz 199

17. Cost analysis of variable fertility management of phosphorus and potassium for potato production in central Washington.
    M. W. Hammond 213

18. Macy Farms - Site specific experiences.
    T. S. Macy 229

19. **Working Group Report**, D. Fairchild, Chair 245

SECTION V. ENVIRONMENT

20. **Keynote Paper** Best management practices for efficient nitrogen use in Minnesota.
    G. W. Randall 257

21. Social issues related to soil specific crop management.
    P. J. Nowak 269

22. Use of soil property data and computer models to minimize agricultural impacts on water quality.
    D. I. Gustafson 287

23.	Nutrient and pesticide threats to water quality. R. S. Marks and J. R. Ward	293
24.	**Working Group Report**, W. E. Larson, Chair	301

SECTION VI. TECHNOLOGY TRANSFER

25.	**Keynote Paper** Prescription farming. W. Holmes	311
26.	Illini FS variable rate technology: Technology transfer needs from a dealer's viewpoint. J. Mann	317
27.	Computerized recordkeeping for variable rate technology. J. S. Ahlrichs	325
28.	**Working Group Report**, D. Buchholz, Chair	335

SECTION VII. POSTER SUMMARIES

29.	Measuring yield on-the-go: The Minnesota experience. K. Ault, J. A. Lamb, J. L. Anderson, and R. H. Dowdy	347
30.	Multi-ISFET sensors for soil nitrate analysis. S. J. Birrell and J. W. Hummel	349
31.	Yield variability in Central Iowa. T. S. Colvin	351
32.	Managing variability of climate and soil characteristics. Characteristics in conservation tillage systems: Effects on field behavior of herbicides. Thanh H. Dao	353
33.	Soil landscape relations and their influence on yield variability in Kent County, Ontario. K. A. Denholm, J. D. Aspinall, E. A. Wilson, and J. L. B. Culley	355

34. A field information system for spatially-prescriptive farming.
    Shufeng Han and C. E. Goering 357

35. Machine vision swath guidance.
    J. W. Hummel and K. E. Von Qualen 359

36. MAPS mailbox - A land and climate information system.
    J. S. Jacobsen, A. E. Plantenberg, G. A. Nielsen, and
    J. M. Caprio 361

37. Leaching and runoff of pesticides under conventional and soil specific management.
    B. R. Khakural, P. C. Robert, and D. J. Fuchs 363

38. Spatial regression analysis of crop and soil variability within an experimental research field.
    D. S. Long, S. D. DeGloria, D. A. Griffith,
    G. R. Carlson, and G. A. Nielsen 365

39. Precision farm management of variable crop land in the Pacific Northwest.
    B. Miller, and R. Veseth 367

40. Management approaches to fertility and biological variation in the inland Pacific Northwest.
    W. Pan, B. Miller, A. Kennedy, T. Fiez, and
    M. Mohammad 371

41. Yield variation across Coastal Plain soil mapping units.
    E. J. Sadler, D. E. Evans, W. J. Busscher, and
    D. L. Karlen 373

42. Sensing for variability management.
    K. A. Sudduth, and S. C. Borgelt 375

43. Nitrogen specific management by soil condition.
    J. A. Vetsch, G. L. Malzer, P. C. Robert, and
    W. W. Nelson 377

List of Participants 379

Conversion Factors for SI and Non-SI Units 391

PREFACE

Historic agronomic practices have been developed with the farm or field as the area of management. The advent of soil conservation began to lead soil management toward topographic and soil-specific features. Even so, agronomic practices and recommendations have largely been made on a field basis rather than on soil-specific properties that might influence tillage, seeding, fertilizing and weed control practices. The near completion of detailed soil surveys nationwide, particularly in the intensive agricultural areas, has provided a database of great magnitude. The advent of computer processed spatial data together with geostatistical analysis enables the display of those soil, hydrologic, and microclimate features relevant to agronomic practices. With the further development of positioning systems suitable to on-site applications, the capability now exists, or can be feasibly developed to deliver real-time, real-space changes in almost any agronomic procedures. There is also much current research in sensor technology applicable to the soil condition or property, such as organic matter content, moisture content, tilth, nitrate content, and crop yields.

Given the capability to assess soil spatial variability and modify agronomic practices accordingly, we now add two other considerations, economic and environmental. Historically, application of inputs, whether seed, fertilizer, or pesticide, has been driven by maximum yields. More recently, emphasis has become maximum economic yields. Soil specific management provides the specific needed inputs on each soil and prevents over and under application of inputs resulting from uniform field applications. The realization of maximizing economic returns will encourage the adoption of this new technology.

If further incentive or justification for soil specific management were needed, the national incentive to reduce the potential for environmental contamination is of concern to all of agriculture. To the extent that application of agri-chemicals can be modified on-the-go according to the potential for retention and transmission of these materials in specific soil conditions, there can be a reduction in ground and surface water contamination and general maintenance of soil quality.

The objectives of this workshop were to: (i) review recent and current knowledge and application technology with respect to soil specific management, (ii) outline the necessary research that will enable adoption of the full range of agronomic practices (tillage to harvest) for soil specific management, and (iii) identify development and technology transfer needs.

The workshop consisted of invited position papers on the topics of soil resources variability, managing variability, engineering technology, profitability, environment, and technology transfer. They were followed by several invited presentations detailing current research and development in each of the six areas. Participants were divided in six working groups corresponding to the same general topics and responded to discussion papers written prior to the workshop.

The workshop also had several poster sessions presenting a variety of specific research and application project results.

This book contains the keynote address papers, session technical papers, working group discussion papers, and recommendations made by the six working groups. It also includes abstracts of most poster presentations.

On behalf of all participants, we wish to express our gratitude to sponsoring organizations for their support and to ASA-CSSA-SSSA for publishing this document. We also wish to express our appreciation to all speakers for their excellent presentations and to all participants who made the workshop a success. We look forward to implementing recommendations, creating an electronic bulletin board system that will facilitate the exchange of information and development of specific management concepts and associated systems, and preparing a second workshop for 1994.

P. C. Robert, co-editor
R. H. Rust, co-editor
W. E. Larson, co-editor

ACKNOWLEDGMENTS

Organizing Committee
  Daryl Bucholz
  Jonathan Chaplin
  H. H. Cheng
  Dean Fairchild
  Thomas Gilding
  C. Steven Holzhey
  Richard Johnson
  William Larson
  Robert Munson
  Gerald Nielsen
  Charles Onstad
  Pierre Robert, Chair
  Richard Rust
  Berlie Schmidt
  John Schueller

Editorial Committee
  Bert Bock
  Paul Fixen
  William Larson
  Pierre Robert, Chair
  Richard Rust
  John Schueller

Co-Sponsors and Contributors
  Deere and Company, Moline, Illinois

  Environmental Protection Agency

  Precision Land and Climate Evaluation Systems (PLACES)
    Bozeman, Montana

  University of Minnesota, Department of Soil Science and
    Minnesota Extension Service

  U. S. Department of Agriculture
    Agricultural Research Service
    Cooperative State Research Service
    Soil Conservation Service

Any opinions, findings, and conclusions or recommendations expressed in this publication are those of the author(s) and do not necessarily reflect the view of the sponsoring organizations.

SECTION I

SOIL RESOURCES VARIABILITY

1 Origin and Nature of Soil Resource Variability

J. Bouma
Department of Soil Science and Geology
Agricultural University
Wageningen, The Netherlands

P. A. Finke
Department of Soil Science and Geology
Agricultural University
Wageningen, The Netherlands

Spatial soil resource variability has resulted from complex geological and pedological processes, which need to be understood before field variability can be successfully characterized. In addition, soil management practices may have caused additional variability to the effect that identical land units from a pedological point of view may act quite differently when subjected to different management. Soil structure descriptions, followed by physical measurements in well-defined structure types, are useful to express variability caused by management. Data obtained can be used in mechanistic simulation models to express temporal variability as well, as is demonstrated with a Dutch case study. Observations and calculations are made for point data to be interpolated to areas of land by geostatistics, which is most effective when applied separately within different soil units of the soil map. Geostatistics can also define the minimum number of observations needed to obtain predictions with a given error. Variability of soil properties as such is of no interest. Attention should be focused on important land qualities for soil specific crop management, such as moisture supply, biocide and nitrogen leaching, trafficability and crop yield, which can only be obtained with simulation modelling, as is illustrated. Lack of basic data for models can be overcome by pedo-transfer functions that relate available soil data from soil survey to parameters needed for simulation.

INTRODUCTION

Soil resource variability can be expressed in terms of static and dynamic properties. Static properties may, for example, relate to variation in soil texture or organic matter content, which are relatively stable soil properties. Dynamic properties are usually of most interest to the soil user as they relate to differences at short distances in water, solute, air, and temperature regimes that

---

Copyright © 1993 ASA-CSSA-SSSA, 677 South Segoe Road, Madison, WI 53711, USA. *Soil Specific Crop Management.*

are considered to be of practical significance. The latter aspect is important when discussing soil variability in an applied context, which is clearly the case in this workshop. Variability, both static and dynamic, is only of interest here when it clearly affects aspects of soil behavior that are considered to be relevant by the user. It is not a purpose in itself that may well be the case in a purely scientific context.

As attractive as this user-focus may appear to be at first sight, it still leaves the question as to which differences among soil properties are significant. The average user of soil information is, for example, interested in crop yields, effectiveness and economy of tillage, spraying of biocides, and rates of fertilization. The average user will have little affinity with differences in texture, water supply capacities, and cation exchange parameters, to just mention some favorite items of concern to soil scientists. Soil scientists, therefore, will always have a major role to play in "translating" their data, be it static or dynamic, into soil properties of practical concern. Close interaction with the user of soil information is therefore crucial, and has, of course, been an essential ingredient of soil survey activities in the past and a major reason for their successes. Over the years, soil survey procedures have been established that produce standardized descriptions of soils that are now input into geographic information systems (GIS) databases making these data more available. The logical question at this point is whether all data gathered is still useful for modern applications or whether we should omit some data and add new items. If so, we may ask: which items?

This question may be more complicated than may appear at first sight. Classic soil surveys in the Netherlands are now being used successfully to produce regional soil vulnerability maps for environmental pollution in a manner that could not have been anticipated by the surveyors 20 years ago. For example, by relating texture data to basic hydraulic properties, iron and aluminum contents to phosphate sorption capacities and organic matter contents to biocide adsorption, effective vulnerability maps can be produced (e.g., Breeuwsma et al., 1986; Wagenet et al., 1991). Not knowing the needs in the year 2020, we may regret omission of data that appear to have little meaning now but may turn out to be quite relevant later. There is good reason to continue to collect standard soil survey data sets, while being alert for new applications.

When discussing variability in a general manner there is yet an important issue to consider, the scale of observation. Significant variability occurs at microlevel in any soil (e.g., Finke et al., 1991) up to horizon, pedon, mapping unit, landscape unit, county and country-level all the way to complete continents. Soil survey has coped with this problem by making soil surveys at different scales. Our workshop is focused on soil management at field level, which corresponds with traditional scales of 1:10000 or 1:5000. Minimum areal units of consideration are perhaps 50 to 100 m^2. Variabilities on a smaller scale cannot be translated into different management procedures from an operational or technical point of view. This restriction of scale is an important consideration for the remainder of this chapter.

In this chapter, the origin and nature of soil resource variability will be briefly discussed, including methodologies of study. Examples from a case study in the Netherlands will be used for illustration purposes. The problem of selection of soil variability data for future databases will be discussed as well.

ORIGIN OF SOIL RESOURCE VARIABILITY

Natural Soil Formation

Natural soil resource variability is caused by geological and pedological processes as has been well documented in numerous papers and textbooks (e.g., Buol et al., 1988; Wilding et al., 1983). The idealized picture of a fresh parent material, be it weathering rock or a sediment, in which a soil profile develops over time applies only to a very limited number of soils. Erosion and deposition of soil materials interrupts the natural processes in many locations. Older soil materials may have been subjected to different climates which each may have left a trace. Understanding both geological and pedological processes and their effects is absolutely crucial when dealing with soil variability in the field.

Occurrence of differences in texture at short distances and different types of stratification can often be explained by an analysis of sedimentation or erosion phenomena. Studies of soil variability in the field with modern (geo) statistical techniques (e.g., Mausbach and Wilding, 1991) are most effective when a stratification is made first in terms of land units that are relatively homogeneous internally or are heterogeneous but in a characteristic manner that can be studied systematically once the "system of heterogeneity" is known. Increasingly, standard soil surveys report internal variability of mapping units of soil maps showing major differences among units (e.g., Brown and Huddleston, 1991). This aspect has not been emphasized much in the past and even though information on variability was contained in soil survey reports it did not really register with the user because the lines on the soil map are the same for homogeneous and heterogeneous units. Moreover, heterogeneity was not systematically reflected in interpretations, creating the impression that it was not taken seriously (Brown and Huddleston, 1991).

Management Induced Variability

We all know that soils belonging to the same soil series can have quite different properties as a result of differences in management. In fact, different soil series subjected to identical management may have rather corresponding properties while different soils belonging to the same soil series, but with different management, may have quite different properties. Soil classification focuses, of course, on more or less permanent soil characteristics and even though management has a major effect on these properties, we expect that each soil series reacts in a rather specific way to different types of management. This "range" of behavior is expected to be characteristic for different soil series and we are principally interested in this range of behavior because this allows us to make predictions about potential soil behavior as a function of soil management. Studies by Van Lanen et al. (1987, 1992) further illustrate this concept for a sandy loam and a clay soil in the Netherlands under grassland and various arable land utilization types. They demonstrate that an analysis of soil structure is useful to define the effects of different management. Just as a clay skin indicates illuviation and possible occurrence of an argillic horizon, types of structure may indicate certain types of management. Wagenet et al. (1991) have therefore suggested to define soil phases of certain soil series, based on well-defined soil structures.

Soil management activities like plowing and levelling may influence the spatial variability of structure-related soil properties. In a field-scale study, Finke et al. (1992) identified a disturbed soil layer in which the thickness showed a clear spatial structure. This spatial structure reflected the surface topography before the levelling took place. Management-induced variability does, however, not only relate to tillage but may also be related to other management features. Fertilizer inputs, for example, are usually assumed to be constant, but experiments report variability due to unequal spreading patterns. Coefficients of variation between 10 and 15% were reported from controlled experiments, depending on the fertilizer-spreading devices used.

Soil management not only affects soil structure, of course. Reduction of organic matter content is an important effect of improper management that does not include addition of organic residue or manure.

Study Methods

Regular soil survey procedures have been published and discussed elsewhere. Using aerial photographs and geomorphological field expertise, soil surveyors delineate areas of land that are expected to be relatively homogeneous. In areas without clear surface features information has to be derived from borings that are also made in all delineated areas to specify soil conditions and their variability. In practice, surveys have been rather strongly focused on the soil map. A rule-of-thumb has been that approximately three borings should be made per square centimeter of map area to allow a reasonable accuracy of soil boundaries. This criterion is, of course, irrelevant because the number of observations should primarily be a function of variability. Fewer observations are needed when variability is low.

Geostatistical techniques have been used successfully to relate the number of observations to variability (Burrough, 1991). Stein et al. (1988) calculated the moisture supply capacity for 600 point observations in a sandy area in the Netherlands. Three major types of soil occurred in the area and geostatistical techniques such as kriging were used within each of the areas and for the area as a whole to interpolate values of moisture supply and the minimum number of observations that were needed to provide data with a specified accuracy. Variograms were characteristically different for the three soil types (Fig.1-1) and when all data were lumped, a variogram was found that was less diagnostic.

ORIGIN AND NATURE OF SOIL VARIABILITY

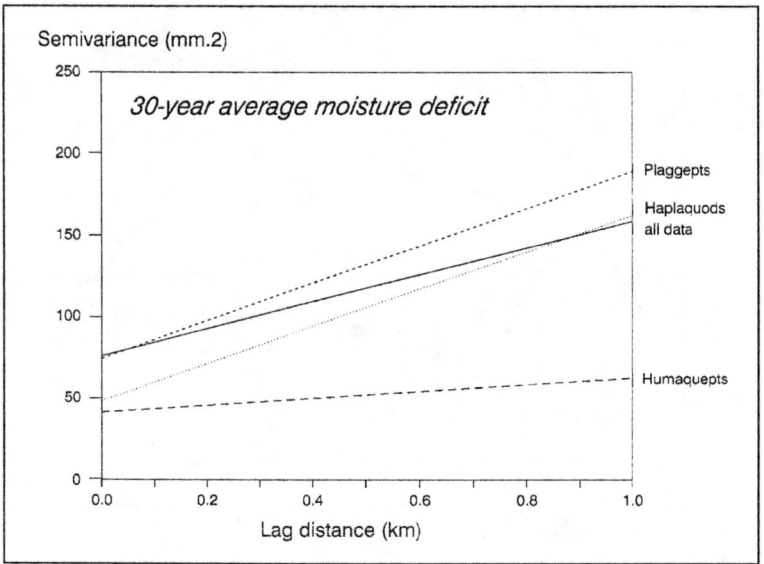

Fig. 1-1. Variograms determined in delineated areas of three soil types showing spatial relations of moisture deficits. (After Stein et al., 1988.)

With some simplification, we can state that variograms can also be used to relate sampling density to the accuracy of estimates after interpolation. It was found that predictions of moisture supply values could have been obtained with a comparable degree of accuracy with only one-third the number of observations which were made according to common practice.

The above example considered moisture supply values that were calculated with a simulation model and interpolated with geostatistical techniques for a large number of points. Thus a method is described which expresses spatial variability patterns, within well-defined land units.

Variability within land units can also be expressed by a method that uses another key ingredient of soil surveys, viz. soil horizons. Soil horizons can be considered as "carriers" of information with specific upper and lower boundaries that are continuous in a three-dimensional landscape. By making multiple measurements in well-defined soil horizons of any property of interest, the average composition of the horizon and its variance can be defined. Next, this information receives a "third dimension" by using geostatistics in interpolating depths below (surface of) upper and lower boundaries of the horizons. Examples for physical soil characteristics were presented by Wösten et al. (1985, 1990). Finke and Bosma (1993) sampled soil horizons in a heterogeneous area with stratified marine soils in the Netherlands with the purpose of obtaining functional layers showing mutually different and internally homogeneous soil physical behavior (Fig. 1-2). Thickness of the functional layers was mapped, in which

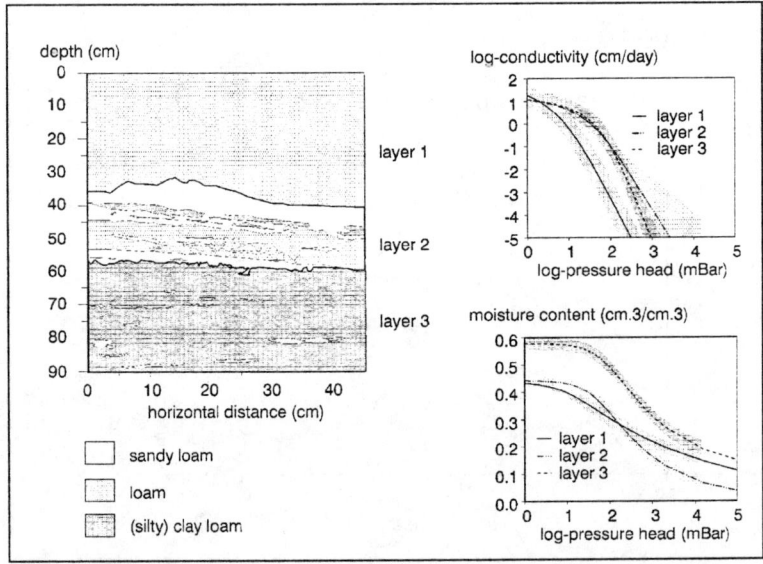

Fig. 1-2. Schematization of stratified marine soils into three different layers, physically characterized by hydraulic conductivity and moisture retention data (Finke and Bosma, 1993).

it was found, that a stratification of the area into four soil mapping units significantly increased the quality of interpolations of various soil properties.

NATURE OF SOIL RESOURCE VARIABILITY

Dynamic Properties Relevant for the User

So far, variability was mainly defined in terms of either natural or management-induced soil properties with a somewhat static character characterized by variability in space only. The example of Stein et al. (1988), however, already touched on what we will consider here as the "nature" of soil variability that includes variability in time. They discussed variability of moisture supply to a crop which is a property of considerable practical interest which covers a growing season and a series of growing seasons when variability among years has to be determined. The moisture supply capacity cannot directly be observed or measured, such as a clay content or a bulk density value, but has to be calculated, in this case with a simulation model. The model can be run on a daily basis for different periods of time. The same holds for many other land qualities in which users have interest. Land qualities are defined, according to Food and Agriculture Organization (FAO), as complex attributes of land that have a distinct impact on its functioning.

Land qualities of interest in the context of soil specific crop management include moisture supply capacity, trafficability, workability, root penetration, crop-yield potential, N dynamics, biocide adsorption and possibly many others. These land qualities are considered for actual conditions of management and, particularly, for new forms of management the potential of which needs to be assessed. The question, then, is how to obtain representative expressions for land qualities for both actual and potential conditions of management.

Study Methods

Two methods exist to obtain values for land qualities relevant for soil specific crop management. The first is by field monitoring for extended periods of time, generally under natural weather conditions because manipulated experiments can only be realized in limited cases. Monitoring procedures are very costly and time consuming even when automated. The second procedure, attractive in principle, involves application of computer simulation modelling for predicting solute movement and crop growth. Several models are available to simulate solute movement (e.g., LEACHM by Hutson and Wagenet, 1990; DAISY by Hansen et al., 1991) and crop growth (e.g., WOFOST by Van Diepen et al., 1988; SWACROP by Feddes et al., 1988; GAPS by Buttler and Riha, 1989). The major problem with running models is to obtain relevant basic data, such as hydraulic conductivity and moisture retention to simulate water fluxes, adsorption coefficients to simulate solute flow and transformations and crop coefficients to simulate crop growth. Besides detailed weather data are needed as well. A discussion of simulation models and their data needs is beyond the scope of this text. The reader is referred to Wagenet et al. (1991) and references therein.

To specifically illustrate use of models to simulate dynamic soil properties that are relevant for soil specific crop management, a case study will be summarized dealing with a N fertilization scenario in a Dutch clay soil. This heterogeneous clay soil was discussed in the section Study Methods and consisted, within one field, of sandy loam and clay loam parts which showed quite different properties in terms of crop growth and response to N fertilization. A sandy loam soil unit (Fig. 1-3) of the area showed significant lower potato yields in 1990, when compared to a clay loam soil unit (8555 vs 9527 kg dry matter/ha). This could be attributed to generally higher moisture stress in the sandy loam area, where capillary rise from the groundwater could not satisfy demand by evapo- transpiration. It was concluded, that the land quality "moisture-availability" showed spatial variability according to soil mapping units, even on a field scale.

Also, these soil units showed significantly different behavior when different fertilizing scenarios were tested in a simulation exercise.

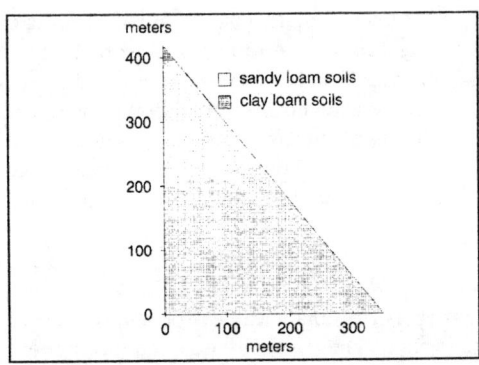

Figure 1-3. Schematized soil map of a farm field in a Dutch polder with two major mapping units.

FOCUSING ON VARIABILITY THROUGH A FUNCTIONAL ANALYSIS

As stated above, soil variability in space and time, as such, is of little interest. We should define those properties of soil that are of particular interest when struggling with questions about soil specific crop management. It was suggested that attention should be focused on important land qualities rather than on static land properties that can be found in soil survey reports. Land qualities have a strong functional focus. They address the issues of concern such as nitrate and biocide leaching to groundwater, nitrogen-use efficiency, and possible crop production. Some important land qualities were mentioned above. The question at hand is then "How can soil variability within a field be characterized in terms of relevant land qualities for soil specific crop management?"

The example for the Dutch clay soil analyzed this question by using a simulation model to calculate the effects of different fertilization scenario (Fig. 1-4). It would have been possible to make estimates of the likely effects by expert knowledge but such data would be unsatisfactory because of its qualitative nature which does not allow quantitative economic calculations. Such calculations are crucial to justify soil specific crop management procedures.

Use of models involves a number of pitfalls. Running a detailed model without having adequate basic data is equivalent to committing scientific fraud. However, by using pedo-transfer functions that relate existing soil survey data to parameters needed for simulation, we can overcome this problem in time. (e.g., Wagenet et al., 1991). On-site monitoring is crucial to allow proper calibration and validation of models for existing conditions. Once validated, models can also be used to predict potential conditions and this represents one

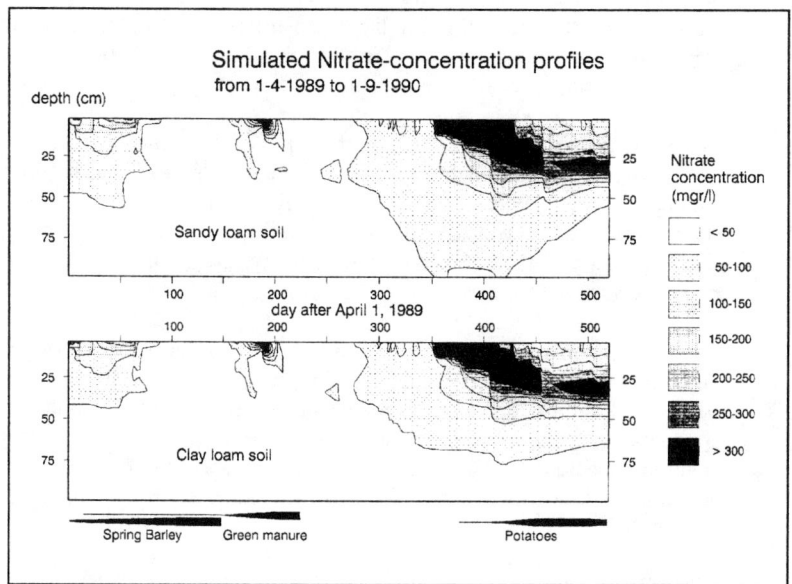

Fig. 1-4. Nitrate concentration profiles in the soil for the period 1 April 1989 to 1 Sept. 1990, for two soil types of which soil physical characteristics were presented in Fig. 1-2.

of the most attractive aspects of model application. Obviously, the alternative would be a long duration and very expensive field experiment, assuming that environmental conditions could be controlled to the extent that potential conditions could be attained.

In summary, we believe that the variability aspect should be translated into a series of specific land qualities which are crucial for, in this case, soil specific crop management. This process is referred to as a functional analysis which also includes an analysis of the calculation procedures to be followed when determining the land qualities being distinguished.

WHICH VARIABILITY ASPECTS SHOULD GO INTO FUTURE DATABASES?

Origin and nature of soil resource variability includes natural and management-induced soil parameters and conditions reflecting variability in space and time. To increase the usefulness of soil survey databases in GIS in future, the question may be raised as to which data should be added. Suggestions are presented in the following points:

1. Point observations should be included in GIS as basic data, because they are the backbone of the system. "Representative profiles" for mapping

units are man-made derived entities, which should be included as well to allow quick estimates of generalized properties. Computer capacities were initially inadequate to contain all point data, but this problem has been erased with ever more powerful computers.

2. More emphasis should be placed on the types of soil structure that result from different land utilization types in different soils. A database should include a unified description of these land utilization types as well as standardized structure descriptions, which should be supported by soil physical analyses in terms of bulk density, hydraulic conductivity, moisture retention, and rates of bypass flow.

3. A systematic effort should be made to derive pedo-transfer functions for physical transport and chemical transformations in soil. Thus, existing soil survey information can be made much more useful for modelling applications. A recent conference in Riverside has resulted in a plan to develop such functions for an international dataset, based on the flow equations of van Genuchten. This effort should be strongly supported.

ACKNOWLEDGMENT

Financial support of the EC-project EV4V*0098-NL "Nitrate in Soils" is gratefully acknowledged.

REFERENCES

Breeuwsma, A., J. H. M. Wösten, J. J. Vleeshouwer, A. M. van Slobbe, and J. Bouma. 1986. Derivation of land qualities to assess environmental problems from soil surveys. Soil Sci. Soc. Am. J. 50 (1):186-190.

Brown, R. B., and J. H. Huddleston. 1991. Presentation of statistical data on map units to the user. p. 127-149. In M. J. Mausbach and L. P. Wilding (ed.) Spatial Variabilities of Soils and Landforms. SSSA Spec. Publ. 28. SSSA, Madison, WI.

Bual, S. W., F. D. Hole, and R. J. McCracken. 1988. Soil genesis and classification. 3rd ed. Iowa State University Press. Ames, IA.

Burrough, P. A. 1991. Sampling designs for quantifying map unit composition. p. 89-127. In M. J. Mausbach and L. P. Wilding (ed.) Spatial Variabilities of Soils and Landforms. SSSA Spec. Publ. 28.

Buttler, I. W., and S. J. Riha. 1989. GAPS: A General Purpose Simulation Model of the Soil-Plant-Atmosphere System.

Feddes, R. A., M. de Graaf, J. Bouma, and C. D. van Loon. 1988. Simulation of water use and production of potatoes as affected by soil compaction. Potato Res. 31:225-239.

Finke, P. A., and W. J. P. Bosma. 1993. Obtaining basic simulation data for a heterogeneous field with stratified marine soils. Hydrol. Processes (in press).

Finke, P. A., J. Bouma, and A. Stein. 1992. Measuring field variability of disturbed soils for simulation purposes. Soil Sci. Soc. Am. J. 56 (1):187-192.

Finke, P. A., H. J. Mucher, and J. V. Witter. 1991. Reliability of point counts of pedological properties on thin sections. Soil Sci. 151 (3):249-253.

Hansen, S., H. E. Jensen, N. E. Nielsen, and H. Svendsen. 1991. Simulation of nitrogen dynamics and biomass production in winter wheat using the Danish simulation model DAISY. Fertilizer Res. 27:245-259.

Hutson, J. L., and R. J. Wagenet. 1990. Simulating nitrogen dynamics in soils using a deterministic model. Soil Use and Manage. 7(2):74-78.

Mausbach, M. J., and L. P. Wilding. 1991. Spatial variabilities of soils and landforms. SSSA Spec. Publ. 28. SSSA, Madison, WI.

Stein, A., M. Hoogerwerf, and J. Bouma. 1988. Use of soil map delineations to improve (Co-)Kriging of point data on moisture deficits. Geoderma 43:163-177.

Van Diepen, C. A., C. Rappoldt, J. Wolf, and H. van Keulen. 1988. CWFS Crop Growth Simulation Model WOFOST documentation version 4.1. Staff working paper SOW-88-01, Centre for World Food Studies, Amsterdam/Wageningen, Netherlands.

Van Lanen, H. A. J., M. H. Bannink, and J. Bouma. 1987. Use of simulation to assess the effects of different tillage practices on land qualities of a sandy loam soil. Soil Tillage Res. 10:347-361.

Van Lanen, H. A. J., G. J. Reinds, O. H. Boersma, and J. Bouma. 1992. Impact of soil management systems on soil structure and physical properties in a clay loam soil and the simulated effects on water deficits, soil aeration and workability. Soil and Tillage Research.

Wagenet, R. J., J. Bouma, and R. B. Grossman. 1991. Minimum data sets for use of soil survey information in soil interpretive models. p. 161-183. In M. J. Mausbach and L. P. Wilding (ed.) Spatial variabilities of soils and landforms. SSSA Spec. Publ. 28. Madison, WI.

Wilding, L. P., N. E. Smeck, and G. F. Hall (ed.) 1983. Pedogenesis and Soil Taxonomy. Developments in Soil Science. 11. Elsevier Publ. Co., New York.

Wösten, J. H. M., J. Bouma, and G. H. Stoffelsen. 1985. Use of soil survey data for regional soil water simulation models. Soil Sci. Am. J. 49:1238-1244.

Wösten, J. H. M., C. H. E. J. Schuren, J. Bouma, and A. Stein. 1990. Functional sensitivity analysis of four methods to generate soil hydraulic functions. Soil Sci. Soc. Am. J. 54:832-836.

2 Mapping and Managing Spatial Patterns In Soil Fertility and Crop Yield

D. J. Mulla
Department of Crop and Soil Sciences
Washington State University
Pullman, WA

There is increasing pressure on commercial agriculture to reduce applications of fertilizer N and minimize nonpoint source N pollution of surface and groundwaters. Spatial variation of soil properties causes uneven patterns in soil fertility and crop growth, and decreases the use efficiency of fertilizer applied uniformly at the field scale (Miller et al., 1988; Bhatti et al., 1991; Larson and Robert, 1991). Application of variable rather than uniform rates of N has been proposed to avoid application of excessive N where it will not be utilized by crops (Carr et al., 1991; Mulla et al., 1992). In order to apply variable rates of fertilizer, a methodology needs to be developed to divide farmlands into management zones that have differences in soil fertility (Mulla, 1991). The present study was conducted to develop an approach for sampling, mapping, and managing soil fertility on a commercial wheat farm located in the Palouse region of eastern Washington.

The Palouse region of eastern Washington is characterized by steep rolling hills (Mulla, 1986) formed from loessial deposits of silt. Erosion may be locally severe, resulting in loss of organic matter rich topsoil and exposure of clay-enriched subsoils. In eroded areas, soil properties such as permeability (Mulla, 1988), water-holding capacity, and fertility (Mulla et al., 1992) are less favorable for crop production than those of soils in non-eroded areas. Frazier and Cheng (1989) showed that patterns in exposure of subsoils could be delineated by remotely sensed estimates of soil organic matter.

The primary objective of this research was to quantitatively measure and assess the magnitude and extent of spatial variability in soil fertility and wheat yields on a locally eroded farm in the Palouse region using geostatistical techniques. Geostatistical methods were used to measure and model the spatial correlation for selected soil properties and wheat yields. The models of spatial correlation were then used along with kriging techniques to develop large-scale maps showing spatial patterns in variability of selected soil properties and wheat yield. These maps were used to divide the field into management zones that

---

Copyright © 1993 ASA-CSSA-SSSA, 677 South Segoe Road, Madison, WI 53711, USA. *Soil Specific Crop Management.*

could be fertilized with variable rates to match existing patterns in soil fertility.

MATERIALS AND METHODS

A study site near St. John, WA receiving an average annual precipitation of 40.6 cm was selected for intensive sampling. The St. John site is located in a region consisting of sharply rolling hills with exposed subsoil or shallow topsoil on eroded hilltops and ridges, and thicker topsoil with a larger organic matter content on lower slopes and bottomlands (SE 1/4 sec. 14 T19N R42E). More than 90% of the St. John site is mapped as a Palouse silt loam (fine-silty, mixed, mesic pachic Ultic Haploxeroll) on 9 to 25% slopes (Donaldson, 1980). About 75% of the site is mapped as having slight to moderate erosion, while 25% is severely eroded.

Four east-west oriented parallel transects 655 m long, each 122 m apart, were established in August 1987 during the fallow portion of the crop rotation. Each transect was sampled to a depth of 1.8 m at intervals of 15.24 m for a total of 172 samples. Soil samples were analyzed in the 0 to 30 cm depth increment for properties such as sodium acetate extractable phosphorus (Olsen and Sommers, 1982), pH, and organic matter. The entire profile was analyzed for nitrate nitrogen and for plant-available water content by determining the difference between measured volumetric water contents and permanent wilting point water contents estimated from soil survey reports.

Fertilizer was applied along each sampled transect in a 20 m wide continuous strip at a uniform rate of 73 kg N ha^{-1} and 6 kg P ha^{-1}. This is the growers typical management practice. "Stephens" winter wheat (*Triticum aestivum* L.) was planted along the fertilized strips in early October. In August 1988, wheat was harvested along the four 655 m long strips at intervals of 15.24 m in 0.6 m wide plots ranging in length from 7 to 10 m. Grain yield in kg ha^{-1} was determined from these samples.

Statistical Procedures

Measured data for soil properties and wheat yield were analyzed using classical statistical techniques (Steel and Torrie, 1980) to obtain values for the mean, standard deviation, coefficient of variation, and correlation coefficients between pairs of properties.

Geostatistical Procedures

Semivariograms (Isaaks and Srivastava, 1989) were used to examine the spatial dependence between measurements at pairs of locations as a function of distance of separation (lag, h). Semivariance ($\gamma(h)$) was computed using the expression:

$$\gamma(h) = \frac{1}{2n(h)} \sum_{i}^{n(h)} [z(x_i) - z(x_{i+h})]^2 \qquad (1)$$

where n(h) is the number of samples separated by a distance h, and z represents the measured value for a soil or crop property.

If sample variation occurs randomly in space, the semivariance will not depend upon the separation distance between the samples. Samples that are correlated in space, however, will have lower semivariance at smaller separation distances than at larger separation distances. At a critical distance known as the range, the sample pairs will cease to be correlated. As separation distance continues to increase, values for the semivariance remain constant at a value known as the sill. Samples separated by distances greater than the range exhibit random variation. Classical statistical methods are best applied to data only if the range has a value which is smaller than the closest sampling distance.

For a quantitative description of these features, it is useful to fit standard semivariogram models to the semivariance data. Typical standard semivariograms include linear, spherical, and exponential models. Model selection is usually based on a criterion of goodness of fit, which in our case, involved fitting the model to data using nonlinear least-squares methods. All of the measured data were well described using a spherical model given by:

$$\gamma(h) = C_0 + C_1 [1.5(\frac{h}{a}) - 0.5(\frac{h}{a})^3] \quad 0 < h \leq a \qquad (2)$$

and

$$\gamma(h) = C_0 + C_1 \qquad h \geq a \qquad (3)$$

where h is the separation distance between observations, a is a model parameter known as the range, C_1 is a model parameter which equals the sill minus the nugget, and C_0 is a model parameter known as the nugget.

Data obtained from the transects were collected in the east-west direction. As a result, the semivariograms based upon this data largely represent spatial correlations in that direction, especially at small separation distances. Given the limitations of how data were collected, it was not possible to fully explore the consequences of anistropy in spatial correlation. At an early stage of the study, semivariograms were computed for yield and organic matter in regions on the landscape having low, medium, or high productivity. Differences between the semivariograms in each of the zones were minimal, so the semivariograms discussed below represent calculations based on crop yields or soil properties at all sample locations regardless of landscape position.

Ideally, the experimental variance should pass through the origin when the distance of sample separation is zero. However, many soil properties have non-zero semivariances as h tends to zero. This non-zero variance is called the "nugget variance" or "nugget effect" (Journel and Huijbregts, 1978). It represents unexplained or "random variance" often caused by measurement errors or variability in the measured property which was not detected at the scale of sampling.

Block kriging (David, 1977; Journel and Huijbregts, 1978) was used to interpolate soil organic matter content and measured grain yields in an area covering 400 x 650 m, or roughly 26 ha. Values for each property were estimated on a regular grid at spacings of 15.2 x 30.5 m using a search radius of 75 m. No attempt was made to optimize parameters of the semivariograms by jack-knifing or cross-validation.

Block kriging is a method for making optimal, unbiased estimates of regionalized variables at unsampled locations using the structural properties of the semivariogram and the initial set of measured data (David, 1977). A useful feature of kriging is that an error term expressing the estimation variance or uncertainty in estimation is calculated for each interpolated value. Kriging differs greatly from linear regression methods for estimation at unsampled locations. Whereas a regression line never passes through all of the measured data points, kriging always produces an estimate equal to the measured value if it is interpolating at a location where a measurement was obtained.

RESULTS AND DISCUSSION

Variability in Soil Properties

Landscape at the study site (Fig. 2-1) exhibits the steeply rolling hills that are characteristic of the Palouse region of eastern Washington. Descriptive statistics of soil properties affecting fertility (Table 2-1) showed coefficients of variation ranging from about 11% for soil pH to about 50% for soil phosphorus. Variability in various soil properties as a function of position on the landscape was not random. For instance, patterns in soil organic matter content measured from surface soil samples tended to be smaller at upper slope positions on the landscape where steep slopes have experienced large historical rates of erosion and subsequent loss of topsoil.

Analysis of Cross-correlation

Correlation coefficients (r) relating soil organic matter to either extractable phosphorus or available profile water content had values of 0.57 and 0.40, respectively. Both of these correlation coefficients were significant at a 1% level of probability. The strong correlation between soil organic matter and phosphorus is important because it suggests that patterns in soil fertility may be related to variations in organic matter content. The correlation between organic matter and profile water content is important because profile water content is the single most important soil property influencing potential yield for winter wheat.

Table 2-1. Mean and coefficient of variation (CV) for measured or kriged properties in the 0-30 cm or 0-2 m depth at the St. John study site.

Property	Mean	CV (%)
Available profile water (cm)	16.9	38.9
Profile nitrate-N (mg kg^{-1})	32.7	39.7
Soil phosphorus (mg kg^{-1})	15.2	50.5
Kriged soil phosphorus (mg kg^{-1})	15.5	33.3
Organic matter content (%)	2.04	41.3
Kriged organic matter (%)	2.06	30.0
Soil pH	6.09	10.8
Grain yield (kg ha^{-1})	4066	29.4
Kriged grain yield (kg ha^{-1})	4086	17.3

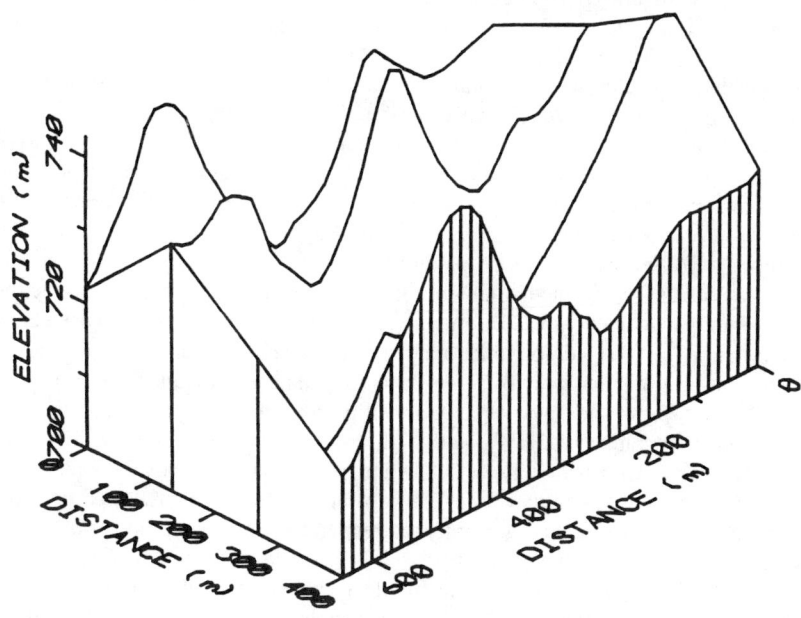

Fig. 2-1. Elevation along the four sampled transects.

Semivariograms

Geostatistical methods are often suitable for analysis of properties that show spatially correlated behavior. Semivariograms were computed for each soil property and parameters for the best fitting spherical models were estimated (Table 2-2). Of particular importance are values for the range. The range is a measure of the maximum distance over which properties remain spatially correlated. The range of influence for soil organic matter and soil P measured on the transects was 114 and 145 m, respectively. At distances shorter than the range, variability is nonrandom, and pairwise sample variation depends upon the distance of separation. All properties in Table 2-2 have sills that are significantly larger than their nuggets. This indicates nonrandom spatial variability in each property. The exception to this pattern was profile nitrate-nitrogen, which exhibited random variability.

Mapping and Management

The classical and geostatistical results presented suggest that spatial patterns in soil fertility at the St. John study site are strongly correlated with patterns in organic matter content. A map of spatial patterns in organic matter (Fig. 2-2) was produced by interpolating from measured values of organic matter using the semivariogram model in Table 2-2 and the method of block kriging. The mean in kriged organic matter is close to the mean of measurements from soil samples (Table 2-1), indicating that the kriging estimates appear to be relatively consistent with measured data. In addition, the broad patterns in kriged organic matter are relatively similar to the measured patterns.

Uniformly fertilizing a farm where soil fertility levels vary with location leads to overfertilization and underfertilizion of large areas. This inefficient use of fertilizer could contribute to degradation of surface and groundwater quality. Fertilizer resources could be better managed by applying variable rates across the landscape to better match broad patterns in soil fertility.

Values for organic matter from kriging (Fig. 2-2) were used to divide the field into management zones having different soil fertility levels. The frequency distribution of kriged organic matter was examined and cutoff values representing the mean plus or minus one-half standard deviation were used to divide the study site into separate fertility management zones having low (<1.5%), moderate (1.5-2.4%), and high (>2.4%) amounts of organic matter (Table 2-3). Approximately 6, 10.5, and 9.5 ha at the study site had low, moderate, or high amounts of surface organic matter. The low, medium, and high organic matter zones generally correspond to top, back, and foot or toe slope landscape positions, respectively.

Profile water content and available P increased significantly from zone 3 to zone 2, and from zone 2 to zone 1 (Table 2-3). Thus, the most highly eroded locations (zone 3), on average, had the lowest profile available water contents and soil test phosphorus levels. Soil pH was significantly lower where organic matter content was highest (zone 1), while profile nitrate-nitrogen was

Table 2-2. Summary of parameters for spherical semivariogram models at St. John.

Property	Nugget	Sill	Range (m)
Available profile water (cm)	10.53	43.88	68
Soil phosphorus (mg kg^{-1})	27.58	63.38	145
Organic matter (%)	0.14	0.72	114
Soil pH	0.17	0.43	132
Grain yield (kg ha^{-1})	8.4 x 10^5	1.14 x 10^6	70

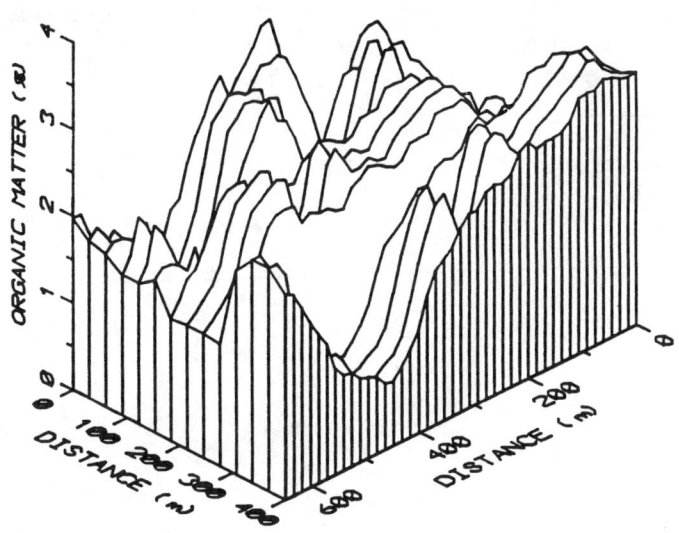

Fig. 2-2. Plot of block kriged organic matter.

significantly higher in zone 1 than in the other zones. Potential yields of winter wheat increased signficantly from zone 3 to zone 1 as a result of increases in profile water content, profile nitrate-nitrogen, and surface organic matter content.

A set of N and P fertilizer recommendations were developed for each management zone based upon the yield potential (Halvorson et al., 1982) and profile N levels (Engle et al., 1975) in each zone (Table 2-3). The spatial pattern in which fertilizer rates would be varied across management zones is illustrated in Fig. 2-3. In this figure, the management zone index corresponds with zones 1, 2, and 3 in Table 2-3. Recommended rates of N fertilizer in zones 1, 2, and 3 are 37, 45, and 28 kg ha^{-1}, respectively, with an overall average of 37 kg ha^{-1}. This represents a significant reduction in N fertilizer relative to the growers typical application rates which were 73 kg N ha^{-1} in the year of this study, but are typically about 95 kg ha^{-1} in most years. Recommended rates of P fertilizer are 0, 0, and 20 kg P ha^{-1}, respectively, in zones 1, 2, and 3 with an average of 7 kg ha^{-1}. This average rate compares closely with the rate applied by the grower (6 kg ha^{-1}).

Table 2-3. Comparison of mean soil properties, potential wheat yields, recommended rates of N and P fertilizer, and measured wheat yields in fertility management zones divided on the basis of organic matter content.

Measured property	Organic matter management zone (%)		
	zone 1 >2.4	zone 2 1.5-2.4	zone 3 <1.5
Available profile water (cm)	19.8 a*	16.4 b	14.1 c
Profile nitrate-nitrogen (kg ha^{-1})	142.4 a	114.8 b	106.0 b
Surface ammonium-nitrogen (kg ha^{-1})	2.2 c	3.6 b	5.3 a
Available P (kg ha^{-1})	21.7 a	18.3 b	10.9 c
Soil pH	5.8 b	6.2 a	6.3 a
Potential grain yield (kg ha^{-1})	5548 a	4918 b	4492 c
Recommended N fertilizer (kg ha^{-1})	37	45	28
Recommended P fertilizer (kg ha^{-1})	0	0	20
Measured grain yield (kg ha^{-1})	4742 a	3933 b	3443 c

*Means followed by similar letter(s) in each row are not significantly different from one another at a 5% level of significance.

Variability in Wheat Yield

Grain yield measured on transects fertilized uniformly with 73 kg N ha^{-1} and 6 kg P ha^{-1} exhibited moderate variability, with a CV of 30% (Table 2-1). Simple correlation coefficients (r) between grain yield and either measured soil organic matter, available profile water, or extractable P had values of 0.52, 0.59, and 0.34, respectively. All of these relations were significant at a level of 1%.

Semivariance values for grain yield were fit using a spherical model. The range of this semivariogram was 70 m (Table 2-2), which is almost identical to the value for the range in available profile water content (68 m). This is not surprising, since research by Leggett (1959) and Lindstrom et al. (1974) has shown that yield of winter wheat is strongly affected by available profile water.

Estimates of grain yield using block kriging and the semivariogram model in Table 2-2 are shown in Fig. 2-4. Field averaged values for mean grain yield at St. John from kriging were comparable in magnitude to mean measured values along the transects (Table 2-1). The coefficients of variability in interpolated grain yield were slightly smaller than those for measured grain yield (Table 2-1). On average, geostatistical interpolation methods accurately represented both mean grain yield as well as the extent of field-scale variability.

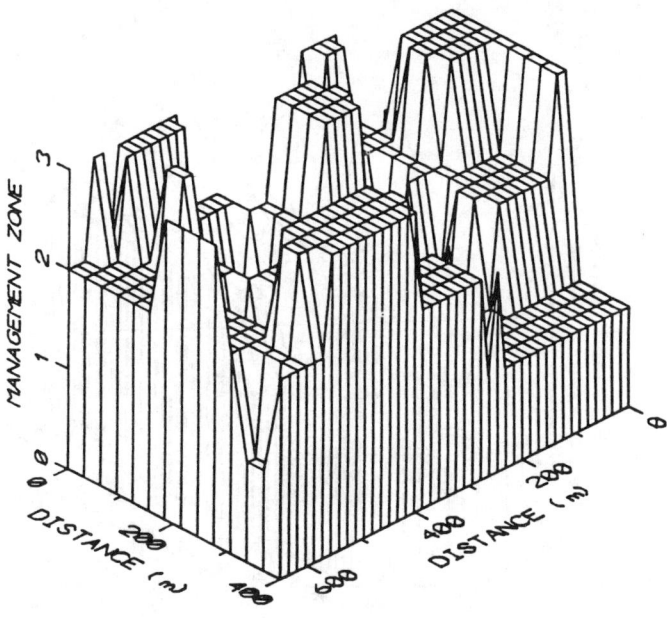

Fig. 2-3. Spatial locations of soil fertility management zones described in Table 2-3.

Patterns in kriged grain yield (Fig. 2-4) closely matched patterns in organic matter (Fig. 2-2). Grain yield varied as a function of landscape position (Fig. 2-1), and significantly higher yields were measured on regions of the landscape having higher organic matter contents than in regions having lower organic matter contents (Table 2-3). This result suggests that the soil fertility management zones not only are a reasonable representation of differences in soil fertility across the landscape, but that they are also a good criteria of variations in crop yield across the landscape. Since the yield of grain and level of soil fertility both depend upon soil organic matter, it seems reasonable to vary rates of fertilizer across the field according to the broad levels of organic matter described in Table 2-3. Mulla et al. (1992) showed that when such a strategy was implemented, there were no significant differences in yield between regions of the field receiving uniform rates of N vs. regions receiving rates that matched the fertility levels and yield goals of specific management zones.

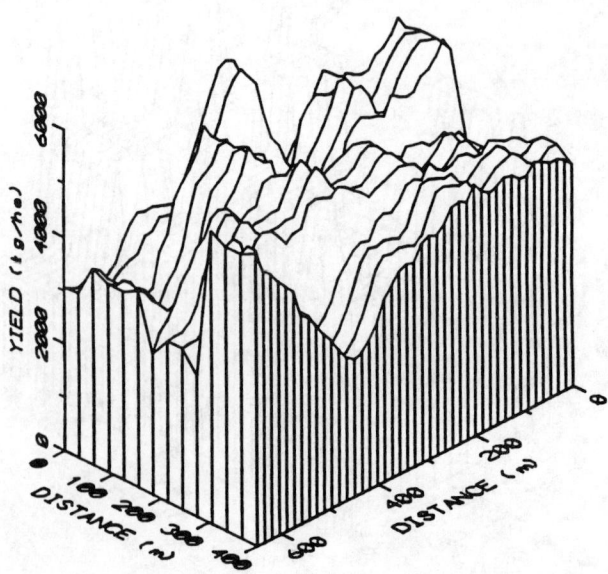

Fig. 2-4. Plot of block kriged grain yield.

SUMMARY

Spatial variability of organic matter, soil P, and wheat yields was studied using classical statistical and geostatistical approaches on a wheat farm in the Palouse region of eastern Washington. The results of this study have significant implications for fertilizer management stategies on farms located in steep rolling topography that have experienced locally heavy rates of erosion and topsoil loss. For such locations, the study shows that:

1. Spatial patterns in soil fertility and wheat yield were nonrandom and were correlated to patterns in soil organic matter.
2. The field could be divided into three fertility management zones associated with differences in organic matter content. Each zone had significantly different levels of soil moisture, residual N, and potential grain yield.
3. Differences in grain yield measured in each management zone were significantly different, with increases in yield corresponding to increases in average organic matter content of a given zone.
4. Nitrogen fertilizer recommendations in each management zone were significantly lower than the uniform rate normally applied by the grower.

Results from Mulla et al. (1992) indicate that there were no significant differences in grain yield for any fertility management zone between locations fertilized at the grower's typical uniform rate vs. a reduced rate that matched the fertility level and potential yield of the management zone. Matching N application rates to fertility levels and yield goals in specific management zones within a farm is a strategy that provides efficient use of fertilizer resources and reduces the potential for nonpoint source pollution of surface and groundwaters (Mulla and Annandale, 1990).

REFERENCES

Bhatti, A. U., D. J. Mulla, and B. E. Frazier. 1991. Estimation of soil properties and wheat yields on complex eroded hills using geostatistics and Thematic Mapper images. Remote Sens. Environ. 37:181-191.

Carr, P. M., G. R. Carlson, J. S. Jacobsen, G. A. Nielsen, and E. O. Skogley. 1991. Farming soil, not fields: A strategy for increasing fertilizer profitability. J. Prod. Agric. 4:57-61.

David, M. 1977. Geostatistical ore reserve estimation. Developments in geomathematics. Elsevier Scientific Publ. Co., New York.

Donaldson, N. C. 1980. Soil survey of Whitman County, Washington. USDA-SCS. U.S. Gov. Print. Office, Washington, DC.

Engle, C. F., F. E. Koehler, K. J. Morrison, and A. R. Halvorson. 1975. Fertilizer guide: Dryland wheat nitrogen needs. Cooperative Extension Service FG-34, Washington State University, Pullman, WA.

Frazier, B. E. and Y. Cheng. 1989. Remote sensing of soils in the Eastern Palouse region with Landsat thematic mapper. Remote Sens. Environ. 28:317-325.

Halvorson, A. R., F. E. Koehler, C. F. Engle, and K. J. Morrison. 1982. Fertilizer guide: Dryland wheat, general recommendations. Cooperative Extension Service FG0019, Washington State University, Pullman, WA.

Isaaks, E. H. and R. M. Srivastava. 1989. Applied geostatistics. Oxford Univ. Press, New York.

Journel, A. G., and C. H. Huijbregts. 1978. Mining geostatistics. Academic Press, New York.

Larson, W. E., and P. C. Robert. 1991. Farming by soil. p. 103-112. In R. Lal and F. J. Pierce (ed.) Soil management for sustainability. Soil Water Conserv. Soc., Ankeny, IA.

Leggett, G. E. 1959. Relationships between wheat yield, available moisture, and available nitrogen in eastern Washington dryland areas. Washington Agric. Exp. Stn. Bull. 609. Washington State Univ., Pullman, WA.

Lindstrom, M. J., F. E. Koehler, and R. I. Papendick. 1974. Tillage effects on fallow water storage in the eastern Washington dryland region. Agron. J. 66:312-316.

Miller, M. P., M. J. Singer, and D. R. Nielsen. 1988. Spatial variability of wheat yield and soil properties on complex hills. Soil Sci. Soc. Am. J. 52:1133-1141.

Mulla, D. J. 1986. Distribution of slope steepness in the Palouse region of Washington. Soil Sci. Soc. Am. J. 50:1401-1405.

Mulla, D. J. 1988. Estimating spatial patterns in water content, matric suction, and hydraulic conductivity. Soil Sci. Soc. Am. J. 52:1547-1553.

Mulla, D. J. 1991. Using geostatistics and GIS to manage spatial patterns in soil fertility. p. 336-345. In G. Kranzler (ed.) Proc. Automated Agric. 21st Century. Am. Soc. Agric. Engineers, St. Joseph, MI.

Mulla, D. J., and J. G. Annandale. 1990. Assessment of field-scale leaching patterns for management of nitrogen fertilizer application. p. 55-63. In K. Roth et al. (ed.) Field-Scale Water and Solute Flux in Soils. Birkhauser Verlag, Basel, Switzerland.

Mulla, D. J., A. U. Bhatti, M. W. Hammond, and J. A. Benson. 1992. A comparison of winter wheat yield and quality under uniform versus spatially variable fertilizer management. Agric. Ecosyst. Environ. 38:301-311.

Olsen, S. R., and L. E. Sommers. 1982. Phosphorus. In A. L. Page (ed.) Methods of Soil Analysis. Part 2. Chemical and Microbiological Properties. 2nd Ed. ASA, Madison, WI.

Steel, R. G. D., and J. H. Torrie. 1980. Principles and procedures of statistics: A biometrical approach. 2nd ed. McGraw Hill, New York.

3 Terrain Analysis for Soil Specific Crop Management

I. D. Moore
Centre for Resource and Environmental Studies
Australian National University
GPO Box 4
Canberra, ACT 2601, Australia

P. E. Gessler
Division of Soils
CSIRO, Canberra, Australia

G. A. Nielsen
Dept. of Plant and Soil Science
Montana State University
Bozeman, MT and
Visiting Fellow, Centre for Resource
 and Environmental Studies
Australian National University
Canberra, Australia

G. A. Peterson
Dept. of Agronomy
Colorado State University
Fort Collins, CO

Soil maps are commonly produced around the world at scales of 1:15000 or smaller. In Montana, digitized soil maps have been linked to a global positioning system (GPS) to control and navigate fertilizer applicators that vary formulations and rates as they move across fields (McEachern et al., 1990; Petersen, 1991). However, conventional soil maps do not delineate all of a field's inherent variability and do not adequately represent specific soil attribute variation. Precise maps (at about 1:6000 scale) of soil water, plant nutrients, and soil microclimate attributes are essential for soil specific crop management and other environmental modelling and monitoring needs.

Users of digital soil maps in the United States often assign attributes to polygons using the Soils-5 database. Ranges given for some attributes, particularly those describing hydraulic properties, vary by an order of magnitude. Furthermore, the nearest sampled pedon used to derive mapping unit attributes could be miles from the point of interest. Therefore, these mapping units are best suited to macro-scale or basin-scale applications where there is a high

---

Copyright © 1993 ASA-CSSA-SSSA, 677 South Segoe Road, Madison, WI 53711, USA. *Soil Specific Crop Management.*

degree of lumping of model parameter values.

Soil survey has played a key role in the advancement of pedological thought (Simonson, 1991) and the usefulness of soil survey maps is unquestioned. But, standard soil surveys were not designed to provide the fine-scale resolution required in detailed environmental modelling applications or soil specific crop management. Creating detailed soil maps of about 1:6000 scale is expensive by conventional methods. Accurate and inexpensive quantitative alternatives are needed. New terrain analysis techniques may allow enhancement of soil maps and other data sources used for soil specific crop management.

The high cost of collecting soil attribute data at many locations across landscapes has created a need for methods of inferring air and water properties of soils using pedo-transfer functions (Bouma, 1989) or economical surrogates derived from soil morphological properties (Rawls et al., 1982; McKeague et al., 1984; McKenzie and MacLeod, 1989; Williams et al., 1990; McKenzie et al., 1991). The most common surrogates used are soil texture, organic matter, soil structure, and bulk density. Methods that also include landform descriptors derived from digital elevation models (DEMs), such as those proposed by Dikau (1989), show potential for improving soil attribute prediction (Moore et al., 1992a; McKenzie and Austin, 1992). In late 1991, the USDA-SCS National Soil Survey Laboratory, in cooperation with the Blackland Research Center at Texas A&M University, released the geo-referenced "Soil Pedons of the United States" database on CDRom. These data when augmented with digital terrain models (DTMs), ground truthing at a range of scales (plot, catchment physiographic region) and suitable spatial interpolation techniques may provide quantitative methods (Mabbutt, 1968) of estimating specific soil attributes. These soil attributes are required for high resolution models and maps of the soil continuum used in applications such as soil specific crop management and macro- and meso-scale models of land surface processes.

Climate, parent material, topography, and biotic factors influence soil formation (Jenny, 1941, 1980), but climate often exerts control at coarser scales than of interest here. Parent material differences are usually differentiated effectively by conventional methods and a large proportion of local soil variation (i.e., within hillslopes) can be attributed to changes in landform. The rationale for this chapter is that in many landscapes, catenary soil development occurs in response to the way water moves through the landscape. Therefore, it may be hypothesized that the spatial distribution of topographic attributes that characterize water flow paths also captures the spatial variabililty of soil attributes at the meso-scale. An exciting new area of research is the attempt to verify such hypotheses by examining the correlation between quantitative topographic attributes, soil horizonation and other soil attributes. The purpose of this chapter is to: (i) describe a geographis information system (GIS)-based terrain analysis system; (ii) compare data from terrain analysis, conventional soil survey sources and extensive soil testing of a field in Colorado; and (iii) suggest potential benefits of terrain analysis for soil specific farming.

SOIL ATTRIBUTES AND LANDSCAPE POSITION: A BRIEF REVIEW

Until recently, soil scientists have emphasized the vertical relationships of soil horizons and soil-forming processes rather than the horizontal relationships that characterize traditional soil survey (Buol et al., 1989). Soil spatial patterns have been captured and displayed as choropleth maps with discrete lines representing the boundaries between map units, which implies homogeneity within map units (Burrough, 1986; Gessler, 1990). Two problems follow from this approach: (i) the lines drawn on the soil survey maps may not accurately depict the boundaries between map units (see Long et al., 1991a); and (ii) the inferred homogeneities do not exist for many physical and chemical attributes needed for environmental modelling and soil specific management.

Since 1970 there have been many attempts to characterize the meso-scale spatial variability of measured soil attributes (Beckett and Webster, 1971; Webster, 1985; Yates and Warrick, 1987; Loague and Gander, 1990). These attempts have concentrated on the characterization of patterns, rather than on the linking of pattern to process. Two techniques are commonly used for spatial predictions: (i) quantitative interpolation methods (i.e., using kriging), that relate the spatial covariance function to the spatial separation of the data, and (ii) methods that relate soil attributes to qualitative measures of landscape position such as toe-slopes and interfluves (an attempt to account for process). Both techniques require large databases and their results are not transferable. Interpolation techniques ignore pedogenesis while methods based on landscape position have lacked a consistent quantitative framework.

Digital terrain modelling methods offer an alternative way of stratifying and extending measured soil attributes based on the way the soil catena develops in response to water movement in the landscape (i.e., process). Semivariograms have shown that spatial correlation lengths of soil attributes, such as saturated hydraulic conductivity, can be on the order of only tens of meters (Webster, 1985; Yates and Warrick, 1987). Another point is that the "high variability" of many variables, and particularly saturated hydraulic conductivity, has much to do with inappropriate measurement methods (Lauren et al., 1988). Better correlations may possibly be obtained using kriging or partial splines if landscape attributes were included as variables or as an initial stratification. McBratney et al. (1991) used topographic information for region partitioning to improve the representation of geostatistically mapped soil attributes. The incorporation of landscape attributes via a parametric submodel of a partial thin plate spline is attractive (Moore and Hutchinson, 1991). In this way, broad changes with position can be accounted for by a smooth dependence on the two spatial variables (x,y) and the parametric submodel can account for more local, process-based effects.

There have been numerous attempts to relate soil properties, soil erosion class and to a lesser extent, productivity to landscape position in the soil science literature (e.g., Walker et al., 1968; Furley, 1976; Daniels et al., 1985; Stone et al., 1985; Kreznor et al., 1989; Carter and Ciolkosz, 1991). For example, organic matter content and A-horizon thickness, B-horizon thickness and degree of development, soil mottling, pH, depth to carbonates and water storage have all been correlated to landscape position (Kreznor et al., 1989). Most of these studies use qualitative mapping units that delineate head slopes, linear slopes, and footslopes rather than quantifiable topographic attributes to map soils. However, Walker et al. (1968) did attempt to correlate a range of depth characteristics, such as thickness of the A-horizon, to slope, aspect, curvature, elevation, and flow path length (distance to hillslope summit).

A recent approach is to organize the land surface according to a formal geomorphological model of landform and inter-landform relations (Speight, 1974; Ruhe, 1975; Weibel and DeLotto, 1988; Dikau, 1989; Lammers and Band, 1990; Gessler, 1990; Mackay et al., 1991). Geomorphological position influences horizonation and soil attributes. Lammers and Band (1990) developed techniques for producing a set of landform files, which they called a "feature model", describing the morphometry, catchment position and surface attributes of hillslopes and stream channels of a catchment. Dikau (1989) demonstrated how digital terrain analysis could be applied to quantitative relief form analysis to define basic relief units for geomorphological and pedological mapping (see also Ruhe, 1975). The main topographic attributes used to define these relief units were slope, plan curvature and profile curvature (Fig. 3-1). This approach provides a systematic basis for derivation of complex relief units. It may be possible to use these relief units to stratify the measured soil attributes and separate the micro- and meso-scale spatial variabilities. Hairston and Grigal (1991) found that topographically stratifying soil-related attributes (organic matter, total N, and soil water) helps reduce the apparent variation of these properties, even in subdued terrain. Odeh et al. (1991) stress the importance of land unit delineation to design optimal sampling patterns that reduce extrapolation error and thus misclassification of soil. They found that slope, plan and profile curvature, upslope distance, and area account for much of the soil variation in their study area.

From a soil science and hydrologic modelling perspective there is merit in using the soil horizon as the basic entry for modelling and quantifying soil attributes in three-dimensional space rather than the map unit or soil series. This stems from the fact that soil horizons are easily identifiable three-dimensional entities that are a result of pedogenic processes (McBratney, 1992). The hydrologically active A-horizon varies greatly in thickness and physical and chemical properties. Gessler et al. (1989) used soil horizon information within a GIS to analyze soil-vegetation-landuse patterns in southwestern Wisconsin. To develop this horizon-based approach for modelling and quantifying soil properties in space requires the development of useful horizon entities by: (i) determining the distribution and arrangement of horizons in space; and (ii) characterizing the chemical, physical, and biological properties of the horizon, for which recent work using fuzzy set theory shows potential (Powell et al., 1992; McBratney and DeGruitjter, 1991; McBratney, 1992). The development of the relationship between horizons and terrain attributes is the subject of ongoing research by the authors.

A 1989 USDA-Soil Conservation Service task force on "The Utility of Soil Landscape Units" concluded that future soil surveys must include more information on the shape of the land surface and that these landform parameters should reflect the combined effects of both the hydrological and erosional processes taking place at different locations in the landscape. The DEMs provide this information via geographic information systems (GIS) (Klingebiel et al., 1987; Moore et al., 1991, 1992a).

TOPOGRAPHIC ATTRIBUTES

There is a general lack of land attribute data, and particularly soil attribute data, for detailed environmental modelling and planning soil specific crop management enterprises. However, we suggest that DEM data at an appropriate scale could be used to calculate terrain attributes that would enhance soil surveys as a source of soil attribute data. As soil test results and additional environmental data become available, they can be used to improve estimates of terrain-based indices. We see different layers of data being developed over time with elevation data and the related topographic attributes constituting a minimum data set needed to enhance soil surveys. This approach has the potential of achieving major cost savings over extensive manual sampling and remapping.

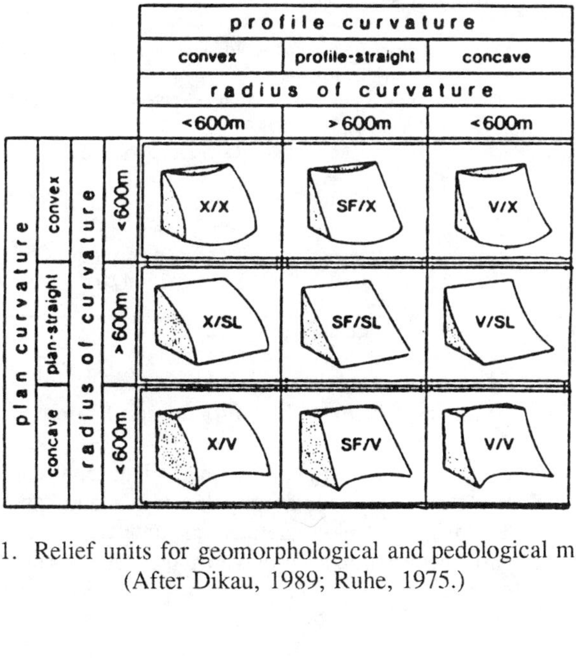

Fig. 3-1. Relief units for geomorphological and pedological mapping. (After Dikau, 1989; Ruhe, 1975.)

Table 3-1. Primary topographic attributes that can be computed by terrain analysis from DEM Data. (Adapted from Speight, 1974, 1980, Moore et al., 1991.)

Attribute	Definition	Significance
Altitude	Elevation	Climate, vegetation, potential energy
Upslope height	Mean height of upslope area	Potential energy
Aspect	Slope azimuth	Solar insolation, evapotranspiration, flora and fauna distribution and abundance
Slope	Gradient	Overland and subsurface flow velocity and runoff rate, precipitation, vegetation, geomorphology, soil water content, land capability class
Upslope slope	Mean slope of upslope area	Runoff velocity
Dispersal slope	Mean slope of dispersal area	Rate of soil drainage
Catchment slope[1]	Average slope over the catchment	Time of concentration
Upslope area	Catchment area above a short length of contour	Runoff volume, steady-state runoff rate
Dispersal area	Area downslope from a short length of contour	Soil drainage rate
Catchment area[1]	Area draining to catchment outlet	Runoff volume
Specific catchment area	Upslope area per unit width of contour	Runoff volume, steady-state runoff rate, soil characteristics, soil water content, geomorphology
Flow path length	Maximum distance of water flow to a point in the catchment	Erosion rates, sediment yield, time of concentration
Upslope length	Mean length of flow paths to a point in the catchment	Flow acceleration, erosion rates
Dispersal length	Distance from a point in the catchment to the outlet	Impedence of soil drainage
Catchment length[1]	Distance from highest point to outlet	Overland flow attenuation
Profile curvature	Slope profile curvature	Flow acceleration, erosion/deposition rate, geomorphology
Plan curvature	Contour curvature	Converging/diverging flow, soil water content, soil characteristics

[1] All attributes except these are defined at points within the catchment.

Primary Data for a Geographic Information System (GIS)

A DEM is an ordered array of numbers that represents the spatial distribution of elevations above some arbitrary datum in a landscape. The DEMs are a subset of DTMs that can be defined as ordered arrays of numbers representing the spatial distribution of terrain attributes (such as elevation, slope, aspect, etc.).

Several hydrologically based, topographically derived indices appear to be particularly powerful and useful for crop management and spatial prediction of soil attributes. They can be used to identify areas of land susceptible to environmental hazards and degradation such as erosion, sedimentation, salinization, nonpoint source pollution and water logging and to assess and manage biological productivity and diversity (Moore et al., 1991). Topographic attributes can be divided into primary and secondary (or compound) attributes. Primary attributes are directly calculated from elevation data and include variables such as elevation and slope. The most important primary topographic attributes for hydrological and soils applications, together with their significance, are listed in Table 3-1. Compound attributes involve combinations of the primary attributes and can be used to characterize the spatial variability of specific processes occurring in the landscape. These attributes may be derived empirically, or by simplifying equations describing the underlying physics of the processes. Topographic indices provide a knowledge-based approach to soil specific crop management and analysis and can be imbedded within the data analysis subsystems that any GIS must have.

Three compound indices that have potential use in predicting the spatial distribution of soil properties and in soil specific crop management are the wetness index, w, the stream power index, Ω, and a sediment transport capacity index, τ. The wetness index has been used to characterize the spatial distribution of zones of surface saturation and soil water content in landscapes (Moore et al., 1988, 1992a) and appears to be useful for mapping forest soils (Skidmore et al., 1991). The stream power index is directly proportional to stream power (=ρgqtanβ, where ρg is the unit weight of water, q is the discharge per unit width and β is the slope angle), which is the time rate of energy expenditure and so is a measure of the erosive power of overland flow. The sediment transport index characterizes erosion and deposition processes and in particular the effects of topography on soil loss (Moore et al., 1992b, Moore and Wilson, 1992). This index is analogous to the length-slope factor in the Universal Soil Loss Equation (USLE), but is applicable to three-dimensional landscapes. The wetness, stream power, and sediment transport indices, in their simplest forms, can be expressed, respectively, as:

$$w = \ln\left(\frac{A_s}{\tan\beta}\right) \tag{1a}$$

$$\Omega = A_s \tan\beta \tag{1b}$$

$$\tau = \left(\frac{A_s}{22.13}\right)^m \left(\frac{\sin\beta}{0.0896}\right)^n \tag{1c}$$

where A_s is the specific catchment area (m^{2}m^{-1}), β is the slope angle (degrees), and m=0.6 and n=1.3. The assumptions in all three equations are that A_s is directly proportional to q, and steady-state conditions prevail (Moore et al., 1991; Moore and Wilson, 1992).

Sources of Digital Elevation Models

There are three principal ways of structuring a DEM; as (i) regular grids, (ii) triangulated irregular networks (TINs); or (iii) as vectors or digitized contour lines (Moore et al., 1991, 1992a). The most widely used data structures consist of square-grid networks because of their ease of computer implementation and computational efficiency (Collins and Moon, 1981). We will restrict ourselves to the use of grid-based DEMs. The data in a grid-network can be stored in a variety of ways, but the most efficient is as z coordinates corresponding to sequential points along a profile with the starting point and grid spacing also specified. Grid-DEMs are ideal for attribute mapping such as is required in soil survey and soil specific crop management. Digital elevation data can be obtained from the Earth Science Information Centers (ESICs) of the U.S. Geological Survey (USGS) as 30 m square grid-DEMs for 7.5-min quadrangle coverage, which corresponds to the 1:24000-scale map series quadrangle. Stereo images available from the French Earth Observation Satellite (SPOT) can be used to produce orthophotos and 10 m grid-DEMs in much the same way as from conventional aerial photography. Satellite data have the advantage that they can be purchased in digital form and directly accessed by computers. However, they also require special software for processing.

The prescribed standard accuracy for delineating the boundaries of soil map units in conventional soil surveys (>1:15000 scale) is 30 m (100 ft). However, significant errors can occur depending on the base map quality, the map reading skill and aptitude of individuals, and terrain conditions (Long et al., 1991a). The GPS (Dixon, 1991) offers a relatively inexpensive, time-efficient, and accurate method of positioning on a base map, navigating to predetermined sites, and field digitizing map units compared to conventional positioning used in soil survey. This is particularly true when surface features are difficult to observe from the ground, aerial photographs, and topographic maps (Long et al., 1991a). It also provides a rapid and relatively inexpensive way of obtaining elevation data for the development of DEMs. In an evaluation of GPS errors, Long et al. (1991a) concluded that positions measured by GPS receivers operating in autonomous mode (single receiver) are comparable with the accuracy of positions from 1:24000 scale topographic maps. In differential mode (two-receivers), the relative accuracy is within centimeters horizontally and vertically. Soon the absolute accuracy will be of the order of 20 mm (Blewitt, 1992). The GPS survey and digital terrain analysis technologies overlap and offer ways of reducing, or at least quantifying, the relational errors in point and vector land surface data in a GIS for soil specific crop management.

Figure 3-2a shows the latitude and longitude (x and y in UTM coordinates) of the x,y,z triplets measured during a kinematic GPS survey of a

20-ha research field at the Northern Agricultural Research Center near Havre in north-central Montana. The survey was conducted in 1989 using Trimble Series 4000 GPS receivers in differential mode. A stationary receiver was located at a benchmark elevation site and a roving receiver was mounted on a vehicle. The goal was to generate a DEM for use in the analysis of autocorrelated terrain, soil and crop variables (Long et al., 1991b). The data presented in Fig. 3-2a were obtained after less than one hour of field work.

Many commercial software packages are available for contouring data and interpolating data onto a grid. Hutchinson (1989) has developed an efficient finite difference method of interpolating grid DEMs from digitized contour lines, scattered surface-specific point elevation data and streamline data. The method, called ANUDEM, has the efficiency of a local method without sacrificing the advantages of global methods. It uses a nested grid strategy that interpolates grid DEMs at successively finer resolutions. ANUDEM is unique in that it has a drainage enforcement algorithm that automatically removes spurious sinks or pits and calculates stream lines and ridge lines from points of locally maximum curvatures on contour lines. It forces the DEM to conform to the drainage pattern, and so is particularly well suited to interpolating topography. Hutchinson and Dowling (1991) describe the application of ANUDEM for developing a 1/40th degree latitude and longitude (approximately 2.5 km), continent-wide DEM of Australia. ANUDEM is probably the most powerful package of its kind available.

The Montana GPS data shown in Fig. 3-2a were interpolated onto a regular 10-m grid using ANUDEM. The x,y,z coordinate triplets were treated as scattered surface-specific points. No contour line or stream line data were available. An isometric projection of the interpolated DEM is presented in Fig. 3-2b (the vertical is distorted to show the variation in the terrain - the total relief is 17 m). These results show that GPS survey techniques in combination with appropriate interpolation algorithms are powerful and effective tools for developing fine-resolution DEMs for use in soil specific crop management.

Calculating Topographic Attributes

Topographic attributes can be easily estimated from a DEM using any one of a number of terrain analysis methods. Most GISs are based on a pixel or cellular structure so that grid-based methods of terrain analysis can provide the primary geographic data for them and can be easily integrated within their analysis subsystems. The following section describes a simple, and computationally efficient method of estimating primary terrain attributes from a grid-based DEM. The attributes that can be calculated using this method include elevation, slope, aspect, curvature, catchment area, flow path length, and others.

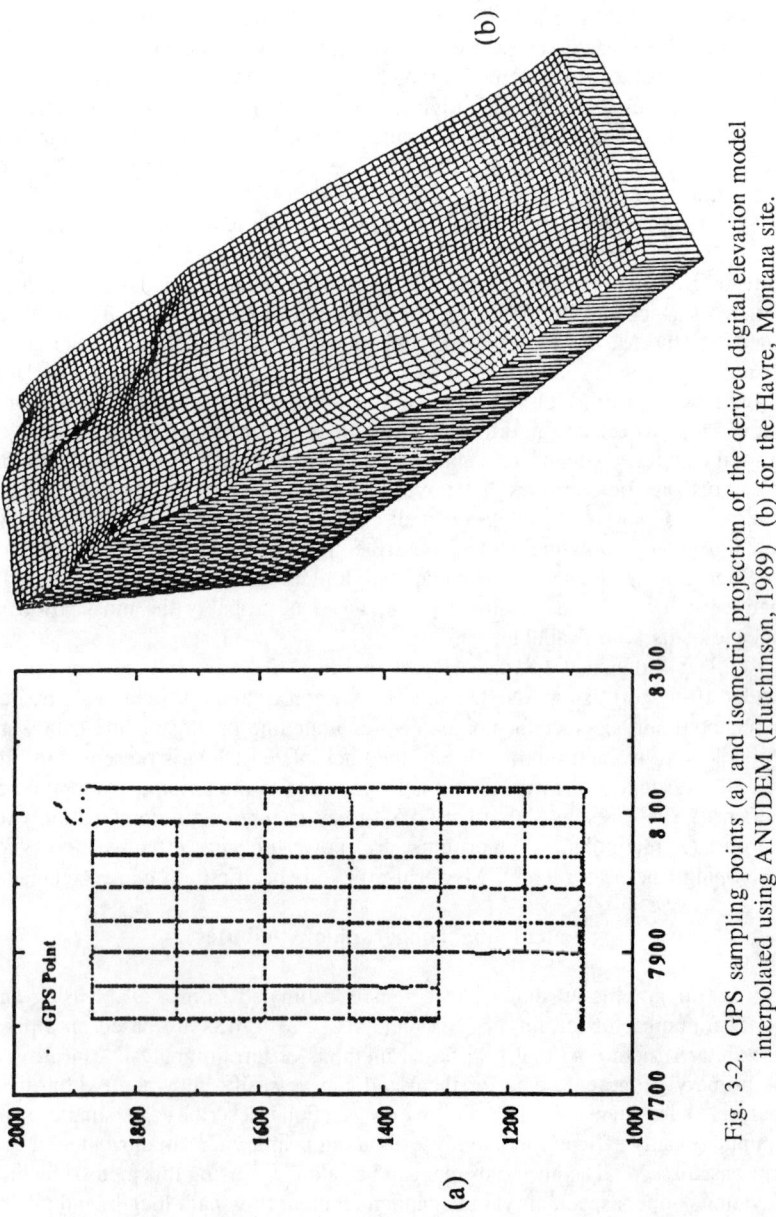

Fig. 3-2. GPS sampling points (a) and isometric projection of the derived digital elevation model interpolated using ANUDEM (Hutchinson, 1989) (b) for the Havre, Montana site.

It is possible to calculate a variety of topographic attributes by applying a second-order, finite-difference scheme centered on the interior node of a moving 3 × 3 square-grid network such as that shown in Fig. 3-3. The grid-spacing of this network is λ. We can simplify the mathematics by using the following notation.

$$f_x = \frac{\partial z}{\partial x}, f_y = \frac{\partial z}{\partial y}, f_{xx} = \frac{\partial^2 z}{\partial x^2}, f_{yy} = \frac{\partial^2 z}{\partial y^2}, f_{xy} = \frac{\partial^2 z}{\partial x \partial y} \quad (2)$$

and

$$p = f_x^2 + f_y^2, \quad q = p + 1 \quad (3)$$

If the Z's are the elevations of the nodes shown in Fig. 3-3, then the central finite difference forms of the partial derivatives for the central node 5 can be written as:

$$f_x = \frac{Z_6 - Z_4}{2\lambda}, f_y = \frac{Z_2 - Z_8}{2\lambda}, f_{xy} = \frac{-Z_1 + Z_3 + Z_7 - Z_9}{4\lambda^2},$$
$$f_{xx} = \frac{Z_4 + Z_6 - 2Z_5}{\lambda^2}, f_{yy} = \frac{Z_2 + Z_8 - 2Z_5}{\lambda^2} \quad (4)$$

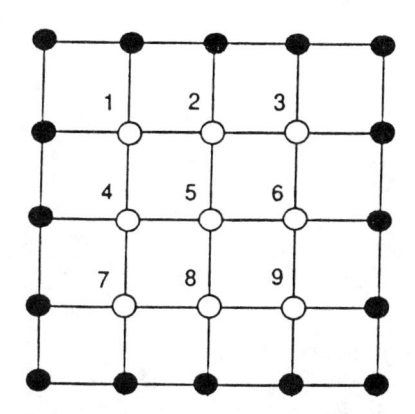

Fig. 3-3. Structure of a grid-based digital elevation model showing a moving 3 × 3 submatrix centered on node 5.

The maximum slope, β (in degrees), aspect, ψ (measured in degrees clockwise from north), and curvature (m^{-1}) of the mid-point in the moving grid can then be calculated using the following relationships.

$$\beta = \arctan(p^{1/2}),$$

$$\psi = 180 - \arctan\left(\frac{f_y}{f_x}\right) + 90\left(\frac{f_x}{|f_x|}\right),$$

and

$$Curvature = \frac{f_{xx}\cos^2\phi + 2f_{xy}\cos\phi\sin\phi + f_{yy}\sin^2\phi}{q^{1/2}\cos\upsilon} \quad (5)$$

where υ is the angle between the normal to the surface and the section plane and ϕ is the angle between the tangent of the given normal section and the x axes (Kepr, 1969; Moore, 1990; Mitasova and Jaroslava, 1992). The two directions of meaningful curvature for hydrological and geomorphological applications are in the direction of maximum slope (profile curvature) and transverse to this slope (plan curvature). Profile curvature is a measure of the rate of change of the potential gradient and is therefore important for water flow and sediment transport processes; plan curvature is a measure of the convergence or divergence, and hence the concentration of water in a landscape. For profile curvature, φ,

$$\cos\upsilon = 1, \quad \cos\phi = \frac{f_x}{(pq)^{1/2}}, \quad \sin\phi = \frac{f_y}{(pq)^{1/2}}$$

so that

$$\varphi = \frac{f_{xx}f_x^2 + 2f_{xy}f_xf_y + f_{yy}f_y^2}{pq^{3/2}} \quad (6)$$

Similarly, for plane curvature, ω,

$$\cos\upsilon = \left(\frac{p}{q}\right)^{1/2}, \quad \cos\phi = \frac{f_y}{p^{1/2}}, \quad \sin\phi = -\frac{f_x}{p^{1/2}}$$

so that

$$\omega = \frac{f_{xx} f_x^2 - 2 f_{xy} f_x f_y + f_{yy} f_x^2}{p^{3/2}} \qquad (7)$$

The plan area in the horizontal plane characterized by each node or gridpoint is $A_h = \lambda^2$. Jenson and Domingue (1988) describe a computationally efficient algorithm for estimating flow directions and hence catchment areas and drainage path lengths for each node in a regular grid DEM based on the concept of a depressionless DEM. They assume that water flows from a given node (say node 5 in Fig. 3-3) to one of eight possible neighboring nodes (nodes 1-4, 6-9 in Fig. 3-3), based on the direction of steepest descent. Upslope flow paths computed using this algorithm tend to zig-zag and therefore are somewhat unrealistic. Moore et al. (1992a) describe a modification of this algorithm that allows flow from one node (say node 5) to be distributed to multiple nearest neighbor elements in upland areas above defined channels on a slope weighted basis. This new algorithm allows flow divergence to be represented in a grid-based method of analysis, which was not possible before now. The algorithm can be combined with the above finite-difference approach to estimate a wide variety of hydrologically significant topographic attributes.

Fig. 3-4 presents shaded class intervals of slope and wetness index calculated by these algorithms and superimposed on an isometric projection of a 15.24 m grid-DEM of a 5.4-ha site located near Sterling in northeastern Colorado. This is a long-term site for studying crop and soil management in dryland agroecosystems (Peterson et al., 1991) in which soil water is a major crop growth determinant. The soils are well-drained, fine-loamy or fine-silty, mesic, Aridic Pachic Argiustolls of mixed mineralogy. They are formed in calcareous alluvial and eolian deposits and slopes range from 0-5% (Amen et al., 1977). Georeferenced elevations and soil properties measured at this site by Peterson et al. (1991) are the basis of the following analysis.

TOPOGRAPHIC AND SOIL ATTRIBUTE RELATIONSHIPS

Relative elevation (m) and A-horizon thickness (A-hor, m) were measured at 231 sites on a regular 15.24 m grid at the Sterling, CO site. In addition, extractable phosphorus (P, ppm), organic matter (OM, %), pH, and percent sand (sand), silt (silt) and clay were measured at these sites at depth increments of 0-0.1, 0.1-0.2, and 0.2-0.3 m. These properties are factors related to water availability and important for soil specific crop management. Figures 3-5a, 3-5c, 3-5e, 3-5g show the measured spatial distribution of A-horizon thickness and the P, organic matter, and pH distributions in the top 0.1-m layer of soil, respectively. Using the grid-based terrain analysis methods described above we calculated the following primary and secondary topographic attributes from the measured grid-based DEM:

Primary - slope (Slope, %), aspect (Aspect, degrees clockwise from west), specific catchment area ($m^2 m^{-1}$), maximum flow path length (m), profile curvature (m^{-1}) and plan curvature (m^{-1})

Secondary - wetness index (Wetind), stream power index (Strpind) and sediment transport capacity index (Sedtind)

The curvature parameters were only calculated at 171 points, as values could not be computed along the boundary nodes of the DEM. The secondary indices are measures of surface and subsurface water and sediment transport processes. We have included these attributes in an attempt to relate pattern to process. Only the relationships between topographic attributes and soil attributes measured in the top 0.1 m of soil profile are explored here.

A-horizons greater than 0.25-m thick are mostly confined to summit and toeslope positions (Fig. 3-5a) where slopes are less than 2% (Fig. 3-4). Where slopes with thick A-horizons do occur on slopes steeper than 2% they are associated with areas that have a wetness index greater than 8.0 (Fig. 3-4b). Here additional subsoil water has apparently allowed greater root activity than in adjacent areas that are just as steep but not as wet. To some extent the A-horizon is a "fossil record" of root activity that reflects the redistribution of water by the terrain. Wetter areas could also have more vegetation cover and consequently less erosion. Soils with organic matter content greater than 1.6% occur mainly where the wetness index is greater than 8 and where slopes are less than 2%. Calcareous BK-horizons, with pH 8 or more, underlie most of the area according to the soil survey (Amen et al., 1977). In parts of the field, material from these horizons is incorporated into A-horizons by tillage. This occurs mostly where slopes are greater than 2% and A-horizons are thin or where calcareous sediment from eroded areas upslope is deposited on thick A-horizons in toe slope positions. Finally, extractable P is spatially related to the distribution of organic matter and possibly to pH. Low P values are associated with low organic matter content, high pH and slopes greater than 2%.

What follows is only a preliminary analysis of the relationships between topography and soil attributes. Because of space limitations we will only present a subset of the results. Fig. 3-6 presents a scatter diagram of each pair of variables below the diagonal and the correlation coefficients of the same pairs of variables above the diagonal. From this matrix it can be seen that the terrain attributes most highly correlated with soil attributes are slope ($|R| = 0.45$-0.64) and wetness index ($|R| = 0.25$-0.61). It is also interesting to note that the sediment transport capacity index, which is a measure of the erosion and transport potential of overland flow is moderately correlated with A-horizon thickness ($R = -0.39$), pH ($R = 0.45$), and sand ($R = 0.42$), and silt ($R = -0.33$) percentages. The correlation between these three terrain attributes and soil attributes supports the hypothesis that the soil catena develops in response to the way water flows through the landscape.

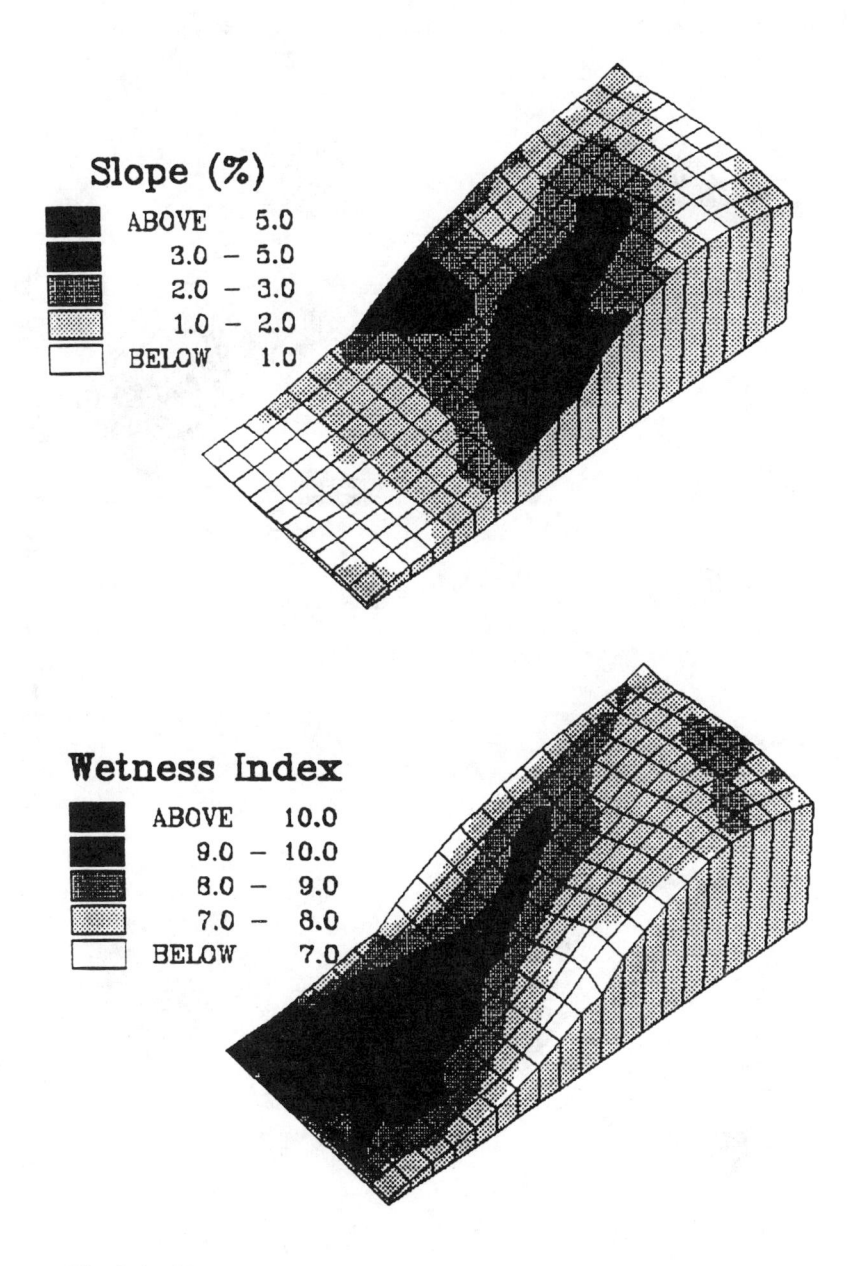

TERRAIN ANALYSIS FOR SOIL SPECIFIC MANAGEMENT

Fig. 3-4. Slope (%) and wetness index computed from the grid-based digital elevation model of the Sterling, CO site.

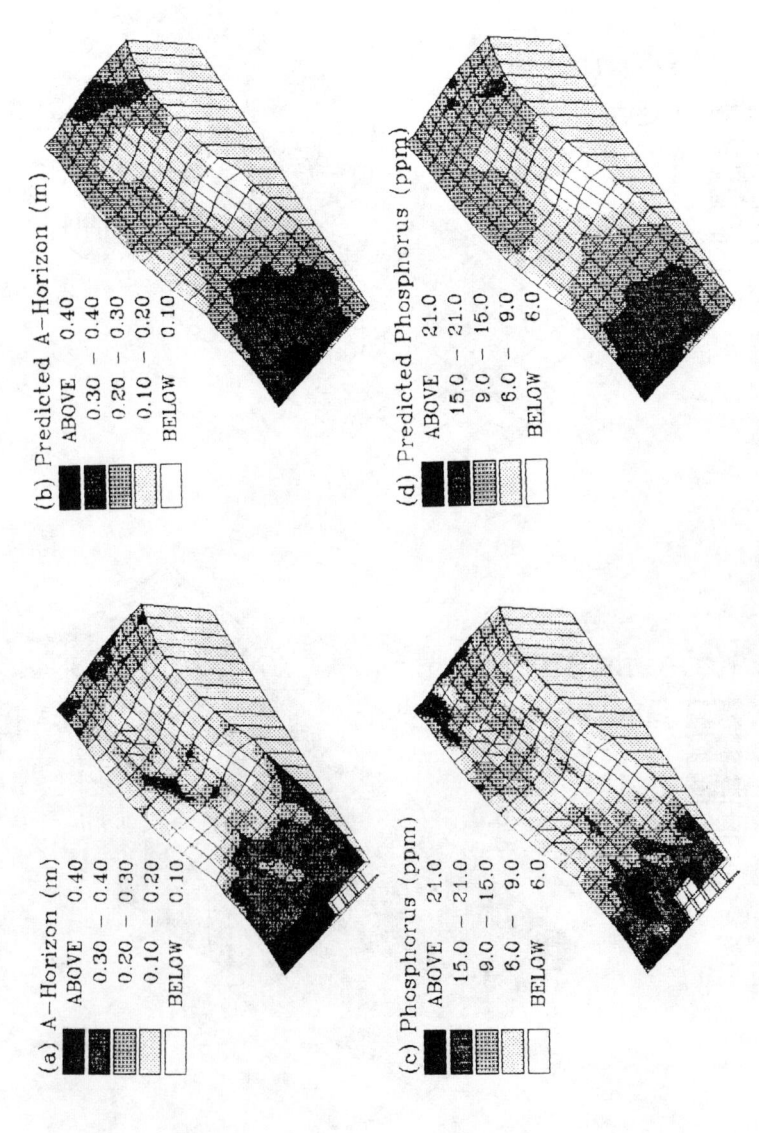

Fig. 3-5. Measured and predicted soil attributes at the Sterling, CO site.

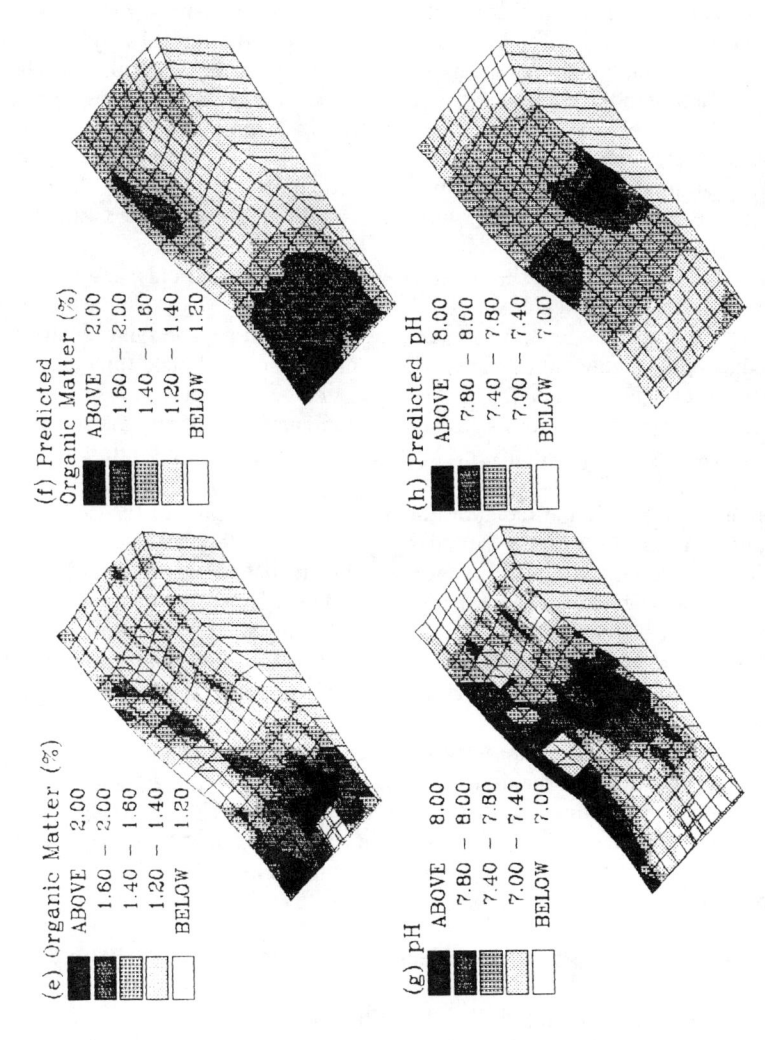

Fig. 3-5. Measured and predicted soil attributes at the Sterling, CO site (continued).

The best combination of terrain variables for explaining the variation of measured soil attributes was explored using stepwise linear regression. Table 3-2 presents the intercepts, terrain attribute coefficients, and R^2's for the best relationships describing the distribution of A-horizon thickness, extractable P, organic matter, pH, and sand and silt percentages in the top 0.1 m of the soil profile. The numbers in parentheses indicate the order in which the variables were brought into the regressions. Only variables that improved the regressions at the 99% level were included. In many cases, profile curvature or plan curvature were significant variables at the 95% level. These regression equations were then used to predict the spatial distribution of A-horizon thickness, extractable P, organic matter, and pH. The predicted and measured values are compared in Fig. 3-5.

The Sterling study site is a relatively small area and only has a limited range of attributes, particularly aspect (mostly northerly). To extend the methods described above to larger and more heterogeneous landscapes would require the introduction of additional descriptors including geology, climate (precipitation, temperature), and the radiation regime. The radiation regime is characterized in part by slope-aspect interactions and is important in modifying the soil water and evaporation distributions in landscapes, and hence soil attributes and agronomic potential. The spatial distribution of radiation can be characterized using a relatively simple radiation index (see Moore et al., 1991, 1992a) and becomes important at higher latitudes. Hutchinson (1991) describes efficient methods of spatially interpolating monthly climatic data on the basis of latitude, longitude, and elevation. These methods could be used to provide the necessary data for a more extensive analysis aimed at predicting soil physical and chemical properties. In more heterogeneous landscapes we believe that the plan and profile curvatures may be more significant than indicated in the analysis reported here.

Surface properties are most modified by land management. Therefore, features of the lower profile may show greater response to topographic attributes (e.g., degree of leaching and distribution of sodium). Also, the topography of the contact between the consolidated and unconsolidated materials or the presence of sedimentary features in the regolith may be significant in some landscapes.

ENHANCED SOIL ATTRIBUTE MAPS FROM SOIL SURVEYS

The estimated range in slope for the Sterling field, based on soil slope classes reported in the country soil survey (0-5%), is essentially identical to the range (0.1-5.2%) of the slope estimates from the DEM of the field. The soil survey here has provided an accurate range of slope estimates for the field but the survey can be enhanced with DEM data than can represent the distribution of terrain and soil attributes within soil survey map delineations.

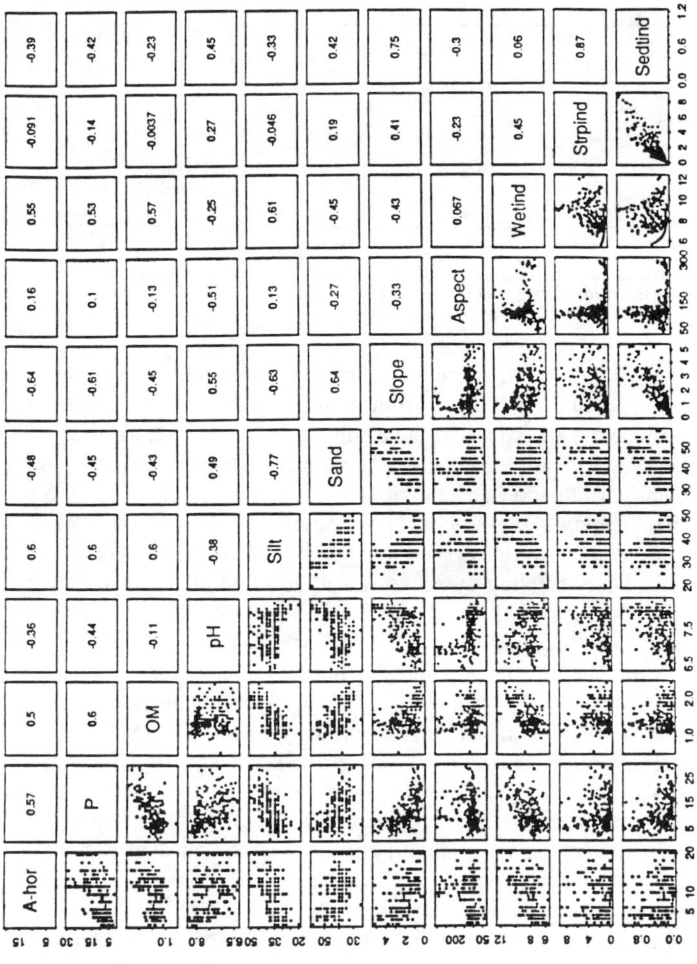

TERRAIN ANALYSIS FOR SOIL SPECIFIC MANAGEMENT

Fig. 3-6. Partial correlation matrix showing the correlation between selected terrain and soil attributes at the Sterling, CO site.

Table 3-2. Regression equations relating soil properties in the top 0.1 m to significant terrain attributes (*P*>99%).

Terrain attribute	A-Hor depth (m)	Organic matter (%)	Extractable P (ppm)	pH	Sand (%)	Silt (%)
Intercept	3.774	0.285	-3.039	7.508	46.417	23.466
Slope	-2.074(1)	-	-1.466(1)	0.190(1)	2.941(1)	-2.009(1)
Wetness index	1.224(2)	0.190(1)	2.311(2)	-	-1.320(2)	2.076(2)
Steam power index	-	-0.070(2)	-0.769(3)	-	-	-
Aspect[1]	-	-0.002(3)	-	-0.003(2)	-	-
Profile curvature	-	-	-	-	27.592(3)	-27.241(3)
R^2	0.503	0.482	0.483	0.409	0.517	0.636

[1] Aspect measured in degrees clockwise from west.

In the following discussion we attempt to illustrate how enhanced soil attribute maps could be derived from conventional soil survey sources using terrain attributes to spatially distribute soil attributes within a field or map unit. The soil survey data are from the most recently published soil survey report for Logan County, Colorado (Amen et al., 1977) and from the USDA-SGS primary characterization date (USDA, 1991). Soil attribute values (Table 3-3) from soil survey sources alone were assigned to map unit delineations. This is a common method of quickly creating inexpensive soil attribute maps but it stretches soil survey data far beyond their intended use. Furthermore, the resulting maps imply that soils are uniform within delineations. Figure 3-7 shows the boundaries of the three mapping units identified in Table 3-3. If we assume that the range of soil attribute values for a field is largely a function of terrain, the assignment of values to raster cells according to slope position, wetness index, or other terrain attributes, could be more realistic than assigning mean values based on the soil survey alone. In other words, we can use the appropriate terrain attributes to scale the reported range of soil attribute values to obtain spatially distributed estimates of soil attributes or soil attribute maps.

In the following analysis we assume that, as a first approximation, soil attributes are linearly related to terrain attributes. With this assumption, the following relationship can be used to predict the spatial distribution of soil properties:

$$\gamma_p = \gamma_{i,min} + (\gamma_{i,max} - \gamma_{i,min}) \sum_{k=1}^{n} \left(\frac{T_k - T_{k,min}}{T_{k,max} - T_{k,min}} \lambda_k \right)_i \qquad (8a)$$

or

$$\gamma_p = \gamma_{i,min} + (\gamma_{i,max} - \gamma_{i,min}) \sum_{k=1}^{n} \left(\left[1 - \frac{T_k - T_{k,min}}{T_{k,max} - T_{k,min}} \right] \lambda_k \right)_i \quad (8b)$$

where γ_p is the predicted soil attribute at the point of interest, $\gamma_{i,max}$ and $\gamma_{i,min}$ are the maximum and minimum values of the soil attribute in the field or mapping unit i (from soil survey), λ_k is a weighting coefficient for terrain attribute k ($\Sigma\lambda = 1$), T_k is the value of the "kth" terrain attribute at the point of interest, $T_{k,max}$ and $T_{k,min}$ are the maximum and minimum values of the terrain attribute in the field or mapping unit i, and n is the number of terrain attributes used in the analysis.

As an example we used this approach to predict the spatial distribution of A-horizon thickness and pH. For A-horizon thickness we assumed $\gamma_{min} = 0.18$, $\gamma_{max} = 0.66$ (Table 3), and slope (negatively correlated) and wetness index (positively correlated) as the topographic attributes with equal weighting coefficients (λ) of 0.5. For pH we assumed $\gamma_{min} = 6.6$, $\gamma_{max} = 8.0$ (Table 3-3) and slope (positively correlated) as the topographic variable. The results are presented in Fig. 3-8a and 3-8b, respectively. The range of values in each mapping unit was available for pH but only one value for each mapping unit was available for A-horizon thickness (Table 3-3). Hence the γ_{min} and γ_{max} values used for pH in the above predictions are probably close to the actual range of values in the field. The γ_{min} and γ_{max} values used initially for the prediction of A-horizon thickness were probably too large (we designated dark colored buried subsoil horizons as A-horizon in Table 3-3). We recomputed the predicted A-horizon thicknesses using $\gamma_{min} = 0.05$, $\gamma_{max} = 0.55$ and the results are presented in Fig. 3-8c.

Another potential method for spatially distributing soil attributes is to use terrain attributes to segment the landscape into essentially stationary process zones where attribute prediction may be done in a more statistically robust fashion. The relief units shown in Fig. 3-1 demonstrate one possible segmentation based on plan and profile curvature.

The mean differences between the measured and predicted A-horizon thicknesses in Fig. 3-8a and 3-8c are 0.40 and 0.31 m, respectively. These compare to the mean difference between the measured and predicted using the regression equation (Fig. 3-5b) of 0.29. Similarly, the mean differences for prediction of pH using Eq. (8) (Fig. 3-8b) and the regression equation (Fig. 3-5h) are 0.45 and 0.31, respectively. These results show that Eq. (8) can effectively spatially distribute observed soil mapping unit properties to produce results that are comparable to direct linear regression analysis (previous section). The major advantages of this method are that the data requirements are much lower and the data can be obtained from the existing soil survey and pedon characteristic databases. The method is, however, highly dependent on obtaining realistic estimates of the range of values of soil attributes.

Table 3-3. Soil attribute data estimated for map units in a cropped field near Sterling, CO.

Source[1]	Attributes	Estimated data			
M	Map unit number	126	90	92	126, 90, 92
M	Location	SUMMIT	SIDE	TOE	WHOLE FIELD
M	Mapped as:	Weld	Plantner	Rango	Weld, Plantner, Rango
S	Sampled as:	Weld	Satanta	Albinas	Weld, Satanta, Albinas
M	Drainage	wd	wd	wd	wd
M	Runoff	slow	medium	slow	slow-medium
M	Slope (%)	1-3	3-5	0-3	0-5
M	A-horizon depth (m)	0.18	0.18	0.66[4]	0.18-0.66
S	OC*1.72,0-0.1 m (%)	1.65	1.55	2.20	1.55-2.20
S	Extr.P, 0-0.1 m (ppm)[2]	27	17	42	17-42
S	Sand, 0-0.1 m (%)	45	54	42	42-54
M,S	pH(1:1),0-0.3 m	6.6-7.8	6.6-8.0	6.6-7.8	6.6-8.0
S	BK-horizon depth (m)	0.51	0.38	0.65	0.38-0.65
M	water supply, rank	intermediate	lowest	highest	-
M	Yield (bu/a)[3]	25-60	26-38	30-55	25-60

[1] M = map unit component data from soil survey report, Logan County, CO (Amen et al., 1977).
S = soil samples from similar soils, Primary Characterization Data (USDA, 1991).
[2] Sodium bicarbonate extractable phosphorus.
[3] Range in yield potential is related to low or high management levels as defined by the USDA-SCS.
[4] Includes buried A-horizon materials with mollic colors.

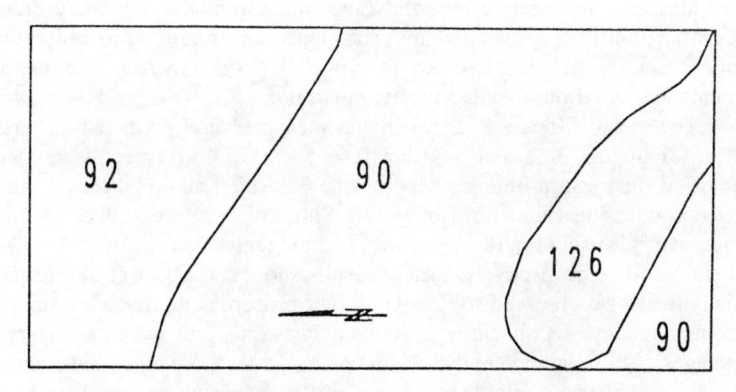

Fig. 3-7. Boundaries of the soil survey mapping units for the Sterling, CO site.

CONCLUSION

Terrain modifies the distribution of hydrologic processes (i.e., soil water content, and runoff) and soil temperature in fields and thereby the distribution of mineral weathering, leaching, erosion, sedimentation, decomposition, horizonation and ultimately those soil and microclimate attributes that determine crop production potential. Soil survey maps and databases are readily available sources of estimated soil attribute data, but the map resolution is generally inadequate for soil specific crop management. We are convinced that data from these sources can be enhanced using terrain attributes (computed from fine-resolution DEMs) to spatially distribute estimated soil attribute data. These methods offer a promising, cost-effective means of creating the fine-resolution maps needed for soil specific crop management. They also allow existing basic datasets (e.g., soil survey primary characterization data) to continue to be used as new techniques and technologies develop.

Fine-resolution grid-based DEMs required by these analysis methods can be quickly developed from GPS survey data. Although specific implementation strategies have not yet been developed, affordable GPS technology encourages the use of these techniques for farm planning and soil specific crop management. Using GIS to organize and build upon these datasets will improve our knowledge of environmental processes and promote economical and sustainable land management.

Initial results from this study indicate useful correlations between quantified terrain attributes and measured soil attributes. Slope and wetness index were the terrain attributes most highly correlated with soil attributes measured at 231 locations in a 5.4-ha field in Colorado. Individually, they accounted for about half of the variability of several soil attributes including A-horizon thickness, organic matter content, pH, extractable P and silt and sand percentages. This represents an incorporation of finer-scale process-based information relating to soil formation patterns in the landscape. One possible method illustrating how relationships between terrain and soil attributes may be used to enhance an existing soil map is presented, even when the exact form of the relationship is unknown. Such techniques may also be applied as a first step in unmapped areas to guide sampling and model development.

In applying Eq. (8) some a priori knowledge of catenary sequences is needed (e.g., pH increases downslope, base status is lower in wet areas, etc.). The underlying pedological or catenary models are usually developed during a soil survey but they are often reported only as verbal models, if at all. If the procedure is to be implemented then survey practice must change so that these verbal models are presented explicitly and consistently. If this could be achieved, then more of the surveyor's knowledge could be transferred than has hitherto been possible. We have attempted to outline one form that this quantitative framework might take.

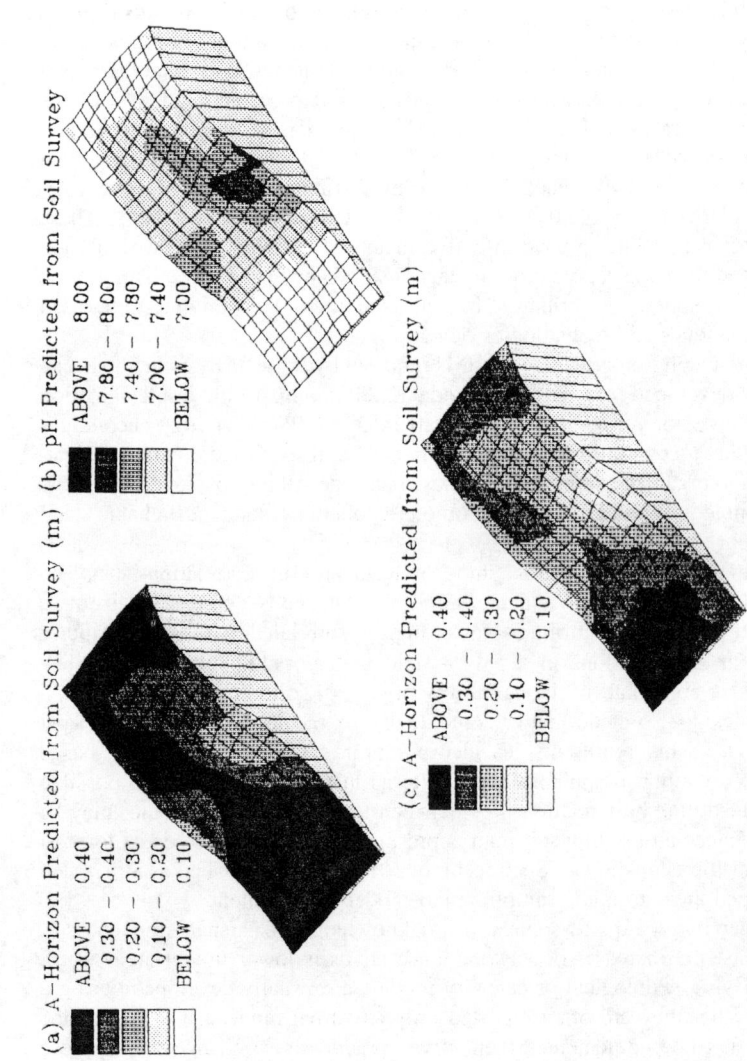

Fig. 3-8. Predicted A-horizon thickness (a and c) and pH (b) using terrain attributes to spatially distribute the range of reported soil attributes available from conventional soil survey and primary characterization databases.

Some key future needs for the advancement of soil specific crop management include: (i) accurate georeferencing of all sites where soil attribute data are acquired, including the sites of all soil profiles observed during soil survey operations; (ii) measurements of soil water, leaching, plant nutrients and crop yields in relation to terrain attributes at georeferenced sites; and (iii) analysis of the economic and environmental costs and benefits of applying the procedures described in this chapter. In addition, soil scientists and agronomists need to begin looking at whole catchments or drainage units, rather than individual plots. The results presented here show that in particular, the hydrology of the catchment area outside of the plot has a significant impact on the soil attributes and therefore crop production potential of the plot.

ACKNOWLEDGMENT

This study was funded in part by Grant no. NRMS-M218 from the Murray-Darling Basin Commission, Grant no. 90/6334 from the Australian Department of Industry, Technology and Commerce, Bilateral Science and Technology Program, the Water Research Foundation of Australia, USDA-CSRS Grant No. 290684, the PLACES Industrial Affiliates Program, and the Great Plains Systems Unit of the USDA-ARS. The authors thank Dr. David Tyler of Montana State University (now at the University of Maine) and Cheryle Quirion of Trimble Navigation, Ltd. for providing the GPS survey data at the Havre, Montana site and Dr. Dwayne G. Westfall of the Department of Agronomy, Colorado State University for the elevation and soil attribute data from the Sterling, Colorado site. Many suggestions and comments from Dr. Neil McKenzie have been incorporated into this chapter and the authors acknowledge this contribution.

REFERENCES

Amen, A. E., D. L. Anderson, T. J. Hughes, and T. J. Weber. 1977. Soil Survey of Logan County, Colorado. USDA-SCS, U.S.Gov. Print. Office, Washington D.C.

Beckett, P. H. T., and R. Webster. 1971. Soil variability: A review. Soils Fertil. 34: 1-15.

Blewitt, G. 1992. Measuring the earth to within an inch using GPS satellites. Trans. Am. Geophys. Union 73: 34.

Bouma, J. 1989. Land qualities in space and time. p. 3-13. In J. Bouma and A. K. Bregt (ed.) Land qualities in space and time. PUDOC, Wageningen.

Boul, S. W., F. D. Hole, and R. J. McCracken. 1989. Soil genesis and classification, 3rd ed. Iowa State Univ. Press, Ames.

Burrough, P. A. 1986. Principles of geographical information systems for land resources assessment. Oxford University Press, Oxford.

Carter, B. J., and E. J. Ciolkosz. 1991. Slope gradient and aspect effects on soil development from sandstone in Pennsylvania. Geoderma 49: 199-213.

Collins, S. H., and G. C. Moon. 1981. Algorithms for dense digital terrain models. Photogram. Engr. Remote Sens. 47: 71-76.

Daniels, R. B., J. W. Gilliam, D. K. Cassel, and L. A. Nelson. 1985. Soil erosion class and landscape position in North Carolina Piedmont. Soil Sci. Soc. Am. J. 49: 991-995.

Dikau, R. 1989. The application of a digital relief model to landform analysis in geomorphology. p. 51-77. In J. Raper (ed.) Three dimensional applications in geographic information systems. Taylor and Francis, New York.

Dixon, T. H. 1991. An introduction to the global positioning system and some geological applications. Rev. Geophys. 29: 249-276.

Furley, P. A. 1976. Soil-slope-plant relationships in the northern Maya mountains, Belize, Central America. J. Biogeography 3: 303-319.

Gessler, P. E. 1990. Geostatistical modeling of soil-landscape variability within a GIS framework. Unpubl. M.Sci. Thesis, Inst. for Environmental Studies, Univ. Wisconsin, Madison, WI.

Gessler, P. E., McSweeney, R. W. Kiefer, and L. M. Morrison. 1989. Analysis of contemporary and historical soil/vegetation/landuse patterns in southwest Wisconsin utilizing GIS and remote sensing technologies. Tech. papers, 1989 ASPRS/ACSM Annual Convention, Baltimore, MD. Vol 4, pp. 85-92.

Hairston, A. B., and D. F. Grigal. 1991. Topographic influences on soils and trees within single mapping units on a sandy outwash landscape. Forest Ecol. Manag. 43: 35-45.

Hutchinson, M. F. 1989. A new procedure for gridding elevation and stream line data with automatic removal of spurious pits. J. Hydrol. 106: 211-232.

Hutchinson, M. F. 1991. Climatic analyses in data sparse regions. p. 55-71. In J.A. Bellamy (ed.) Climate risks in crop production. CAB International, Wallingford.

Hutchinson, M. F., and T. I. Dowling. 1991. A continental hydrological assessment of a new grid-based digital elevation model of Australia. Hydrol. Processes 5: 45-58.

Jenny, H. 1941. Factors of soil formation: A system of quantitative pedology. McGraw-Hill, New York.

Jenny, H. 1980. The soil resource: Origin and behaviour. Ecological Studies 37, Springer-Verlag, New York.

Jenson, S. K., and J. O. Domingue. 1988. Extracting topographic structure from digital elevation model data for geographic information system analysis. Photogram. Engr. Remote Sens. 54: 1593-1600.

Kepr, B. 1969. Differential geometry. p. 298-396. In K. Rektorys (ed.) Survey of applicable mathematics. The MIT Press, Cambridge, MA.

Klingebiel, A. A., E. H. Horvath, D. G. Moore, and W. U. Reybold. 1987. Use of slope, aspect, and elevation maps derived from digital elevation model data in making soil surveys. p. 77-90. In Soil survey techniques. Soil Sci. Soc. Am. Spec. Publ. 20. SSSA, Madison, WI.

Kreznor, W. R., K. R. Olson, W. L. Banwart, and D. L. Johnson. 1989. Soil, landscape, and erosion relationships in a northwest Illinois watershed. Soil

Sci. Soc. Am. J. 53: 1763-1771.
Lammers, R. B., and L. E. Band. 1990. Automated object description of drainage basin. Computers Geosci. 16: 787-810.
Lauren, J. G., R. J. Wagenet, J. Bouma, and J. H. M. Wosten. 1988. Variability of saturated hydraulic conductivity in a glossaquic hapludalf with macropores. Soil Sci. 145: 20-28.
Loague, K., and G. A. Gander. 1990. R-5 revisited, 1: Spatial variability of infiltration on a small rangeland catchment. Water Resour. Res. 26: 957-972.
Long, D. S., S. D. DeGloria and J. M. Galbraith. 1991a. Use of the global positioning system in soil survey. J. Soil Water Conserv. 46: 293-297.
Long, D. S., S. D. DeGloria, D. A. Griffin, G. R. Carlson, and G. A. Nielsen. 1991b. Spatial statistics for survey and analysis of autocorrelated soil and crop variables. p. 142. In Agronomy abstract. Denver, CO. October 1991.
Mabbutt, J.aA. 1968. Review of concepts of land classification. In G.A. Stewart (ed.) Land evaluation. MacMillan, Melbourne.
Mackay, D. S., L. E. Band, and V. B. Robinson. 1991. An object-oriented system for the organization and representation of terrain knowledge for forested ecosystems. In Proc. LIS/GIS 1991.
McBratney, A. B. 1992. Some remarks on soil horizon classes. Catena, Special Issue on Soil Horizon Classes (in press).
McBratney, A. B., and J. J. DeGruijter. 1991. A continuum approach to soil classification and mapping: Classification by modified fuzzy k-means with extragrades. J. Soil Sci. (in press).
McBratney, A. G., G. A. Hart, and D. Mcgarry. 1991. The use of region partitioning to improve the representation of geostatistically mapped soil attributes. J. Soil Sci. 3: 513-533.
McEachern, K. L., G. A. Nielsen, J. S. Jacobsen, and P. A. McDaniel. 1990. Application of global positioning system technology for a field-scale geographic information system. p. 250. In Agronomy Abstract. San Antonio, TX. October 1990. ASA, Madison, WI.
McKeague, J. A., R. G. Eilers, A. J. Thomasson, M. J. Reeve, J. Bouma, R. B. Grossman, J. C. Favrot, M. Renger, and O. Strebel. 1984. Tentative assessment of soil survey approaches to the characterization and interpretation of air-water properties of soils. Geoderma 34: 69-100.
McKenzie, N. J., and M. P. Austin. 1992. Developing parametric soil survey for land evaluation: An Australian example. Geoderma (submitted).
McKenzie, N. J., and D. A. MacLeod. 1989. Relationships between soil morphology and soil properties relevant to irrigated and dryland agriculture. Aust. J. Soil Res. 27: 235-258.
McKenzie, N. J., K. R. J. Smettem, and A. J. Ringrose-Voase. 1991. Evaluation of methods for inferring air and water properties of soils from field morphology. Aust. J. Soil Res. 29: 587-602.
Mitasova, H., and H. Jaroslava. 1992. Interpolation by regularized spline with tension: II. Application to terrain modeling and surface geometry analysis. Math. Geol. (submitted).

Moore, I. D. 1990. Geographic information systems for land and water management. In Land use and declining water quality. Proc. Annu. Conf., Soil & Water Assoc. of Aust., Canberra, 5-6 November 1990.

Moore, I. D., G. J. Burch, and D. H. Mackenzie. 1988. Topographic effects on the distribution of surface soil water and the location of ephemeral gullies. Trans. Am. Soc. Agric. Eng. 31: 1383-1395.

Moore, I. D., R. B. Grayson, and A. R. Ladson. 1991. Digital terrain modelling: A review of hydrological, geomorphological and biological applications. Hydrol. Processes 5: 3-30.

Moore, I. D., and M. F. Hutchinson. 1991. Spatial extension of hydrologic process modelling. Proc. Int. Hydrology and Water Resources Symposium, Institution of Engineers Australia 91/22, pp. 803-808.

Moore, I. D., A. K. Turner, J. P. Wilson, S. K. Jenson, and L. E. Band. 1992a. GIS and land surface-subsurface process modelling. In M. F. Goodchild et al. (ed.) Geographic information systems and environmental modeling. Oxford University Press (in press).

Moore, I. D., and J. P. Wilson. 1992. Length-slope factors in the revised universal soil loss equation: Simplified method of estimation. J. Soil Water Conserv. (in press).

Moore, I. D., J. P. Wilson, and C. A. Ciesiolka. 1992b. Soil erosion prediction and GIS: Linking theory and practice. Proc. Geographic Information System for Soil Erosion Management, Taiyuan, China, 2-11 June, 1992.

Odeh, I. O. A., D. J. Chittleborough, and A. B. McBratney. 1991. Elucidation of soil-landform interrelationships by canonical ordination analysis. Geoderma 49: 1-32.

Petersen, C. 1991. Precision GPS navigation for improving agricultural productivity. GPS World 2(1): 38-44.

Peterson, G. A., D. G. Westfall, L. Sherrod, E. Mcgee, and R. Kolberg. 1991. Crop and soil management in dryland agroecosystems. Tech. Bull. 91-1, Agric. Exp. Stn., Colorado State Univ., Fort Collins.

Powell, B., A. B. McBratney, and D. A. MacLeod. 1992. Fuzzy classification of soil profiles and horizons from the Lockyer Valley, Queensland, Australia. Geoderma 52: 173-197.

Rawls, W. J., D. L. Brakensiek, and K. E. Saxton. 1982. Estimating soil water properties. Trans. Am. Soc. Agric. Engr. 25: 1316-1320, 1328.

Ruhe, R. V. 1975. Geomorphology. Houghton Mifflin Co., Atlanta, Georgia.

Simonson, R. W. 1991. The U.S. soil survey - contributions to soil science and its application. Geoderma 48: 1-16.

Skidmore, A. K., P. J. Ryan, W. Dawes, D. Short, and E. O'Loughlin. 1991. Use of an expert system to map forest soils from a geographical information system. Int. J. Geograph. Inf. Sys., 5: 431-445.

Speight, J. G. 1974. A parametric approach to landform regions. Spec. Publ., Inst. Br. Geogr. 7: 213-230.

Speight, J. G. 1980. The role of topography in controlling throughflow generation: A discussion. Earth Surf. Processes Landforms 5: 187-191.

Stone, J. R., J. W. Gilliam, D. K. Cassel, R. B. Daniels, L. A. Nelson, and H. J. Kleiss. 1985. Effect of erosion and landscape position on the productivity of Piedmont soils. Soil Sci. Soc. Am. J. 49: 987-991.

U.S. Department of Agriculture. 1991. Primary characterization data (Logan County, Colorado). USDA-Soil Conservation Service, Natl. Soil Surv. Lab., Lincoln, NE.

Walker, P. H., F. F. Hall, and R. Protz. 1968. Relation between landform parameters and soil properties. Soil Sci. Soc. Am. Proc. 32: 101-104.

Webster, R. 1985. Quantitative spatial analysis of soil in the field. Advances in Soil Sci., Vol 3, Springer-Verlag New York, New York.

Weibel, R., and J. S. DeLotto. 1988. Automated terrain classification for GIS modeling. GIS/LIS 99. Proc. ASPRS/ACSM, San Antonio, Texas, Dec. 1988, p. 618-627.

Williams, J., P. Ross, and K. Bristow. 1990. Prediction of Campbell water retention function from texture, structure and organic matter. In M. Th. van Genuchten (ed.) Proc. Int. Workshop on Indirect Methods for Estimating the Hydraulic Properties of Unsatursated Soils. Univ. California, Riverside, 11-13 October 1990.

Yates, S. R., and A. W. Warrick. 1987. Estimating soil water content using cokriging. Soil Sci. Soc. Am. J. 51: 23-30.

4 Application of Soil Survey Information to Soil Specific Farming

Maurice J. Mausbach
Soil Survey Division
Soil Conservation Service
Washington, DC

Dennis J. Lytle
Soil Survey Division
Soil Conservation Service
Lincoln, NE

Lawson D. Spivey
Soil Survey Division
Soil Conservation Service
Lincoln, NE

Whether it is soil specific farming, farming on the grid, farming by soil type, farming by the foot, farming soils not fields, variable rate technology, or prescription farming, it is clear that users of the soil survey information need more specificity. In this chapter, we define the increase in precision required by soil specific farming as farming by phases of soil series or by specific components of a map unit.

One of the main objectives of the National Cooperative Soil Survey (NCSS) is to provide a sound soil resource inventory for making land use decisions. This inventory together with soil attribute data and soil interpretations is the basis for resource planning in the Soil Conservation Service (SCS). Soil information includes interpretations or predictions of behavior for a specific use of the soil based on soil properties (phase of a soil series). Soil interpretations make soil surveys useful to people. They are not, however, recommendations for specific areas within a tract of land (Kellogg, 1961).

Soil surveys in the NCSS are multipurpose in design to serve users ranging from state and regional planners, soil resource planners for sustainable farming systems, tax assessors, wetland delineators, state highway departments, pipeline and buried cable companies, and others that use the soil or build on it. The multipurpose soil survey, however, is not designed for the specificity to make onsite or small tract evaluations. The site specific surveys often become single purpose since they are site specific, often require management or time dependent soil properties, and often are aimed at a specific use.

Copyright © 1993 ASA-CSSA-SSSA, 677 South Segoe Road, Madison, WI 53711, USA. *Soil Specific Crop Management.*

Charles Kellogg (1955), as a vision for soil survey, discussed soil surveys with respect to modern farming. Even at this early date, he recognized the need for soil survey to define kinds of soil in ways that knowledge about soil behavior could be related to farming systems and be organized and presented to farmers so that they could fit the best management practices to specific soils. He was talking about general or resource management systems, not fertility applications. Kellogg (1961) was very clear that the soil survey and its interpretations are not site specific.

We will discuss soil survey information with respect to soil specific farming, soil survey databases and geographic information systems (GIS), and future considerations for using soil survey information in soil specific farming.

BACKGROUND

Soil Survey

Soil maps are made in the field. Properties used to map are those that can be sensed in the field. Hence, soil fertility is inferred, but levels of fertility are not mapped. Soil scientists formulate landscape models, collect data on the soils and their behavior under various uses, and create process models in the form of soil interpretations to transfer their knowledge about how soils perform under land uses. They judiciously sample carefully chosen sites and extend the data to other soils based on interrelationships among analytical and field collected properties. Properties are recorded that infer how the soil can be used, and not those that vary so frequently that the status has to be tested annually or on a regular basis. Human-induced activities such as land leveling, ripping, severe erosion, or extremes in additions of manure abruptly render any soil map out-of-date.

Soil Series

Phases of soil series are used to describe the kinds of soil in map unit delineations (Soil Survey Staff, 1983, 1992). Soil Series, the lowest category in Soil Taxonomy (Soil Survey Staff, 1975), are the most homogeneous classes in Soil Taxonomy, and are closely related to interpretations for uses of the soil (Soil Survey Staff, 1983, 1992). The pedon (Soil Survey Staff, 1975) represents the sampling unit for soil series.

Information on the soil series in the United States is stored in the Official Series Description (OSED) and Soil Interpretations Record. Supporting analytical data are in the National Soil Characterization Database.

Map Units

Map units are specially tailored to each soil survey in the United States. They are designed to meet the specific needs of users with respect to common land uses and the complexity of soils and landscapes in the area. Map units are named by phases of soil series. Phases include surface texture, organic surface layers, rock fragment, slope, erosion, depositional, depth, wetness, saline, physiographic, and other phases. The phases narrow the range of properties of a series and are specific to the various soil components that occur in delineations of a map unit. We are presently limited to three major components (named phases). Minor components are mentioned in the map unit description.

Most detailed soil surveys are at a scale of 1:12000, 1:15840, 1:20000, or 1:24000. At these scales the minimum size of the delineation ranges from about 2 to 10 acres. Most map delineations, even in simple landscapes, have inclusions of similar to contrasting soils. In areas of complex landscapes, map delineations usually have more than one component. Neither the major components of a map unit nor the minor inclusions are specifically located within a map delineation. They are described in the map unit description according to landforms or landscape position.

Information on the map unit composition, and phase information of the components is stored in the Map Unit Use File (MUUF) and specific information on the soil components of the map units are stored in the Map Unit Interpretations Record (MUIR) in the State Soil Survey Database.

Soil Specific Farming

Several articles in the popular farm literature indicate an increase in profits from applying plant nutrients by soil vs. a general application for the entire field (Bechman, 1992; Reichenberger and Russnogle, 1989; Richter, 1991). The principal savings is from reduced fertilizer costs that more than offset the additional costs of application.

Munson and Runge (1990) discuss high precision farming and propose farming by soils as a tool. They suggest that farming by soils requires integration of soil survey information systems and research databases for site specific testing for nutrient and other chemical additions. The integrated soil information helps identify soils sensitive to nutrient leaching or erosion and runoff.

Carr et al.(1991) show through yield and fertility trials that fertilizing by soil is profitable. In their study, they supplemented the soil survey by using remote sensing photography to fine tune soil boundaries. They also took composite fertilizer samples for each of the soil units.

Most articles report fertility testing of the map units to obtain optimum fertilization rates for specific soils. However, the Richter (1991) article, on a Minnesota application, was unclear as to the use of the original soil survey in their study. They cite grid sampling and fertility analysis but do not mention soil surveys directly. Holmes (1992) suggests that variable rate technology requires sophisticated systems to handle the massive site data need to assess site specific rates. They found little correlation to specific soils, however, soils in their study were poorly drained and wetness factors may have an overriding affect. The Reichenberger and Russnogle (1989) article suggests that farming by grid is better in some areas because it considers both the soil type and field history.

It is unclear from many of these studies how the previous management and fertility practices were factored into the research results. Previous over or under fertilization practices could easily mask the benefits from soil specific farming and certainly would confound the results. It is clear that the many interactions among fertilization history, soil drainage, available water capacity, compaction, and other soil qualities in addition to climate need to be recognized in soil specific farming research.

Soil Survey Databases

Soil Series

Information on the soil series of the USA is stored in the Official Soil Series Description, National Soil Characterization Database (Pedon Data), and Soil Interpretations Record. The Official Soil Series Description is a narrative technical description of the typical pedon (profile) of the series. It includes morphological properties and their ranges for each soil horizon, and information to differentiate similar soil series.

The soil characterization database contains analytical data on more than 17,000 pedons in the USA. Most of the data are representative of soil series and are used to develop the Soil Interpretations Record for each soil series. The data are also useful in developing interrelationships among properties to predict soil qualities and some of the more difficult-to-measure soil properties.

The Soil Interpretations Record (SIR) contains the range in estimated properties of the soil series and soil interpretations for engineering, water management, recreation, agronomic, woodland, range, and wildlife uses of the soil. The estimated properties include mean annual air temperature, precipitation, elevation, frost-free days, flooding frequency and duration, depth to high water table, depth to bedrock, and drainage for site characteristics; and particle-size distribution, bulk density, permeability, organic matter, available water capacity, soil reaction, salinity, cation exchange capacity, sodium adsorption ratio, gypsum, and calcium carbonate equivalent for layers. Estimates of potential subsidence, shrink-swell, corrosivity, and frost action are also listed on the SIR. The SIR lists the parameters for soil erosion equations including the soil erodibility factor K and Soil Loss Tolerance T for the Universal Soil Loss Equation and the Soil Erodibility Index 'I' for the Wind Erosion Equation.

Interpretations specific to crop management include pesticide leaching and runoff potentials; drainage; terraces and diversions; grassed waterways; capability class and subclass; crop yields; woodland, windbreak, and rangeland suitability; and potential range production.

The kinds of soil property, soil quality, and soil interpretation information relate to the multipurpose nature of the NCSS. Certain properties relate to road construction while others relate to wildlife uses of the soil. Many of the properties may also be directly or indirectly useful in soil specific farming.

Map Units

The MUUF contains the map symbol, map unit name, counties where mapped, acreage of each unit, percent composition of multitaxa units, texture of the surface layer, slope of the map unit and map unit components, and SIR numbers and phase information that tie map unit components (phases of series) to the estimated properties of the soils. The MUUF file when merged with the SIR provides a single phase record for each component (phase of a series) in a map unit. This merged file forms the basis of the MUIR for map units in a soil survey area. The MUIR is part of the State Soil Survey Database (SSSD). The SSSD is a compilation of data for all survey areas in a state and is maintained in the SCS State Office.

The MUIR information is specific to the map unit components of a soil survey area. This is the most detailed information on map unit components in our soil survey databases and is the data that is linked with the digitized soil map of a survey. The MUIR information is used by our resource conservationists for developing resource management systems and is the information most pertinent to soil specific farming. Interpretations related to natural soil fertility are given in the map unit description and are often stored as narrative technical descriptions in the MUIR.

Soil Geographic Databases

The NCSS supports three geographic databases that represent different scales of soil maps. They are the Soil Survey Geographic Database (SSURGO), the State Soil Survey Database (STATSGO), and the National Soil Geographic Database (NATSGO) (Cartographic Staff, 1980; Reybold and TeSelle, 1989). SSURGO is the digital data for the detailed soil surveys. National specifications require the survey area maps to be on orthophotography in a 1:24,000 quadrangle or 1:12,000 quarter quadrangle format (Cartographic Staff, 1980; Mausbach et al., 1989).

The STATSGO database consists of general soil maps on a 1:250,000 United States Geologic Service (USGS) quadrangle format for each state. They are an abstraction of the SSURGO data. The STATSGO database is useful for state and regional planning but is not applicable to farming by specific soil.

The NATSGO database is a 1:7,500,000 scale general map of the USA. It is used for broad national planning purposes.

FUTURE CONSIDERATIONS IN SOIL SPECIFIC FARMING

What is the Role of Soil Survey?

Soil surveys are the basic building blocks for developing the comprehensive data set needed to drive specific soil farming. Our detailed soil survey (order 2) is a multi-purpose survey used by SCS to plan operational resource management systems. The key words are planning and operational.

Planning means evaluating the resource condition, formulating options, and selecting the best options. Operational means that there is a firm intention to initiate or change a resource management system but it stops short of implementation or installation of related works, structures, and practices.

The implementation or installation of works, structures, and practices such as variable fertilizer application, requires a rather precise fit to the soil. The more detailed order 1 procedures would be more site specific and could be used to guide the design of a close interval grid sampling plan. Then site specific data for the particular practice or chemical application could be collected for the specific cropping system.

Our map unit information on the composition and location of the various components within a delineation is crucial to soil specific farming. Percent composition is not very useful. We must develop techniques to locate the components on landforms or landscape positions within a delineation. Most users can see landforms or landscapes and thus can more specifically locate the specific soil components within a delineation. Boundaries between adjacent delineations may be sharp or broad and adjacent delineations may contain strongly contrasting soils while others contain similar soil. This information is not presently provided in the NCSS but may be very useful in designing sampling schemes and chemical application. Our field soil scientists know many of these relationships, but we need to develop information systems that will easily store the information and display it to the user. Brown and Huddleston, (1991) succinctly state that as soil scientist, we need to collect what we (i) know about the landscape, (ii) what more we need to know, and (iii) how to go from i to ii efficiently. Fertilizer operators could adjust application rates based on this knowledge --- the trick being how do we convey the information easily to user in the fertilizer rig.

Our basic soil surveys serve as a general guide for determining if an area is suited to specific cropping systems. Once the basic decision that a cropping system will be profitable, a farmer can proceed to collect the detailed information for soil specific farming. The detailed soil survey is also useful for identifying areas of concern whether it is the identification of highly erodible land, soils that are susceptible to pesticide leaching or surface runoff loss, wetlands, or soils that are suitable for the disposal of agricultural or municipal wastes. Once the area of concern or the suitable areas are identified, more specific information will drive the application of chemicals or waste by soil.

Kellogg (1955) mentions that at one time in soil survey we did not make interpretations. It was a grave error since no one filled the void to develop interpretations from the soil survey information. Soil scientists who make the surveys are also the scientists that observe how soils behave under different uses and thus are integral in developing interpretations for their use. We believe this principle is still valid. The partners in NCSS must continue to research new interpretative uses of soil survey to include soil specific farming. We are the experts in the techniques of soil survey and must continue to expand the science.

Soil surveys are excellent technology transfer vehicles to extend limited and often expensive analytical data from the sampled area to other areas of similar soils. State highway departments have been able to significantly decrease sampling for road construction by using soil survey maps to extend borehole data. Producers can extend their data for soil specific farming in a similar fashion. Fertility data, however, is related to cropping systems and fertility practices in addition to soil properties. But the scheduling of irrigation, seeding rate, and application of pesticides rely mostly on relatively stable soil properties in our databases.

Onsite determinations have traditionally been a private sector activity. We believe that collecting the information for specific soil farming should remain a private sector activity. The partners of the NCSS have the responsibility to develop the science for collecting site specific information and the procedures for linking this information to soil survey data. The display of the site specific data (technical transfer) to the user is crucial to soil specific farming.

The interpretative soil properties in the SIR and MUIR provide basic data on physical properties such as particle size, bulk density, and available water capacity; chemistry such as pH, calcium carbonate, salinity, cation exchange capacity, gypsum, and SAR; and agronomic properties such as expected yields. This information is useful in determining seeding rates, wetness hazards, general fertility problems such as iron chlorosis or zinc and other micronutrient deficiencies. Farmers should be able to adjust seeding rates, pesticide rates, as well as fertilizer rates by using basic soil information. The catch is how to transfer the soil specific information to the farmer. Soil scientists must be able to bridge the gap between the scientific community and the producer (Brown and Huddleston, 1991).

Improving Soil Survey Information and Its Delivery

Soil Survey Information

The NCSS, in order to survive into the next century, must be adaptable to new technologies and user needs for information. The basic data in our databases is among the best in the world but, as technology progresses, users need more kinds of- and more specific- information. Soil property data on such things as phosphorus sorption, buffering capacity of the soil, surface area, and other relatively static or basic soil properties are needed to adequately evaluate rates of chemicals to apply to specific soils. We can model these properties and need to continue to perfect these models, but the models are more precise if we supplement them with analytical data. We also need to assimilate the site specific information into our information systems and correlate it to specific management systems and soils.

Soil tests that are time and management system dependent have traditionally been excluded from our databases. Soil specific farming requires this information and information systems are needed that capture and analyze the data for maximizing profits using specific management systems. Some of these data are user generated and may be available only if users provide it. However, limited data for specific management systems and soils can be used to develop process models for extending the information. If we can convince the user to provide the data, eventually the models will help reduce the onsite data collection costs of the producer.

Reliability estimates of our soil maps, attribute data, and interpretations will be crucial in developing risk assessment scenarios for agricultural chemicals or application of waste materials to the soil. We have avoided providing this information largely due to the complexity of the problem and the amount of resources needed to obtain the basic data to estimate reliability. Fuzzy set theory in mathematics was developed for biological systems (Zadeh, 1968; Burrough, 1989). It provides a means for both capturing the heuristic knowledge of our soil scientists who make all of the field observations in mapping and providing estimates of reliability. We are excited about using fuzzy sets to more precisely define our attribute data and interpretations. It will be crucial for developing statements of risk associated with our interpretations for such things as potential for pesticide leaching and runoff.

Munson and Runge (1990) suggest that farming by soils requires integration of soil survey information and analytical databases. Geostatistics is a technology for relating information on point data such as fertility samples to a larger area. It was developed for use in mining applications which are not much different from soil specific farming. We need to continue to develop geostatistical techniques to expand information from point samples using soil surveys. These methods help us understand the variability and order of soils in nature and to alter application of chemicals via this understanding.

Stein et al. (1988a,b) demonstrated that soil map units (delineations) are effective strata for geostatistical analysis. Theoretically, soil maps can reduce sampling needs and facilitate the extension of point data. Burrough (1991), in a review of sampling designs for map units, states that there is no one optimum sampling design for quantifying map unit composition. He also mentions use of cokriging and regression or other models that have use for extending expensive-to-measure data. The NCSS needs to research new and innovative approaches for representing and extending the soil information.

Geographic Information Systems

Geographic Information Systems are a prerequisite for managing the massive data required for soil specific farming. The GIS not only are necessary for fertilizer application, but are the crux for analyzing fields for the most economic and environmentally sustainable farming system. The technology for geographic information systems continues to develop with respect to analysis of digital imagery data and display of reliability information. We now have, in the NCSS, technology for digitizing soil maps and integrating soil attribute data with the map unit delineation. However, we are not able to easily show landscapes or landforms overlaid with the digital imagery information, nor are we able, at the scales needed for soil specific farming, to easily display information about more than one component in a polygon. We must continue to develop technology to locate components of map units based on our knowledge of where they occur in the landscape and display this on a GIS. Perhaps digital elevation models will help but most of this data is at 30-m intervals and the resolution is not adequate for soil specific farming.

Another possibility is to show a miniature block diagram along with the digital soil maps to array soil components in a polygon to specific landscape positions. The landscape diagram would be displayed for each delineation as the applicator travels across a field. Fertilizer and chemical applicators could adjust rates as they visually identify these landforms or landscape positions similar to what farmers have done for years. This information could be digitized when developing the detailed onsite survey and would be instrumental in designing the amount and placement of grid sampling sites.

Technology is available (hardware and software) to overlay the digital imagery data with digital soil map and attribute data. We need to continue research in remote sensing techniques that will allow correlation of digital imagery data to specific components within a delineation or polygon in a GIS. Perhaps using digital elevation models (DEMs), digital imagery, and landform/landscape models together will eliminate or greatly decrease the need to make grid soil maps. One of the ironies is that soil scientists want leaf-off imagery, while leaf-on imagery may be most useful in soil specific farming.

Emerging technology in GIS, known as the 'S' language, may be useful in extending point data, such as fertility analyses. The process makes complex analysis transparent to the user. This technology may be extremely useful in extending point data and may be an alternative to geostatistical approaches or congruent with them.

Environmental Uses

It is only logical that if farmers can use soil specific information, the public ought to be able to use soil specific information to regulate or provide guidance for the use of agricultural chemicals and soils in the disposal of agricultural or nonagricultural wastes. The philosophy of soil specific farming can also be the philosophy of an environmentally sustainable farming system. Sustainable with respect to the soil resource and the environment. Regulation is shunned by most of us and abhorred by the farming community. We prefer voluntary programs driven by the stewardship ethic. However, we must also deal with the realism of the dwindling patience of the environmentally minded public.

Soil survey information is now being used to evaluate the bids received from farmers for the Conservation Reserve Program (CRP). Productivity of the soil(s) in the bid is used to evaluate the environmental benefit relative to program cost (Resources and Technology Staff, 1991). We are working with a group of scientists to develop a soil productivity index driven by the basic properties of the soil. The Agricultural Stabilization and Conservation Service (ASCS) is very interested in using this index to evaluate productivity in all of their programs. This group has set three conditions that the index must satisfy;

it must be defensible, doable, and durable. This may not be soil specific farming per se but it may be useful in an economic analysis to maximize profits based on the potential of the soil to produce vs. cost of chemical amendments.

Hydric soils, are one of three criteria used to identify wetlands. Location of hydric soils must be soil specific, which often means onsite visits to locate inclusions of hydric soils in some delineations. Our agency is using remote sensing techniques along with soil surveys to make offsite location of wetlands. Offsite techniques include the use of the ASCS compliance slides to locate areas of surface water. The slides are not as useful for saturated hydric soils. The ASCS compliance slides may represent a data layer for soil specific farming if they are available to the producers.

Water quality and risk assessment of potential nonpoint source pollution is an active consideration. Erosion from cropland is recognized as a serious source and carrier of nonpoint source pollutants. Soil survey information supplemented with information on the vadose zone, regional aquifers, and GIS technology of overlaying soil information on DEM projections of landscape will be invaluable in assessing potential nonpoint pollution problems. The landscape projections are especially helpful in screening hot spots, such as sloping areas next to streams that may be particularly susceptible to runoff pollution problems. Membership functions and possibility statements from fuzzy set theory will be needed to express reliability of our data and for use in environmental risk assessment.

Recent legislative proposals for application of municipal wastes on agricultural land rely heavily on soil properties. These proposals also specify that soil scientists will determine the rates of application and areas suitable for application. If this legislation is enacted, we will rely on soil specific criteria, perhaps similar to the hydric soil criteria, for suitable sites. Application rates must also be tied to soil properties.

SUMMARY

Soil surveys in the NCSS are multipurpose inventories of the soil resource. They are useful for planning on-farm resource management systems, locating least cost routes for highway construction, and urban and other planning activities. Soil surveys are not designed for site specific applications such as soil specific farming.

Soil survey information is useful in determining the potential for various cropping systems and provides the basic information needed to determine optimum seeding rates, irrigation scheduling, and pesticide application rates. The basic soil surveys are also technology transfer mechanisms for extending expensive-to-measure soil properties and designing efficient sampling schemes for soil fertility testing.

Cooperators in the NCSS need to continue to research methods of capturing information on occurrence of soils within map unit delineations and for transferring the information to users. They must develop reliability statements on the information via fuzzy set theory or other appropriate means.

The NCSS is the focal group for conducting research on sampling schemes for soil specific farming and transferring the information to the user. The actual collection of soil specific information should continue to be a private sector activity using technology developed by the partners of the NCSS.

REFERENCES

Bechman, Tom J. 1992. Farming by soil type cuts costs, boosts profits. Prairie Farmer (Jan. 7) 1992:12-13.

Brown, R. B., and J. H. Huddleston. 1991. Presentation of statistical data on map units to the user. p. 127-147. In M. J. Mausbach and L. P. Wilding (ed). Spatial variabilities of soils and landforms. SSSA Spec. Publ. 28. SSSA, Madison, WI.

Burrough, P. A. 1989. Fuzzy mathematical methods for soil survey and land evaluation. J. Soil Sci. 40:477-492.

Burrough, P. A. 1991. Sampling designs for quantifying map unit composition. p. 89-125. In M. J. Mausbach and L. P. Wilding (ed.) Spatial variabilities of soils and landforms. SSSA Spec. Publ. 28. SSSA, Madison, WI.

Carr, P. M., G. R. Carlson, J. S. Jacobsen, G. A. Nielsen, and E. O. Skogley. 1991. Farming soils, not fields: A strategy for increasing fertilizer profitability. J. Prod. Agric. 4:57-61.

Cartographic Staff. 1980. National cartographic manual. USDA Soil Conservation Service, Washington, DC.

Holmes, Bill. 1992. Prescription farming in Missouri. Oral presentation at SCS State Soil Scientist Workshop, Mobile, Alabama. 10-14 Feb. 1992.

Kellogg, Charles, E. 1955. Soil surveys in modern farming. J. Soil Water Conserv. 10:271-277

Kellogg, Charles, E. 1961. Soil interpretation in the soil survey. In service publ. USDA, Soil Conservation Service, Washington, DC.

Mausbach, Maurice, J., David L. Anderson, and Richard W. Arnold. 1989. Soil survey databases and their uses. p. 659-664. In Joe K. Clema (ed.) Proc. 1989 Summer Computer Simulation Conf. Austin, TX. 24-27 July 1989. Sponsored by the Soc. for Computer Simulation.

Munson, Robert D., and C. Ford Runge. 1990. Improving fertilizer and chemical efficiency through "high precision farming." Center for International Food and Agric. Policy, Univ. of Minnesota, St. Paul.

Reichenberger, Larry, and John Russnogle. 1989. Farm by the foot. p. 11-15. Farm J. (Mid-March) 1989.

Resources and Technology Staff. 1991. Agricultural resources - Cropland, water and conservation - Situation and outlook report. USDA Economic Research Service, Report AR-23 Agricultural Resources and Situation and Outlook, Sept., 1991, Rockville, MD.

Reybold, William U., and Gale W. TeSelle. 1989. Soil geographic data bases. J. Soil Water Conserv. 44:28-29.

Richter, Steve. 1991. Applying plant food by bits and bytes. Cooperative Partners (March/April) 1991:1-4.

Soil Survey Staff. 1975. Soil taxonomy: A basic system of soil classification for making and interpreting soil surveys. USDA Soil Conservation Service Agric. Handb. 296. U.S. Gov. Print. Office, Washington, DC.

Soil Survey Staff. 1983. National soils handbook. USDA Soil Conservation Service, Washington, DC.

Soil Survey Staff. 1992. Soil survey manual. USDA Soil Conservation Service, U.S. Gov. Print. Office, Washington, DC (in press).

Stein, A., M. Hooderwerf, and J. Bouma. 1988a. Use of soil-map delineations to improve (Co-)Kriging of point data on moisture deficits. Geoderma 43:163-177.

Stein, A., W. van Dooremolen, J. Bouma, and A. K. Bregt. 1988b. Cokriging point data on moisture deficit. Soil Sci. Soc. Am. J. 52:1418-1423.

Zadeh, L. A. 1968. Fuzzy sets. Information Control 8:338-353.

5 WORKING GROUP REPORT

C. S. Holzhey, Chair
H. R. Sinclair, Recorder
E. G. Knox and B. Hudson, Discussion Paper

Other participants: J. Bouma   S. Rawlins
                     I. Moore   P. Robert
                     D. Mulla   J. Wilson

IMPORTANCE OF SOIL VARIABILITY

The practical objective is to identify/delineate meaningful management zones within fields in order to optimize resource use. The purposes for optimization will necessarily incorporate economic, efficiency and environmental considerations in balances that will differ with the specific management decisions and physical settings.

Natural resource variability originates from differences originating from geological and pedological processes, as indicated on existing soil survey maps. In addition, management procedures may have a lasting effect on soil structure and on nutrient contents to the effect that identical soils within a given phase of a soil series can have quite contrasting properties. Morphological descriptions of soil structure can provide important clues as to the physical properties that have resulted from different management procedures. Chemical properties, however, have to be deduced from laboratory measurements. Aside from natural and management factors, variability of soil properties in time is determined by weather conditions which dictate water and air contents and soil temperatures at any given time.

To explain observed soil variabilities in the field, an understanding of the effects of natural, management and weather induced processes on soil variability is necessary.

The analysis of variability should be focused on parameters that are of prime concern to the user. Here, use of land qualities (defined by FAO as attributes of land which have a distinct function on the performance of land for a defined type of land utilization) is recommended. For agricultural land utilization, important land qualities are soil moisture availability, soil nutrient status, nitrate leaching potential. Process-oriented simulation models which integrate nutrient fluxes in the soils and crop growth, are suitable to determine

---

Copyright © 1993 ASA-CSSA-SSSA, 677 South Segoe Road, Madison, WI 53711, USA. *Soil Specific Crop Management.*

these land qualities. Calculations should be made for points in the landscape followed by interpolation to areas of land as represented by delineated mapping units on the soil map.

The group at once recognized the holistic nature of the topic, pointing out that, even excluding topics assigned to other groups, the responses of crops to soils, and to management practices, are dependent on interactions of plant, soil, other physiographic factors, climate, and variable weather events. There was recognition of the necessity to isolate soil specific issues, both in the planning of approaches, and in experimentation, but that it must be done in the context of the whole system.

Although much of the current soil specific farming is calibrated to nutrient status, organic matter, and pH, significant soil variables also include other chemical properties, and an array of physical properties (especially those that effect water storage, and transmission). It will be important in many settings to incorporate such information into information bases.

Soil Survey and Grid Sampling Approaches

With respect to site specific crop management, the fundamental concern with soil variability is to partition it into relatively homogeneous soil management zones of size and shape appropriate for application of differential management within a field.

Soil survey maps provide information on surface features and the properties of the soil profile to a depth of 2 meters based on the correlation between profile properties and surface features that can be observed in the field and detected by remote sensing.

Map unit delineations of the soil survey range down in size to about 1, 2, or 3 hectares, depending (Knox and Hudson, discussion paper) on the map scale. Attribute data provided for components of map units (commonly based on phases of soil series) include soil texture, bulk density, Atterberg limits, cation exchange capacity, estimated saturated hydraulic conductivity, computed erodibility, and others, and continues to be augmented to supply information to drive new models for erosion and water quality applications.

These attributes give little or no direct information about nutrient availability and no direct information on seasonal changes (in bulk density or crusting, for example) or in conditions resulting from management (a tillage pan, for example).

Attribute data for soil map units include properties that relate to the land qualities that control the movement of water in the soil and on the soil surface, the supply of water to plants, and the general chemical environment of nutrients. Hence, **they are used to estimate yield potentials for common crops. Soil map unit delineations can therefore be used to establish management zones within a field.**

Soil tests for nutrients, at least in surface layers, show large variability within soil map unit delineations or even without respect to map units.

Variability of this sort can be detected by systematic grid sampling. Papers presented at this workshop show strong variability at whatever the scale of the grid with some irregular tendency for clumping.

Where grid sample results for a nutrient are spatially aggregated into field segments of appropriate size and shape, these segments can be used as management zones for application of that nutrient for one or more years. **Optimal grid size can be investigated with geostatistical techniques (Mulla, this proceedings). Grid size and management area may differ with both kinds of nutrient and management practices.**

The advantage of soil map unit delineations as the basis for management zones is that soil maps are readily available for most of the country and they show the location of soils defined in terms of soil properties of the whole profile, that, interpreted into important land qualities, are the basis for the selection of crops and crop rotations and a whole range of land and crop management practices. Soil map delineations can be fine-tuned by more detailed integration of airphoto interpretation or remote sensing. The disadvantage of soil map units is their internal variability in surface soil nutrient test results.

The advantage of grid sampling as the basis for management zones is that it identifies areas of uniform nutrient test results as the basis for selection of fertilizer rates from response curves. If grid sampling is restricted to nutrient tests, then the management zones are applicable only to fertilizer applications for one or a few years. If grid sampling includes observation of the whole profile, sampling of more than the surface layer, and analysis of more than nutrient levels, then it becomes quite expensive.

OPERATIONAL LEVELS IN DEALING WITH MANAGEMENT ZONES WITHIN FIELDS

Five operational levels were identified to focus discussion of issues at the appropriate level for each (Fig. 5-1). The levels are based on operational criteria. The low levels use available data, while increasingly sophisticated methodology is added at higher levels to allow better delineation of management zones. The group recognized sizeable differences between: (i) procedures that would be technologically feasible in the future, or even in the present for some operations, and (ii) those that might be practical in the present for the majority of producers. Perhaps largest among these are the issues of computerized information storage systems, records of soil tests, and production, familiarity of the producer with the observable spatial variabilities of fields, the capacity to invest in grid sampling, and the degrees of technological complexity used by consultants. There are several issues at each level of complexity, ranging from relatively simple compilations of existing information to long-term research and development to build tools for the future. The five levels of this report encompass the full range, recognizing that information flows from level to level, and that the ultimate purpose is to enhance soil (or site) specific decisions.

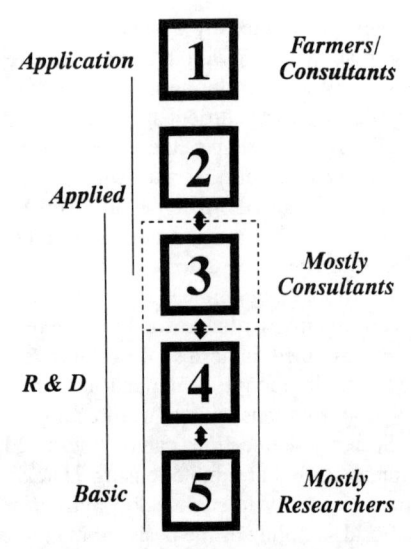

Fig. 5-1. Operational Levels in Dealing with Management Zones Within Fields

Level 1

Step 1 is to use existing knowledge to delineate management zones, then to treat each management zone within the field differently. This approach has long been used by farmers who would spread extra manure on the least fertile parts of fields, would fall plow only the more clayey bottom lands, drain only wet areas, restrict corn to the soils with adequate water storage for the climate, or terrace the more erosive slopes.

The expectations at this level are that the farmer knowledge and soil survey maps is the basis for establishing management zones, and that sampling for fertility tests is done within zones. At the least expensive level, samples from each zone would be composited, or with greater cost, locations would be recorded and samples analyzed separately.

Statistically, 12-15 samples per management zone has been recommended as adequate in some settings.

Issues

The optimal number of samples within management zones in different landscapes needs further investigation.

A number of studies show that fertility management zones are often different from the zones delineated in this fashion. An issue is which predeterminable soil differences are significant from a fertility standpoint. Level 3 research and development would be expected to address this issue.

Level 2

This level involves grid sampling by field to establish management zones. This is the operational level that is growing rapidly among consulting groups. The delineated management zones for nutrients are transitory and are constructed whenever necessary for application decisions. Soil surveys and other knowledge sources are used by producers as supplemental information for management decisions, but maps created by consultants, from the grid information, are the sole or chief basis for establishing levels of chemical applications in the corresponding management zones for each nutrient.

Issue

The appropriate sampling strategy, in terms of observation density, tailored to specific soil associations and landscapes for the various nutrients.

Level 3

This level integrates spatial information from a variety of sources in a decision support system, including soil survey maps, remote sensing images, air photos. Management zones are created from combinations of the spatial data, and data from grid or other statistically reliable sampling and testing procedures.

Soil surveys and other knowledge sources are used by producers for estimating yield potentials, for determining appropriate fertilization at specific nutrient levels, and for establishing a variety of management practices. Maps created by consultants, from the grid information, are the sole or chief basis for establishing management zones for each plant nutrient.

Leading edge consultants already have much of this capability, and that in the future, regional mapping centers might provide the management zone information to other consulting groups and producers.

Geographically rectified (georeferenced) maps are recommended. It was noted that information such as soil survey maps will be digitized in the future, along with soil attribute data, but that few georeferenced soil survey observations will be available for some years, whereas the soil test results from grid or other sampling will be georeferenced. By contrast, soil survey data provide a broader array of soil attribute information over a greater soil depth. The group favored the continuation of data storage by soil horizon wherever possible, and georeferencing of observations. Several participants favored research to optimize sampling and management strategies, using combinations of ground truth, existing knowledge, remote sensing, and existing data bases. This would be done separately for a variety of soil associations, physiography, and climate. Such information might be used in part in Level 2, but would generally require more information than listed for that level.

We recognized that sampling strategies can be adjusted to the different kinds of patterns and contrasts, as effected by management. It was noted that management variations sometimes create larger variations within one kind of soil than occur naturally between some soils. **Research and development from Level 4 could provide such generalizations to be incorporated at this level.**

Issues

Development of decision support systems capable of utilizing the various kinds of information (soils are only part of the story) and overcoming problems of different levels of detail among different information sources. Levels of detail become a concern when information from a variety of sources is overlain in a support system. The group felt that scale problems are manageable, but that levels of reliability related to detail should be recorded with the information.

Ways to create improved combinations of sampling, ground truth, expert knowledge, existing maps, and remote sensing.

Level 4

This level is the research and development, using existing tools and long-term observations, to develop management strategies that take into account basic soil variability as modified by management and the ephemeral variations driven by climate and weather. The knowledge gained helps in the development of models (Level 5), and in the strategies applied by consultants and others at the practical level. It is significant to both farm operations and environmental considerations.

This level covers strategic research and development, using existing research tools and long-term observations, either on location or by remote sensing. The objective is to take into account basic soil variability as modified by management and the ephemeral variations driven by climate and weather. Long-term observations of an area by remote sensing allow distinction of subareas with a consistent optical response. These areas may or may not correspond with mapping units of the soil map. Sampling in the field and exploratory use of models is needed to explain the behavior of crops in the subareas and to determine whether these areas qualify as management units. The knowledge gained helps in the application of models (Level 5), and in the strategies applied by consultants and others at the practical level. It is significant to both farm operations and environmental considerations.

The group again emphasized a holistic approach to the study of the systems in the field, and the potentials of existing technologies, such as geographic information systems, digital elevation models, repeated observations through remote sensing, field observations of soil conditions relating to hydraulic and hydrologic soil attributes, Weather Service, USGS, and SCS data, accumulations of georeferenced, chemical and physical point observations. Also mentioned was the importance of understanding and modelling variability in soil properties and processes, using landscape models.

Issues

Many of the unexplained variations in fields are artifacts of past management. **Some are consequences of random variation, whereas others would be predictable with better understanding of the long-term responses of soils, and the influences of those responses on such things as insect populations, crop diseases, the root environment and so on. Although the variations require holistic approaches to explain completely, it is possible to evaluate variations in specific soil attributes through changes in weather and management, and** risks associated with a given decision are not known until long-term responses of the soil in its specific environment are known. This level feeds information to Level 5 that assists in modelling. It feeds information to management strategies for Level 3. The key is long-term observations to document ephemeral and management influences on variability and on soil responses.

Enhancements to the decision support system of Level 3 will be required as knowledge is gained and new types of information become operational.

Level 5

This level covers research and development to build new tools, but only when tools tested at lower levels prove to be inadequate. It includes integrated modelling of crop growth and nutrient fluxes, measurement of physical properties, and influences of management on soil variation. Simulation models and statistical tests are used to determine sensitivity of crops and environmental considerations to soil attributes.

New tools besides improved simulation models and sensitivity analysis include improved statistical techniques to create inferences, to interpolate between observations, new remote sensing technologies, application of such techniques such as ground-penetrating radar, electromagnetic and nuclear magnetic resonance sensors, alternative fertilizer application techniques, better modelling of pesticide/herbicide interactions with soils, establishment of intelligent information bases, and equipment on farm machinery to sense key soil attributes as equipment crosses fields.

Issues

Yield information is an important factor in determining the economic success in the selection of management zones and optimization of management decisions. Yet, yield information is in short supply and not easy to obtain in a management-by-soil setting. When it is obtained, short-term monitoring is subject to vagaries of weather, plant diseases, management, and other factors. Hence, quality simulation models, calibrated to the long-term observations of Level 4, are important in evaluating alternatives.

SECTION II

MANAGING VARIABILITY

6 Some Practical Field Applications

Robert E. Ascheman
Ascheman Associates
2921 Beverly Drive
Des Moines, IA

We are at the threshold of a major breakthrough in the utilization of fertilizer, pesticides, and other inputs in U. S. agriculture.

Just a few years ago the dream of using equipment that would automatically apply fertilizer or pesticides at varying rates in farmers' fields was considered futuristic. How to get these machines to profitably apply inputs according to some predetermined need is the subject of this chapter. This technology is no longer based on a Buck Rogers or Star Wars idea.

The equipment may still be Mercedes or Cadillac rather than Chevy or Ford priced at present. However, the computerized components of these systems are more like the pricing of Apple and IBM PCs and are likely to soon be reduced substantially even while performance is being enhanced. The really encouraging aspect of this technology is that agriculture alone will not be paying all of the costs of development. The systems that control these devices are most likely to become standardized. They will use generic or "off-the-shelf" software where available and usually not require agriculture-specific programming.

SITE-SPECIFIC MANAGEMENT

This chapter is presented from the viewpoint of an agricultural consultant and particularly from the aspect of providing advice for a fee directly to growers about crop and soils management on their farms. Our livelihood depends directly on providing "site-specific," cost-effective information for their farm(s), their individual fields, and where feasible at sites within those fields. Management recommendations for the different sites within fields are often soil-type related, but almost always include other parameters as well. These other parameters usually are the result of human activities causing differences in past management; for example, manure or fertilizer applications, drainage, crops grown, or weed and pest control practices.

Copyright © 1993 ASA-CSSA-SSSA, 677 South Segoe Road, Madison, WI 53711, USA. *Soil Specific Crop Management.*

These differences within soil types are often greater than the differences between soil types. Thus when addressing the topic of "Managing Variability" in crop production, we prefer to use the term "site specific" rather than "soil specific."

The first prerequisite for managing variability in crop production is to determine what the variability is for some agronomically important characteristic.

In order to be a useful tool in managing variability, that agronomic characteristic referred to above must be one that is cost effective, reliable, reproducible, and specific or detailed enough to make decisions.

Two good examples of tools already available are soil fertility testing and crop scouting or monitoring data. Discrete, definitive data can be obtained with these tools that are the basis for generally reproducible and reliable recommendations. The more site specific the information, the greater the probability of favorable results. When information obtained through use of these and other tools is collected in a suitable format, it can become the database upon which computerized recommendations are made. If suitable and readily available formats are used, then the application of fertilizer and pesticides can to a large degree be automated and documented. The hardware and most of the software is already available or being adapted to these uses.

In the past, the cost, timeliness, and massive amount of data generated made site-specific management difficult. Recently however, mostly in the last 6 to 12 months, some substantial developments in technology have become available to agriculture that overcome or at least minimize three limiting factors: cost, timeliness, and the massive data handling requirements. They include:

GPS	-	Navstar Global Positioning System
GIS	-	Geographical Information System
VRT	-	Variable Rate Technology, (pioneered by Soil Teq Inc.) and several computer programs to use this information.

This conference is devoted to studying the various aspects of this new technology and its application to crop production.

The rest of this segment on "Managing Variability" will address yield goals (mapping), tillage, and pest variability.

The rest of this chapter will deal with some personal observations and a discussion of the applications of this concept in our own consulting practice and some recommendations for further research.

GRID SAMPLING

Our firm's concept of site-specific crop management began with "Grid Sampling" of soils in 1979. During the intervening 12 years some of the details were modified slightly, but the concept has remained essentially the same and is described. (For details, write to the author.)

It is important to recognize that the new satellite technology or even the use of computers is not essential to "get started." Massive amounts of data accumulate and soon become a burden and of much less value if not readily available and preferably in a format convenient for use.

Our own experience with computerizing soil fertility data during the past 10 years illustrates the desirability of developing a format that permits rapid retrieval and analysis of data. By developing a grower, farm, field, and block numbering system that also includes township, range, and section numbers, we were able to assemble, store, retrieve, and sort soil fertility data on computers even before we had the capability of producing computerized maps.

Linking these functions, database, and mapping, however, proved to be a real challenge. Through the use of a commercially available software program, data can now be sorted, stored, and collected from numerous sources and through an integrated auto-cad function, produce computerized detailed colored maps that show visually numerous relationships not otherwise readily detected. Fortunately, the program stores data in a GIS format that is also compatible with the Navstar GPS system. Thus data is imported, integrated with our own soil fertility or crop scouting data, analyzed, and provided to the grower client with rapid, accurate, and detailed information and recommendations.

This information on where to apply the fertilizer or pesticide product could also be provided to the applicator in the format compatible for use on variable rate technology (VRT) applicator equipment.

Hopefully, VRT equipment for fertilizer and pesticide application and that being developed for yield evaluation "on-the-go" will be compatible and a uniform standard developed. A module that can be moved from one implement to another has been proposed and is a likely development.

When farmers are presented with colored, computerized maps, they have much greater capacity to visualize and understand the severity and location of areas in their fields that need special attention. Farmers intuitively know that it is profitable to put fertilizer or pesticides where they are needed and to not put them where they are not needed. This usually results in greater yields in areas (or blocks) that were previously undertreated and savings in areas that were overtreated. "Made and Save" calculations on thousands of acres indicate in excess of $10/acre per year above the cost of the service.

By using information from multiple databases (i) soil types from a GIS file, (ii) insect infestation levels from crop scouting data in a Paradox-generated and compatible file, and (iii) the leaching and runoff potential of candidate insecticides from a D-Base compatible file, it is relatively quick and easy to make plain English queries in Paradox and to select the safest and most effective insecticides on numerous soil types and pesticide combinations.

This is sustainable agriculture on a practical field scale basis, not the reduction of inputs only, but rather the choice of the safest and most efficacious product and use only where needed based on all available information.

Grid sampling for soil fertility was the basis for Ascheman Associates "getting started," making possible a much more effective soil consulting practice and continues to be the best developed component of our consulting efforts.

Some examples of soil fertility management follow:

Phosphorus Management
- On high pH soils, especially in Clarion, Nicollet, Webster soil association with wide pH variability use Bray P1 test on soils of pH 7.2 or less and change to Olsen Bicarbonate test when pH exceeds 7.2.
- On soils with pH more than 7.8, omit broadcast P treatments and change to deep placement bands or dribble bands of high phosphate starter fertilizer — especially where several blocks have high pH soils.
- Apply broadcast P correction treatments where indicated once every four years.
- Consider no P (except possibly starter) on very high P soils.
- Consider no P (except possibly starter) on high P soils for one of two application dates.
- Use high P starter only if most of blocks are very high or high P (if correction treatments were made).
- Generally apply starter if feasible and at low rates in northern Iowa and cooler soil areas. Increase rate if many blocks are medium or lower.

Potassium Management
- Apply K broadcast correction treatments where indicated once every four years.
- Do not exceed 200 lb/A of K_2O on sandy soils with a cation exchange capacity less than 10 meq/100 g.
- Generally apply P and K broadcast every other year in a corn-soybean (*Zea mays* L. - [*Glycine max.* (L.) Merr.]) rotation.
- Apply K broadcast only if adequate starter is used. Apply only once every four years if K test levels are mostly high or very high and corrections treatments were made.
- Use higher K application rates for no-till or ridge-till planting.

pH Management
- Apply lime where indicated, usually at 2000lb/A of calcium carbonate equivalent (CCE) assuming conservation tillage.
- Lime fields for corn or soybeans only when pH is less than pH 6.
- For alfalfa use greater amounts of lime and correct to pH 6.8. Select fields for alfalfa one or two years in advance and apply lime as needed.
- Generally avoid liming blocks that are above pH 6.5 and especially avoid liming high pH alkali spots (pH 7.8 or greater). Apply lime only when little or no soil compaction will result due to wet soils. Incorporate lime to speed up pH neutralization

process but avoid or minimize tillage on 5% or greater slopes.

Manure Management
- On high priority livestock farms use manure to build P and K values but due to lack of uniform application and risk of substantial N loss from surface applications, be cautious about optimistic credits for N due to high cost of downside risk of yield loss. Consider late spring N test where appropriate.
- Use manure as principal source of P and K maintenance and also for correction treatments if feasible.
- Consider cost of spreading manure as a charge against livestock operations rather than against the crops.
- Avoid soil compaction by not spreading when soil is wet; use set-aside land where feasible to permit manure spreading on drier soils.

Nitrogen Management
- Apply N-P combination fertilizers (especially starter) for corn to maximize P uptake.
- When broadcasting diammonium or monoammonium phosphate (DAP or MAP) or other ammonium or urea containing N fertilizers, incorporate into the soil to minimize losses of N. Avoid or minimize tillage on soils with 5% or greater slopes.
- Use late spring nitrate test where applicable.
- Side dress part of the N if equipment and conditions are feasible.
- Be realistic in establishing yield goals especially to avoid excessive N use on blocks within fields that have lower yield potential.
- Give credit for legume N being conscious of uniformity and extent of legume in the previous crop.

Secondary or Micronutrient Management
- Make pH corrections before spending money for micronutrients.
- Consider S soil test on blocks with sandy soil.
- Consider Zn soil test when pH and P levels are both high.
- Consider B soil test on sandy soils especially for alfalfa.
- Utilize micronutrients in starter if feasible and needed rather than broadcast. Be aware of crop injury risk with pop-up fertilizers with or without micronutrients.

MODELING AND SIMULATION

Through computerization of this soil fertility data, it has been possible to make some computerized recommendations for certain frequently occurring situations, thus allowing more time for special problems and ensuring that the grower has a written record of the recommendations.

These computerized responses to frequently occurring situations, either automatically or through "boiler plate" statements, in effect become simple models.

Modeling and simulation of crop monitoring and crop scouting data would appear to be likely methods of anticipating the results of given insect or weed infestations when left untreated or given specific pesticide treatments with different timings, rates, and additives.

Assuming that these crop monitoring records are uniform in format, it also appears likely that they could become the basis for new and improved recommendations. Incorporating the influence of weather from an additional database could further refine the process of selecting the most cost effective and environmentally safe pesticides for particular situations.

Differences in costs of the pesticide application, value of the crop at harvest, etc., could be entered and the effect readily be determined. Then through the mapping function, visual evaluation of the projected outcome could help determine the feasibility of treatments. Common sense will surely almost always be needed to verify the feasibility of treatment (and how to best accomplish it), but the time and effort in doing so should be greatly reduced.

RESEARCH NEEDS

Several recommendations for increased research needs, especially in soil fertility follow. It becomes apparent after some experience with soil sampling on the GRID SYSTEM that some of the old parameters about soil sampling techniques need to be modified or at least reevaluated. They are itemized as follows:

Soil Sampling Techniques

A. Monthly variation
   Study amount of variation of P, K, pH, N, and organic matter (OM).
   Study influence of, and time of, previous fertilizer or lime applications.
   Study influence of soil type.
   Study influence of fertility levels.
   Study influence of crop.
   Study influence of crop growth.
   Study influence of weather (especially rain-leaching K from soybean leaves).
   Are adjustments to a given standard month or date feasible?
   Best time of year for sampling based on (i) accuracy (precision), (ii) reproducibility, (iii) economics, (iv) usefulness to end user, (v) usefulness for others (consultant, dealer/applicator, and researcher).

B. Number of cores needed (and size of cores).
   Whole sample processing.
   Entire sample submitted to lab (accuracy and efficiency).

C. Sampling procedures
   Especially with band applications
   - Deep band, dribble, row-banded starter
   - Ridge-till
     - interval since last application
     - method of application (in band or broadcast)
     - Cultivation/ridging effect on broadcast application increasing test levels in ridge
     - Cultivation/ridging effect on "in-the-ridge" banding decreasing test levels in ridge
     - Knocking down ridges

   Nitrogen application - especially anhydrous between rows effect on pH (after multiple applications); not uncommon with continuous corn on erodible or highly erodible land.

D. Potash tests
   - Comparability and usefulness of several tests.

E. Micronutrient testing
   - Trouble shooting only? or
   - Develop a practical/cost-effective way to incorporate micronutrient testing into the specific soils management.

CROP SCOUT

A. Develop uniform format to facilitate data handling.
   - Feasibility for electronic data transfer from remote locations.
   - Perhaps establish regional standards for crop/pest relationships.
   - Establish uniform severity code (??)
   - Develop common terminology for crop growth

B. Evaluate critically weed population dynamics.
   - Improved recordkeeping capabilities makes more precision recommendations feasible.

   - Study and develop terminology for "patches" of weeds - annuals and perennials.
   - Competition studies - Establish criteria for measuring variability of weed populations: single species, groups of species, and combinations with like and unlike growth habits. Consider also relationships of weed and crop growth stages.

C. Reexamine thresholds for insect control when used in site specific situations.
- Consider insect movement and escapes.
- Study cost of spot treatment vs. benefits of broadcast including evaluation of risk.
- Integrate weather predictions for period(s) following pesticide application.
- Consider specific insect control recommendations for site specific management.

D. Study and prepare for using pesticide application techniques that permit change of materials, additives, rates, carrier volume, and application methods on-the-go. This kind of capability makes precision scouting more necessary and valuable.

E. Use the data obtained from scouting, compare the control obtained from pesticide application vs. untreated strips, document yields for on-the-go harvesting equipment, and establish procedures for collecting and analyzing data from these field experiments. This could greatly increase the amount and usefulness of on-farm research.

CONCLUSION

Dramatic breakthroughs in operating efficiency in agriculture and related agribusiness industries can be expected, as a result of using this new technology. At the same time it will contribute substantially to the sustainability of agriculture, both economically and environmentally.

This is an exciting time in the history of agriculture, similar to the introduction of power machinery, the discovery of hybrid corn, the discovery of 2, 4-D, and the general widespread introduction of agricultural chemicals. Site specific crop management is now at the cutting edge of applied science in food and fiber production. We are approaching or are now at the critical mass stage where logarithmic growth of this technology is about to start.

In my opinion, this is a great time to be an agronomist and agricultural consultant. I hope you share my enthusiasm for the concept.

7 Yield Mapping and Application of Yield Maps to Computer-Aided Local Resource Management

Ewald Schnug
Donal Murphy
Eric Evans
The University of Newcastle upon Tyne
Department of Agriculture
King George VI Building
Newcastle upon Tyne
NE1 7RU, Great Britain

Silvia Haneklaus
ILLIT - Institute for Agricultural and
  Ecological Innovations and Technologies
Luedemannstrasse 72
D-2300 Kiel 1, Germany

Juergen Lamp
Institute for Soil Science and Plant Nutrition
Christian-Albrechts-University
Olshausenstrasse 40
D-2300 Kiel 1, Germany

The need for coordination of economical and ecological demands require new perspectives in future plant production. At Kiel University a modular organized local resource management system called computer-aided farming (CAF) has been developed that enables farmers to consider the spatial variability of soil fertility parameters when designing pest control and fertilization as well as any other field work in their production technique (Fig. 7-1).

In practice, the management of local resources in agriculture commences with yield mapping. Thus yield maps are an essential part of CAF. They provide basic information for the setup of nutrient balances, the evaluation of equifertiles (which are defined as areas of identical or comparable soil fertility or productivity see Schnug et al., 1992) in the field and last but not least enable control of the efficiency of the whole system.

Copyright © 1993 ASA-CSSA-SSSA, 677 South Segoe Road, Madison, WI 53711, USA. *Soil Specific Crop Management.*

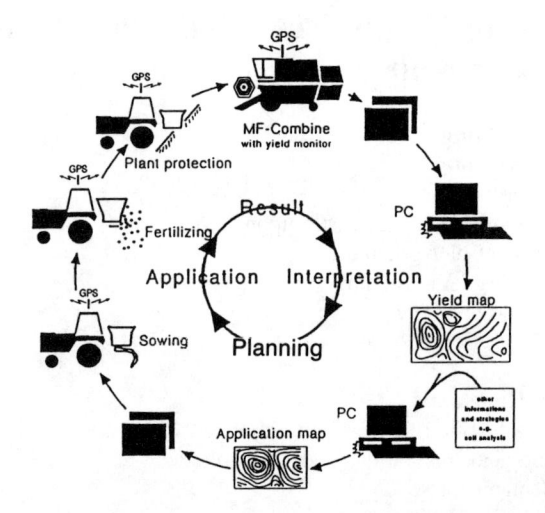

Fig. 7-1. Flow chart for computer-aided local resource management.

MATERIAL and METHODS

The results presented in this chapter derived from the 1991 harvest of wheat (*Triticum aestivum* L.) and oilseed rape (*Brassica napus* L.) fields belonging to farms situated in the German federal state Schleswig-Holstein and the British county Northumberland.

The combines were equipped with two different types of yield monitors. One was the Claydon yield-o-meter system that works principally on a volumetric base (for technical details one may refer to Searcy et al., 1989). The other one was the "Flow Control" system relying on the absorption of gamma rays by the grain flow (Thoustrup, 1991).

To localize the actual geographic position of the combine SHIPMATE and SEL-GLOBOS 2000 receivers for the GPS service were used (Global Positioning System, a free satellite service provided by the U.S. Department of Defense). The systems operated in more than 85% of the run time of the combines with an estimated positioning error of less than 20 m. Figure 7-2 displays an example of GPS monitored combine tracks in a wheat field of approximately 80-ha size.

The raw data coming from the yield monitors and the GPS system were recorded on board and transferred by chipcards to conventional computing systems (Fig. 7-1). Localized samples for grain moisture corrections were taken during combining.

By means of a specially designed software system data sets with erroneous yield or positioning values were screened out and by KRIGING, a geostatistical interpolation procedure (Matheron, 1971), the yield and moisture values for a regular spaced grid were calculated.

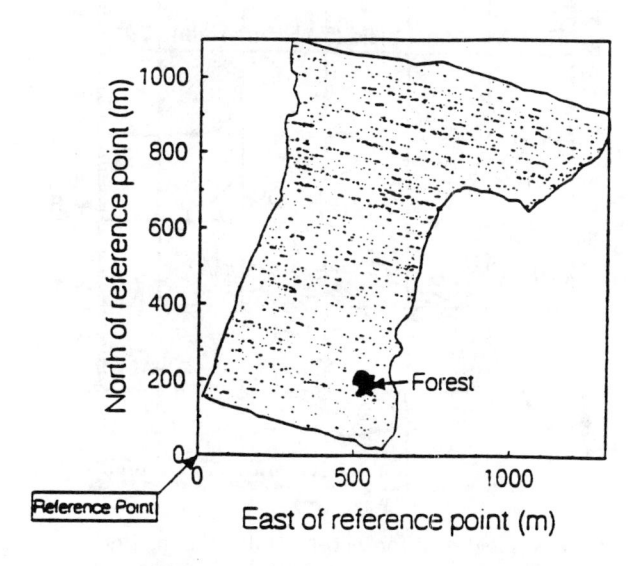

Two combines operating in tandem mode
5,000 recorded points plotted out of 30,000

Fig. 7-2. The GPS monitored combine tracks during wheat harvest - Pronstorf farm 1991.

YIELD MAPS AND THEIR APPLICATION TO COMPUTER-AIDED LOCAL RESOURCE MANAGEMENT

An example for a yield map is given in Fig. 7-3. Although fields in the landscape where the data were captured look fairly even and the growing crop appears uniform, the variability of the grain yield within the field is surprisingly high. When yield mapping has been carried out over several years, it is envisaged that the sampling of soils and crop can be simplified by establishing equifertiles that are defined as areas that have been found to have identical or similar productive capacity in space and time. Within equifertiles permanent monitoring points can be established to limit the sampling for relevant data.

To establish the technology for mapping spatially variable yield during combining seems to be fairly easy compared to the efforts needed to sort out the causes of the observed variability. For that task the yield maps need to be converted into Digital Agro Resource Maps (DARMs) which are part of the Local Resource Information System (LORIS) (Schnug and Junge, 1991). LORIS is a software specially designed for CAF and enables the processing of the yield maps (Fig. 7-4) together with already existing spatial information of a certain field (Fig. 7-5).

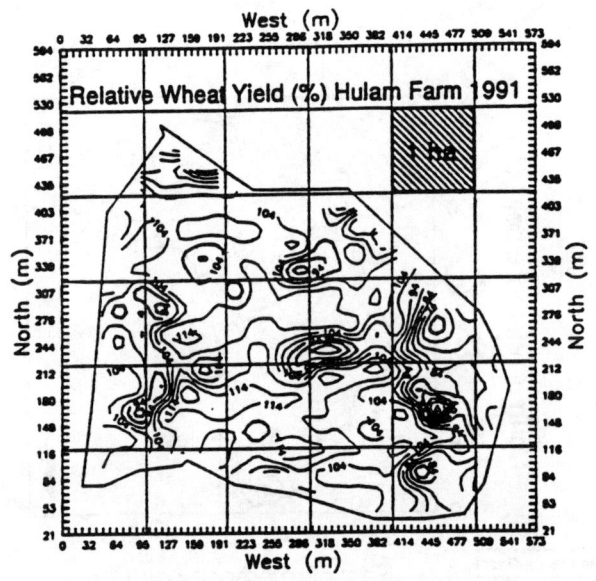

Fig. 7-3. Yield map for wheat - Hulam Farm, 1991.

Depending on the nature of the factors there are two main strategies to improve the efficiency of the production technique: In the case of naturally limiting factors (e.g., water stress) which cannot be managed by production technique, the optimum strategy will be to adapt the inputs as close as possible to the local productivity. If, however, the factors can be technically influenced (e.g., nutrient supply) adaption of inputs to the crops demand for highest productivity is recommended.

An example how the combination of yield data with only one extra information layer can be used for the optimisation of production is given in Fig. 7-4 to 7-7 where the evaluation of the optimum pH value for a field is demonstrated. The map in Fig. 7-4 shows the relative yield data of an oilseed rape crop (upper half) and a wheat crop together (mean yield for each crop = 100%).

Border line analysis as shown in Fig. 7-6 is an extremely powerful tool in determining the optimum of any factor, especially plant nutrients (for details refer to Schnug and Haneklaus, 1991). After spatial synchronization of yield and soil pH data, with this analysis, the values for the factor concerned, in this case soil pH, are plotted against the relative yield. The top line that fits to the uppermost boundary of the points plotted marks the maximum yield obtained at the site for each pH level. This line therefore marks the response to soil pH where all other factors are as near as possible within the limitations of the site. This type of analysis seems commonly to be more valid than the results of classical field trials where the response to increasing levels of a crop nutrient may be limited by other factors such as water stress resulting in under estimation of optimum requirements. The results given in Fig. 7-6 show that the highest

YIELD MAPPING 91

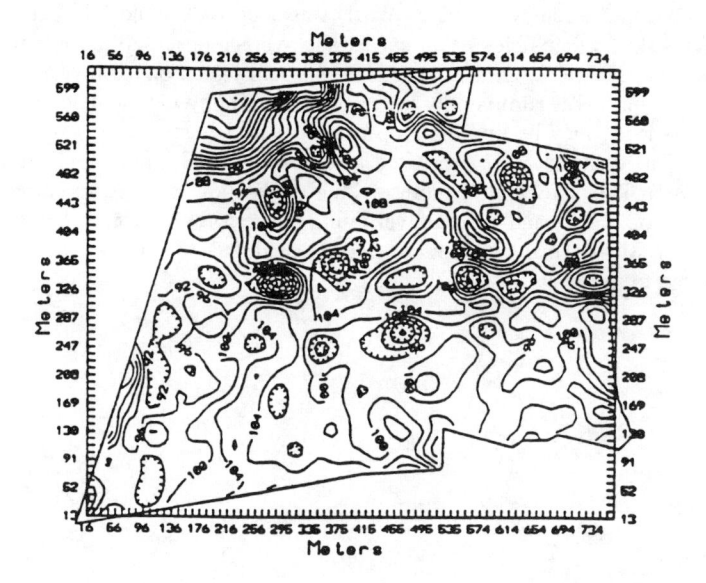

Fig. 7-4. Yield map for oilseed rape and wheat - The Shaw 1991.

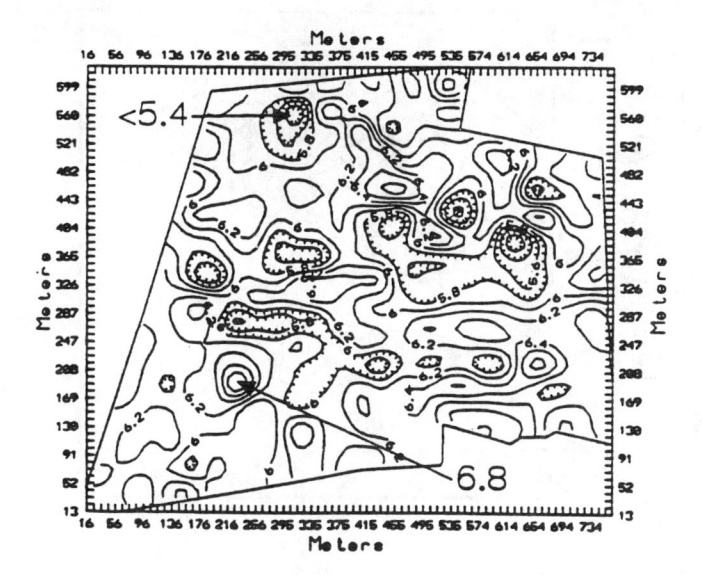

Fig. 7-5. Soil pH values - The Shaw 1991.

yields were only achieved where soil pH was between 5.5 and 6.1. The top line analyses shows that yields never exceed the average or are 20% lower than the highest yields where the pH exceeds 7 and below 5. For the investigated site this optimum pH is surprisingly low, and therefore provides some evidence that Mn deficiency may be limiting crop performance where soil pH exceeds 6.1.

The data sets were processed further to produce a lime application map (Fig. 7-7). This shows the pattern of lime application required to correct yield limiting soil acidity and prevent overliming where soil pH exceeds the optimum. The map shows that applications that exceed 0.5 t/ha CaO equivalent are confined to less than 20% of the area.

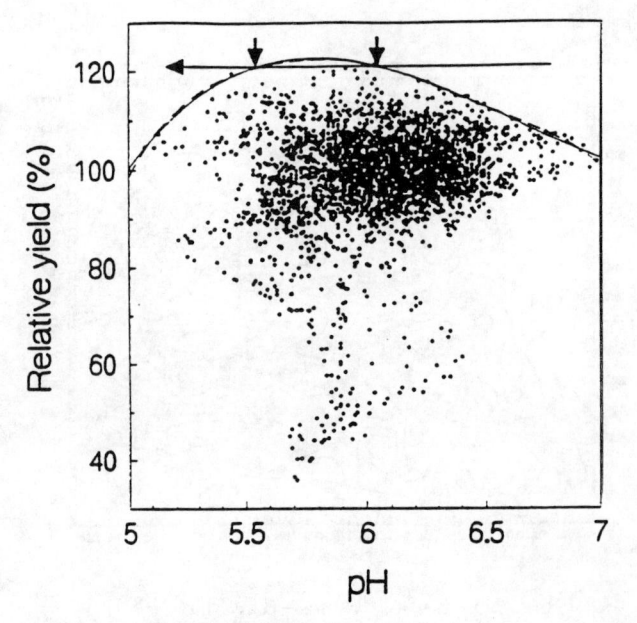

Fig. 7-6. Border line approach to evaluate optimum pH values - The Shaw 1991.

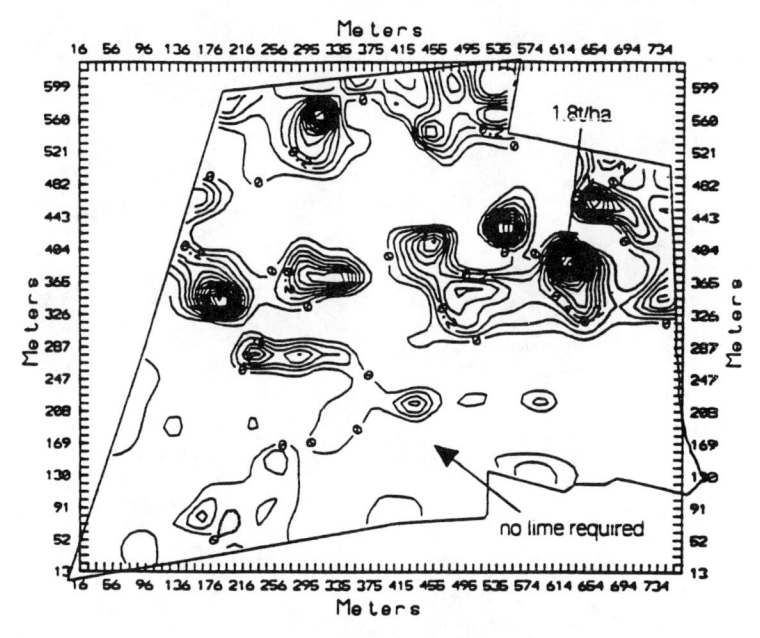

Fig. 7-7. Liming map - The Shaw 1991.

REFERENCES

Matheron, G. 1971. The theory of regionalized variables and its applications. Cahiers du Centre de Morphologie Mathematique de Fontainbleau No. 5. France.

Schnug, E., and S. Haneklaus. 1991. Evaluation of the nutritional status of oilseed rape plants by leaf analysis. Proc. 8th Int. Rapeseed Congress Saskatoon 1991 (in press). Saskatchewan, Canada.

Schnug, E., and R. Junge. 1991. Strukturierung des Interpretationsmoduls und Konzeption des LORIS (Local Resource Information System) für die Anwendung in 'Computer Aided Farming' (CAF). KTBL-Schriftenreihe 1991. Germany. (in press).

Schnug, E., D. Murphy, and S. Haneklaus. 1992. Importance, evaluation and application of equifertiles to CAF (Computer Aided Farming). Proc. VDI/MEG-Colloquium Agrartechnik "Ortung und Navigation landwirtschaftlicher Fahrzeuge". Munich, Germany.

Searcy, S. W., J. K. Schueller, Y. H. Bae, S. C. Borgelt, and B. A. Stout. 1989. Mapping of spatial variable yield during grain combining. Trans. ASAE 32: 826-829.

Thoustrup, A. 1991. Bestimmung von Ertraegen mit dem dronningborg flow control. KTBL-Schriftenreihe 1991. Germany. (in press).

8 Tillage Considerations in Managing Soil Variability

W. B. Voorhees
USDA-Agricultural Research Service
University of Minnesota
Morris, MN

R. R. Allmaras
USDA - Agricultural Research Service
University of Minnesota
St. Paul, MN

M. J. Lindstrom
USDA-Agricultural Research Service
University of Minnesota
Morris, MN

Recent developments in technology that allow "on-the-go" sensing of various soil characteristics have stimulated a whole new approach in agricultural production, often referred to as farming by soils (FBS). Most of the applications to date of FBS have been in the application of chemicals such as fertilizer and herbicides. The most obvious benefits from FBS are higher net income or less environmental pollution (due to better matching of inputs to the productivity potential of a soil.).

Another area of agricultural production that might benefit from FBS, but that has been almost totally ignored, is tillage and its associated changes in soil and plant environment. This chapter will address how knowledge of within-field soil variability can be matched with various options for controlling tillage-induced soil changes to provide for a better seed and root environment, and improved conservation of soil and water resources. The impact of tillage machine design and operation will also be discussed.

An often ignored aspect of tillage, but not restricted just to tillage operations, is soil compaction associated with the wheel traffic of almost every field operation.

The objectives of this chapter are to:

1. Review information available from soil mapping unit databases regarding soil physical properties that can affect tillage performance, or that can be altered by tillage to affect erosion or plant growth. Such factors include drainage, erosion potential, and soil structural properties.

Copyright © 1993 ASA-CSSA-SSSA, 677 South Segoe Road, Madison, WI 53711, USA. *Soil Specific Crop Management.*

2. Relate effects of various combinations of tillage-soil properties to plant residue burial, soil structure, and surface roughness, and how these relate to erosion control and plant growth.

3. Review interactions between tillage and wheel traffic- soil compaction, and their spatial variability impact on erosion and plant growth.

4. Discuss practical application of using the FBS concept in design and operation of tillage systems to sustain agriculture and conserve natural resources.

SCALE OF TILLAGE

The concept of managing tillage operations in response to soil variability is not new. The many different designs over time of the simple moldboard plow reflect the diversity of individual efforts to develop a tillage tool unique to some farm field or group of fields. Often these special efforts were subjective. But never-the- less, there was a recognition that all soils were not the same, and all tillage tools did not perform equally well in all soils. One can speculate that this same awareness applied to the selection of a simple stick to scratch a shallow seedbed in the earliest of times.

In a sense, agricultural mechanization dampened our awareness of soil variability as tillage and planting equipment became larger. Fields became larger in an effort to take advantage of the larger capacity of modern machinery. Along with that movement (which, incidentally, is still progressing), farm operators shifted their scale of thinking from meters to kilometers. One result has been farming systems built around the concept of uniform tillage and planting within a field, and often within a farm, both of which are now much larger, and therefore more variable with respect to soil type.

The shrinking profit margin in farming, increased environmental concerns as a result of all-out production, and increased knowledge of machine-soil-plant interactions have combined to focus recent attention on reversing the scale of thinking away from kilometers back to meters. Detailed soil surveys now encourage the farmer to again look at areas within a field. Technological advances allow real-time monitoring of soil variability. Thus, agriculture is again faced with the challenge and opportunity to find ways to use a stick to till the soil "that works better" than the current practices.

MAPPING UNITS - - TILLAGE RELATIONSHIPS

While farm managers have a general knowledge of what kinds of data are available from a detailed soil survey, these data are not commonly used to make decisions regarding tillage management. For purposes of this discussion, it will be assumed that tillage is performed mainly for incorporation or placement of soil amendments, pesticides and plant residues, or for modification of surface soil structure. Other tillage objectives, such as compaction alleviation and weed control, are no less important but are beyond the scope of this chapter.

Information contained in a detailed soil survey data base (such as the SCS Soils 5) can be used to predict or estimate soil response to a given tillage operation on a given soil. Soil characteristics such as particle-size distribution and water-holding capacity, or drainage classification can be especially helpful as shown in the following example.

In a field study comparing a loam and clay loam in western Minnesota, soils were moldboard plowed at various soil water contents and resulting surface roughness and change in total porosity were measured (Allmaras et al., 1967). The clay loam soil showed a nearly linear decrease in surface roughness due to plowing as the soil water content at time of plowing increased. This soil also exhibited a decreasing response in total porosity due to plowing as the water content increased. The loam soil, however, showed a curvilinear response to plowing, with the roughest soil surface and maximum porosity change occurring at either end of the soil water range. The inflection point occurred at a soil water content near the lower plastic limit. Thus, in terms of creating a certain soil condition for erosion control or water infiltration, both soil texture and soil water content need to be considered when performing a tillage operation. Both factors can be obtained or inferred from soil survey data.

Pidgeon and Soane (1978) reviewed the effects of different tillage methods on bulk density for a well drained sandy loam and a poorly drained sandy clay loam. As shown in Fig. 8-1, a given tillage operation resulted in a higher bulk density on the poorly drained sandy clay loam than on the well-drained sandy loam. Within a soil, different tillage operations produced different bulk densities.

Subsoil properties linked to soil mapping units have variable influences on tillage relationships; two of these are water content of the surface layer and the reservoir of plant available water. Internal drainage of soils in a drainage catena significantly influences surface water contents. Although soils are routinely rated into one of four permeability classes, saturated hydraulic conductivity is a critical parameter controlling infiltration and subsequent redistribution of water. Logsdon et al. (1990) and Coen and Wang (1989) have shown that saturated hydraulic conductivity of the subsoil can readily be estimated from soil survey knowledge of soil texture, area occupied by biopores, and ped characteristics using the pedotransfer function of McKeague et al. (1982). Wang et al. (1985) used this method to estimate changes in saturated hydraulic conductivity of the upper subsoil brought about by tillage and traffic. Rawls and Brakensiek (1988) used a range of soil information (in TAXONOMY database) to simulate infiltration using the Green-Ampt equation -- both surface and subsoil layers were included in the simulation. Many other algorithms are available to use simple soil properties for estimating the soil water content at various pressures to remove water from the soil (Rawls et al., 1991; Wosten and van Genuchten, 1988). These can be used with soil survey databases to estimate infiltration and drainage controls for tillage variations in a drainage catena.

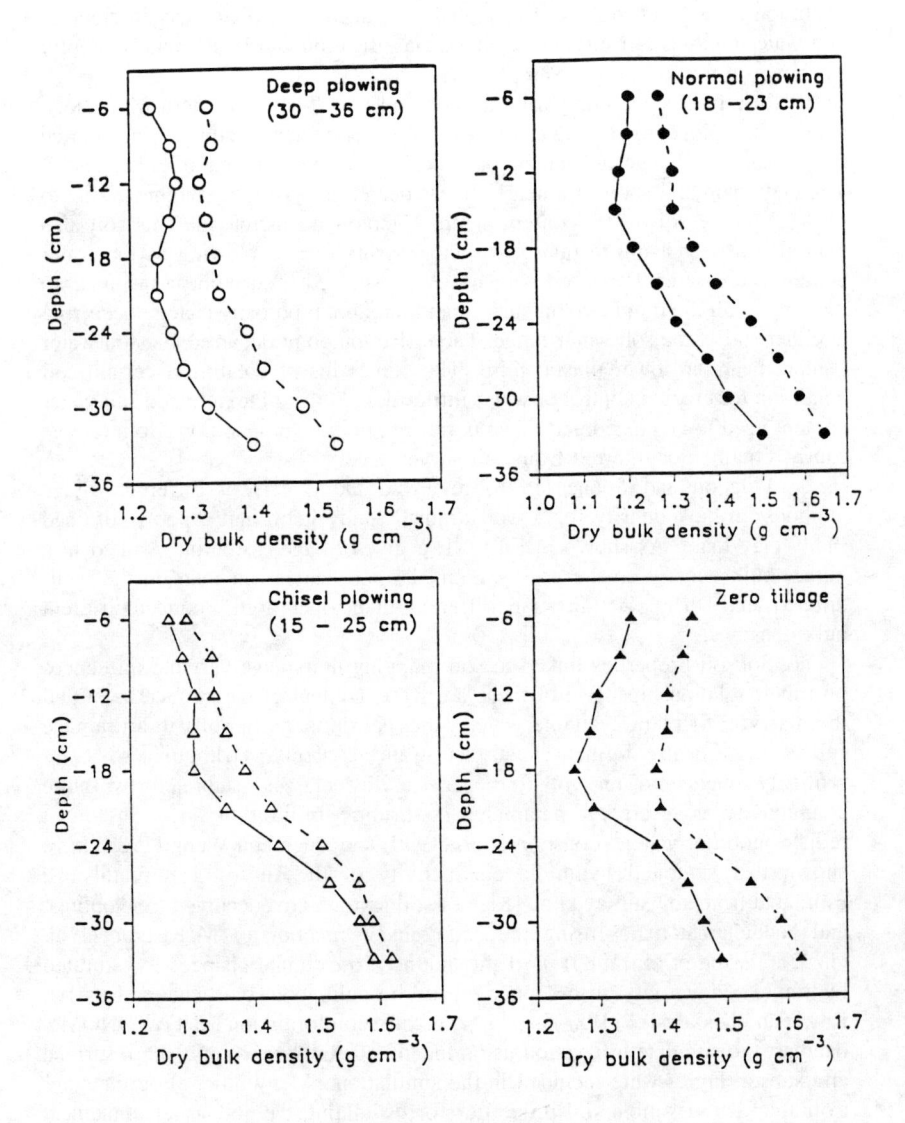

Fig. 8-1. Equilibrium bulk density (1973-1975) of two soils under four tillage systems. (From Pidgeon and Soane, 1978.)
———— well-drained sandy loam
- - - - poorly drained sandy clay loam

While soil survey properties are quite useful to predict infiltration and internal drainage, Ratliff et al. (1983) found that the standard -33kPa and -1.5 MPa water retention did not estimate plant available water when compared with field measured water contents; further investigations showed only limited value of adding other standard database soil parameters into the prediction (Cassel et al., 1983). Pierce et al. (1984) had a similar experience while developing their model for soil erosion - productivity. It appears that variations of FBS in landscapes where stored soil water is important may require direct field measures of the reservoir for soil water storage, as well as the associated root zone, which itself can vary by plant species.

Another set of soil properties necessary for varying tillage in a FBS plan is the soil consistence and compression index. These relate to the timeliness for workability and response to traffic loads. McBride and Bober (1990) predicted the lower plastic limit (water content) using clay and organic matter content, while the upper plastic limit (water content) only varied with clay content. Larson and Gupta (1980) demonstrated that increased clay content up to 35% increased the index of susceptibility to compaction. Kassa (1992) suggests that organic matter may reduce and sand content may increase the susceptibility to compaction, but more measurements are needed on a wider range of companion surface (Ap) and underlying subsoil.

TILLAGE - - EROSION RELATIONSHIPS

Tillage systems, ranging from full width tillage with complete residue incorporation to no-till, can strongly influence soil erosion. But this influence is not necessarily a straight-forward relationship. Tillage influence will be modified by soil type, crop rotation, soil moisture content at tillage, and climate (Benoit and Lindstrom, 1987). If erosion is not a potential problem, then any tillage system would be acceptable provided it met the requirements for subsequent crop growth. However, when erosion (either wind or water) becomes a problem then tillage becomes an important management tool.

The effectiveness of any tillage method for controlling erosion ultimately depends upon the amount of crop residues left on the soil surface (Duley and Russell, 1941; Laflen et al., 1985). Erosion and sedimentation by water involves the processes of detachment, transport, and deposition of soil particles. The major forces are from raindrop impact and water flowing over the land surface. Crop residues will effectively intercept and dissipate energy in raindrop impact, which reduces rainfall erosivity. Surface residues also slow runoff flow rates, which increases infiltration and reduces runoff erosivity. Surface residues effectively control wind erosion by intercepting and diverting wind energy away from the soil surface. Soil roughness has also been demonstrated to effectively reduce water runoff through both surface storage capacity and increased infiltration (Burwell and Larson, 1969). Soil roughness also is effective in reducing wind forces on erodible soil particles and acts as a trap for eroding soil particles, but soil roughness generated by tillage is not stable and reverts to pre-till roughness through weathering (Onstad et al., 1984; Zobeck and Onstad,

1987). Soil roughness becomes a less effective erosion control device as slope gradients increase.

An estimate of tillage on soil erosion reduction by water is possible with the Universal Soil Loss Equation (USLE) (Wischmeier and Smith, 1978). The cover and management factor (C factor) can vary from 0.027 in a typical corn-soybean (*Zea mays* L.)-[*Glycine max* (L.) Merr.] rotation in which all residues are left on the surface with a winter cover crop after the soybean crop to 0.390 for a fall moldboard plow based tillage system. In a sensitivity analysis showing the effect on soil erosion of each factor in the USLE, (Fig. 8-2), the combined landscape factors of slope length "L" and slope gradient "S" were shown to be tremendously important in determining soil erosion (Franzmeier, 1990). Cosper (1983) reports that soil taxonomy offers one means of predicting how soils might control plant response under various forms of conservation tillage. Using this same reasoning but including the landscape factors of slope gradient (often reported as a phase in the taxonomic system) and slope length, then soil taxonomic classification should be a possible tool in determining tillage systems or surface residue amounts required to control erosion within acceptable limits.

Many soil series are often present on individual farms and within the same fields, and represent differences in soil formation primarily due to topography, and microclimate. In rolling landscape, where many soil series are in close association ranging from upland soil series to depressional soil series, then it may be impracticable to till the soils as separate entities. Khakural et al. (1992) measured soil properties across two discrete but closely associated landscape positions and found that the improvement in soil properties on an upland soil with conservation tillage outweighed the effects of these same high residue tillage on a depressional soil. When soil series are more distinct, then individually designed tillage systems can be developed for specific soil series or landscape positions.

Surface residue amounts can be controlled to some degree with type of soil engaging tool. Johnson (1988) measured remaining spring surface residue amounts from 44 to 59% after corn and 22 to 54% after soybean with fall chisel-type primary tillage implements. Residue remaining on the surface ranked from high to low as follows: low crown sweep ≥ medium crown sweep > chisel point > high crown sweep > concave twisted shovel. Johnson (1988) also found that field operating speed between 5 to 9 km/h did not affect quantity of residue remaining on the surface but reducing field operating speed to 2.4 km/h significantly increased surface residue with a disk chisel with twisted shovels. Also operating the disk chisel at 20- to 25-cm depths incorporated more residue than when operated at 10 or 15 cm. Increased residue incorporation at the deeper setting was at least partially attributed to the more aggressive action of the disk. These data suggest that both speed and tillage depth could be changed when tilling across soil series and landscape positions for erosion control when necessary and residue incorporation when desired. Other options that could be considered are changing angles of disk blades or ground engaging tool to leave or incorporate residue as desired by soil type.

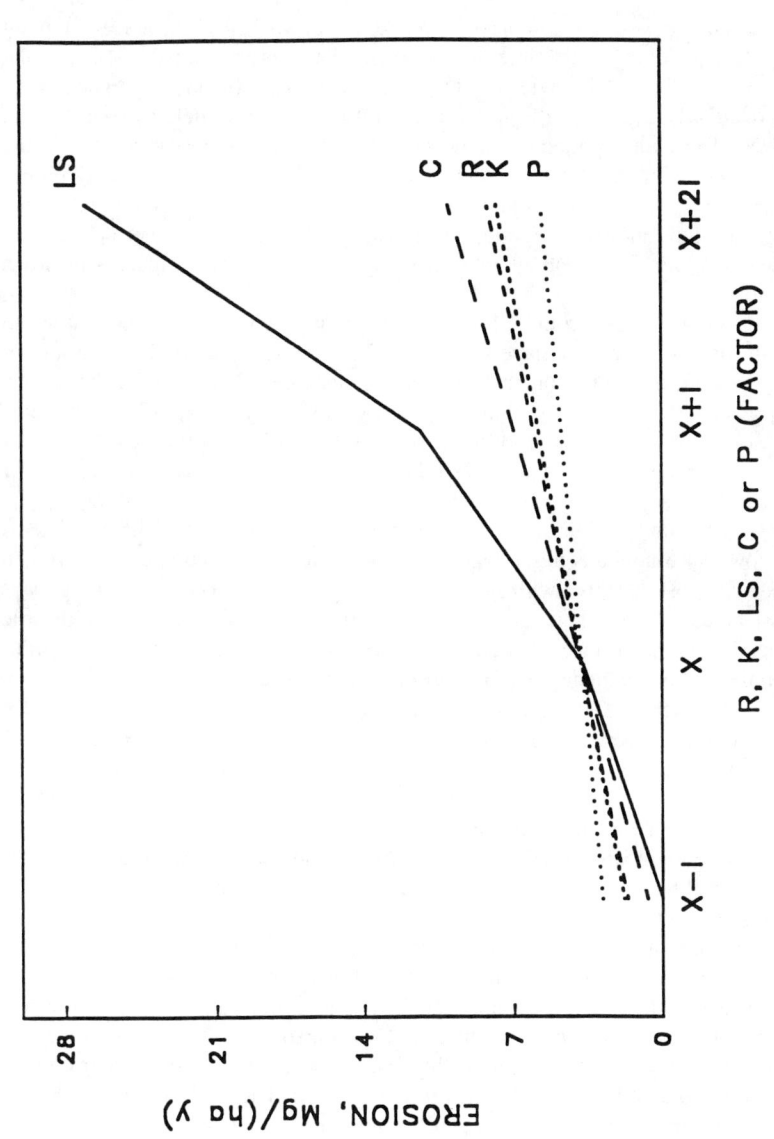

Fig. 8-2. Sensitivity diagram showing effects of factors in the Universal Soil Loss Equation on erosion for field crops in the midwestern USA. (Franzmeier, 1990).

TILLAGE - - PLANT GROWTH RELATIONSHIPS

Tillage systems can have a marked, but variable, influence on plant growth depending on the internal drainage characteristics of the soil profile. Allmaras et al., (1991) reviewed the effects of tillage systems on several soils, and related the resulting tillage depth and residue cover to yield of soybean and maize. The data reported in Table 8-1 are from several research sites in the midwestern United States on soils that ranged from well drained to poorly drained and in a subhumid to humid climate. The soils belong mostly within the aquic and udic moisture regime, and mostly within the mesic thermal regime. Research plots were planted to uniform plant density with successful weed control.

As one ranges from moldboard plowing through chiseling and disking to no-till forms of primary tillage, the depth of tillage decreases and the amount of plant residue remaining on the soil surface increases. Note that in the poorly drained soils, both maize and soybean yields declined as tillage systems increased surface residues and decreased depth of tillage (Table 8-1). In well-drained soils, yields increased as tillage depth decreased and residue cover increased.

The depth of burial of plant residues is also affected by tillage systems as shown by Staricka et al. (1991) in Fig. 8-3. Moldboard plowing incorporated residues most deeply with chiseling being the shallowest; disking was intermediate. Note that disking was most efficient for shallow burial with little remaining at the surface; this could be a desirable feature when linking multi-purpose tool attachments to a single implement frame.

The depth of maximum tool operation shown in Fig. 8-3 is typical for primary tillage operations in northern soils. When the chisel and disk are used as secondary tillage tools, they reproduce the same residue depth distribution as seen in Fig. 8-3, provided the residue was initially on the soil surface.

Soil bulk density measurements made in midsummer on a clay loam soil show that moldboard plowing typically loosens soil to a depth of about 27 cm while chiseling (as a primary tillage) penetrates to only about 15 cm (Fig. 8-4). Subsequent secondary tillage did not completely repack the 15 to 27 cm layer. Similar observations have been reported by Logsdon et al. (1990, 1992).

In the situation shown in Fig. 8-4, moldboard plowing not only penetrated deeper than chisel tillage but also effected some relief of a slightly compacted layer at 15 to 27 cm. One can envision from Larson and Gupta (1980) that the 10 to 25 cm layer is more likely to be compacted due to a higher water content in a poorly drained compared to a well-drained soil. Tile drainage should reduce these differences somewhat, especially if the tile spacing is closer in the poorly drained portions of the field.

Table 8-1. Effects of soil properties, cropping practice, and tillage on soybean and corn grain yield [soybean data from Van Doren and Reicosky (1987)].

Soil property		Cropping	Moldboard plow[4]		Chisel[4] Normal		Chisel[4] Deep		Disc or field	Ridges[4]	No-tillage[4]	
Drainage[1]	Texture[2,3]	sequence[3]	Fall	Spring	Fall	Spring	Fall	Spring	cultivate[4]		<6 year	≥6 year
			——————————— Soybean grain yield (kg/ha) ———————————									
Poor	Fine	Continuous	3,440	- 20	- 110	--	--	--	- 170	--	- 380	- 290
		Rotation	2,980	- 30	- 70	- 120	--	--	- 140	--	- 230	- 420
	Medium	All	- 120	2,870	+100	- 190	+ 20	- 30	- 170	--	+ 30	--
	Fine	DC	--	1,380	--	--	--	- 100	+ 360	--	+ 210	--
Imperfect	All	All	2,810	--	0	+ 200	--	+ 160	+ 100	--	+ 60	--
	Fine	DC	--	1,500	--	--	--	--	--	--	+ 80	--
Well-1	Medium	All	+ 320	2,190	+ 260	+ 80	--	+ 70	+ 20	--	+ 40	+ 240
Well-2	Coarse	All	--	2,230	--	+ 320	--	+ 380	+ 20	--	--	--
			——————————— Corn grain yield (kg/ha) ———————————									
Poor	Fine	Continuous	8,020	+ 80	- 470	- 940	- 320	--	- 510	- 290	- 650	-1,020
		Rotation	8,120	- 230	- 220	- 350	+ 20	--	- 190	- 250	- 310	- 410
	Medium	All	8,340	- 280	- 360	+ 50	- 10	- 180	- 300	- 470	- 710	- 650
Imperfect	All	Continuous	7,510	- 350	- 210	+ 40	- 200	--	- 140	- 440	- 360	--
		Rotation	7,700	- 670	+ 30	+ 200	+ 490	--	- 90	- 60	- 100	--
Well-1	All	Continuous	+1,090	6,750	+1,110	+1,990	--	+ 870	- 970	+ 870	+ 650	+1,700
		Rotation	--	6,810	--	--	--	+ 890	+ 80	+ 700	+ 760	+1,210
Well-2	Coarse	All	+ 260	6,120	--	+1,080	--	--	+ 20	+ 480	+ 780	+1,470

Footnotes are listed on page 104.

[1]Poor drainage are soils in the aquic suborder; imperfect drainage are soils from aquic and nonaquic subgroups; well-1 drainage are soils from nonaquic subgroups except Paleudults; and well-2 are Paleudults.

[2]Fine textures are surface horizons of silty clay loam, clay loam, or clay; medium textures are from sandy to silt loam weighted toward silt loam; coarse textures are from sandy loam to silt loam weighted toward the sandy loam.

[3]All indicates insufficient data for separating texture or cropping systems into separate categories. DC is double cropping soybean with wheat.

[4]Moldboard plow depth = 15 to 25 cm, normal chisel depth = 15 to 20 cm, deep chisel depth = 30 to 50 cm, disc or field cultivate depth = 8 to 12 cm, number of years consecutive no-tillage on the same plot; the italicized yield is the base yield [see Van Doren and Reicosky (1987) for method to interpret statistically] from which all others are compared either as an increase (+) or decrease (-).

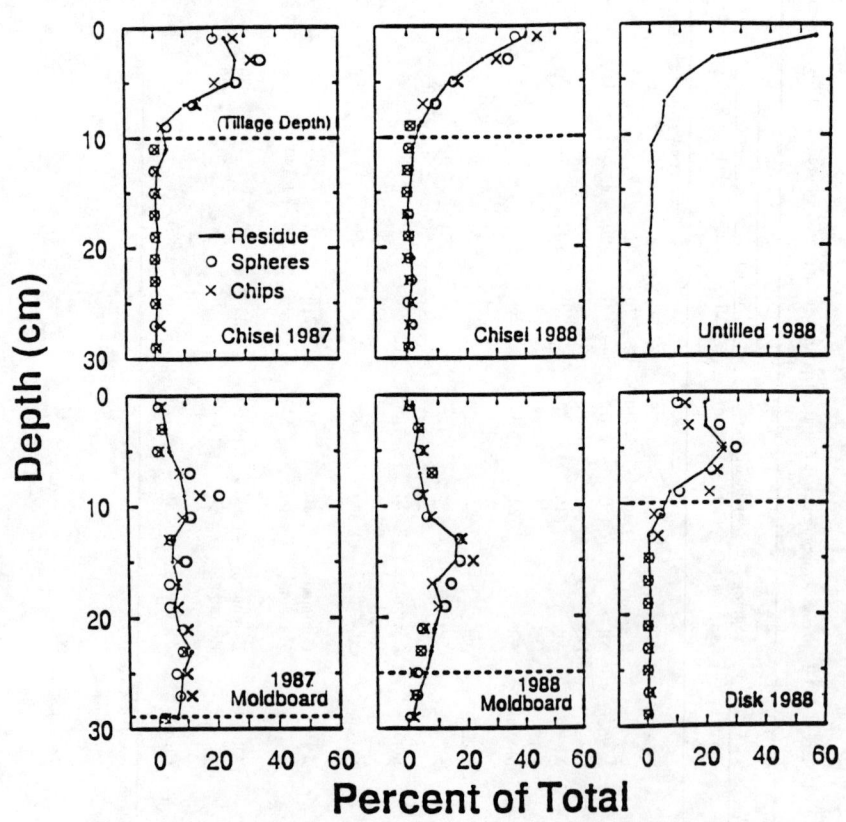

Fig. 8-3. Vertical distribution of crop residues as affected by tillage. Dotted lines indicate depth of tillage. (From Staricka et al., 1991.)

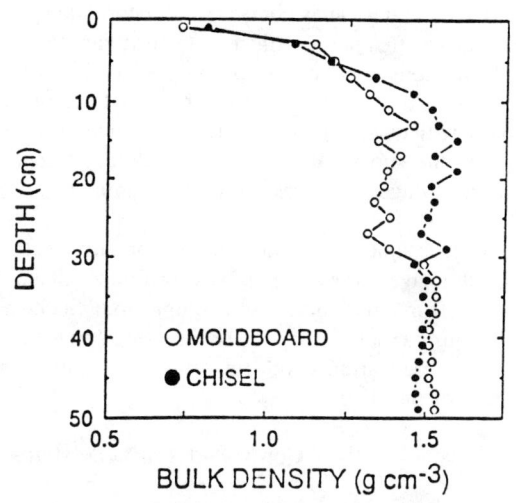

Fig. 8-4. Depth distribution of soil bulk density related to primary tillage treatment. (From Staricka et al., 1990.)

The above discussion has established the potential yield variations due to tillage, i.e., positive response from reduced tillage systems in the well-drained sites and from deeper tillage in the poorly drained sites. Primary tillage has tool-specific control of residue burial and tillage depth, and also exerts a strong control in loosening the 10- to 25-cm zone. In a field with mixed soil drainage characteristics, the ideal primary tillage tool may be a tool frame that supports a front tool rank (disk gang) with on-the-go variable angle and depth control to handle residue incorporation, followed by a chisel gang for deeper tillage (ie., 25 to 30 cm) in the poorly drained sites, and very shallow but horizontally complete tillage in the well-drained sites.

For secondary tillage, the same multi-tool rank arrangement might be used except that the second (third and fourth) gang would range to work at depths from 5 to 10 cm to give shallow but complete coverage as needed for incorporation of fertilizers or pesticides. Some variations on this overall tillage scheme specific to drainage criteria and taxonomy is already incorporated into tillage systems now used in the Corn Belt of the USA.

SOIL VARIABILITY AND WHEEL TRAFFIC

Conventional Systems

Wheel traffic is the most common source of current soil compaction concerns on soils in the western world (Voorhees, 1991). With most conventional tillage systems, wheel traffic is generally random spatially and therefore may not seem germane to the topic of managing soil variability. There are some important exceptions, however, that need to be mentioned.

Certain landscapes contain small depressional areas that normally do not require special treatment with respect to tillage. But in abnormally wet years, these depressions may be susceptible to excessive soil compaction due to wheel traffic. These areas may not be sufficiently wet to prevent field operations, especially with current high-powered equipment, and therefore will be trafficked along with the surrounding drier area. Although conditions may not warrant a change in tillage practices, efforts to minimize soil compaction may be beneficial.

Depressional areas that are historically wet, or soils that have poor internal drainage, warrant not only different tillage considerations, but also additional efforts to minimize soil compaction. Such areas may exhibit a gradual reduction in internal drainage over time, accompanied by decreasing crop yield. Such a response warrants adjusting tillage or planting practices separate from the remainder of the field.

Controlled Traffic Systems

After a row crop is planted, most wheel traffic for the remainder of the growing season is confined to certain interrow areas of the field. Even with close seeded crops, post-planting wheel traffic generally follows established traffic lanes or tramlines. This is a source of soil variability, albeit human-induced and in a predictable pattern. Lateral root growth patterns are often very sensitive to interrow soil compaction as discussed by Voorhees (1989, 1992). Such rooting patterns suggest a preferential placement of banded fertilizers relative to interrow wheel traffic.

In no-till or ridge-till planting systems, interrow wheel traffic patterns are usually constant over time. Since natural forces do not ameliorate these compacted interrows (Voorhees, 1983), they become, in effect, permanent and well-defined sources of soil variability, lending themselves to special treatment. A good example of this is reduced yield and K uptake in rows of maize with wheel traffic on both sides under a ridge-till system.

Subsoil Compaction

Irrespective of the type of tillage, or lack there-of, the greatest threat to long-term soil productivity (next to soil erosion) may easily be subsoil compaction from heavy harvest equipment. Depending on the type of planting system and extent of crop rotation, this subsoil compaction may be random or patterned.

Soils in Minnesota that are annually subjected to freezing and thawing to depths exceeding 60 cm, have exhibited long-term responses to a one-time application of wheel traffic equivalent to that of harvest equipment. Figure 8-5 shows significant maize yield reductions due to subsoil compaction up to 9 years after heavy trafficking (20 ton axle load compared with 5 ton). The average long-term yield reduction was 6%. While this subsoil compaction may not be a readily detectable source of soil variability, it can easily alter the characteristics

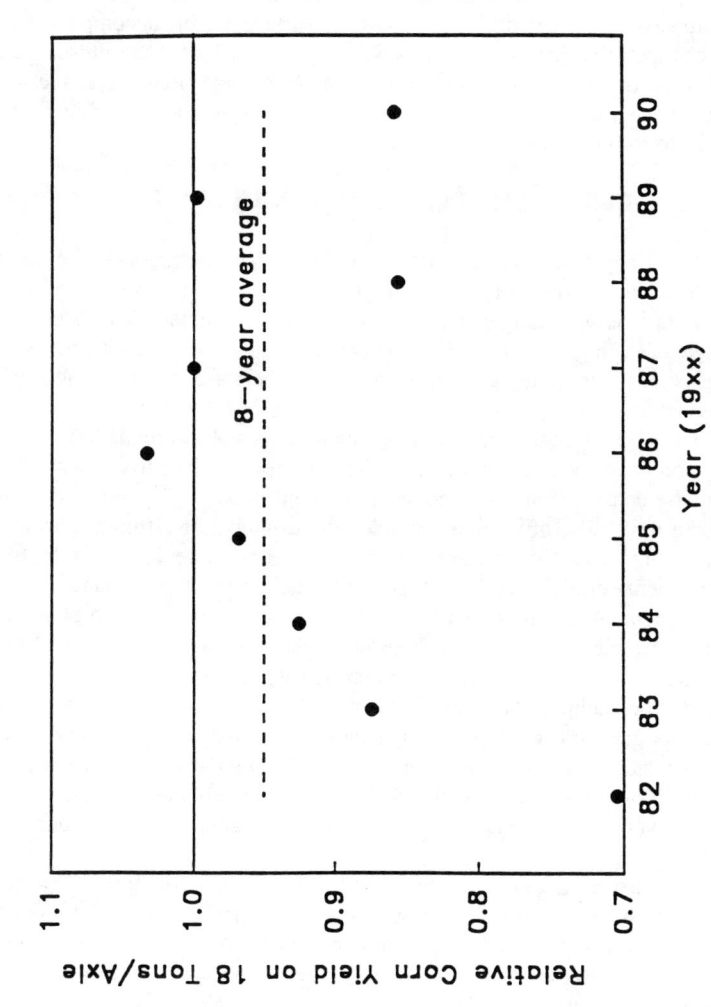

Fig. 8-5. Maize yield over time as affected by subsoil compaction from 20 ton axle load wheel traffic (W. B. Voorhees, 1993 unpublished data).

of surface soil parameters that are being monitored to identify soil variability. Lodgson et al. (1992) and Voorhees et al. (1986) showed significant reductions in hydraulic conductivity with evidence of decreased rooting in subsoils compacted by heavy farm equipment. This decreased subsoil conductivity may be manifested in slower drainage of excess surface water, accompanied by a drastic change in surface soil color (Voorhees et al., 1986). Care must be taken when interpreting technology that assesses color change because, as the above example shows, the color change may be due to varying soil water but assumed to be due to varying organic matter content.

RESEARCH NEEDS AND OPPORTUNITIES

As in any small plot research, it is difficult to extrapolate site specific data to another site, or to infer sweeping generalities. Furthermore, managing tillage operations according to spatially varying soil characteristics has the added challenge of trying to satisfy multiple, and often opposing, objectives; soil conditions best for plant growth may not be best for erosion concerns or pollution impact.

The biggest challenge is not so much being able to recognize or detect varying soil characteristics on-the-go. Technology is in place to do this, or is on the drawing boards. Similarly, tillage equipment is becoming more adjustable, with some units now offering independent adjustments of different portions of the implement to satisfy different objectives. It is possible, for example, to use one implement across several soil types or landscape positions, changing output of that implement to minimize plant residue burial for one portion of the field while maximizing some other variable in another portion of the field. But are such changes economical, either in terms of conserved natural resource, increased crop yield, or environmental impact? Ultimately, the farm operator must ask the question "at what point does it become cost-<u>ineffective</u> to adjust my tillage to match detectable variations in soil characteristics?" Can agriculture justify the cost of developing the technology to continually adjust a tillage operation every few meters to obtain a certain degree of residue burial, for example?

We also need to consider accumulative effects over time. A negative response to changing or adjusting soil tillage the first 1 to 2 years may turn positive if the tillage practice is continued for 4 to 5 years. This is where decision-aid technology such as computer models and GIS can be used to help make some decisions.

REFERENCES

Allmaras, R. R., R. E. Burwell, and R. F. Holt. 1967. Plow-layer porosity and surface roughness from tillage as affected by initial porosity and soil moisture at tillage time. Soil Sci. Soc. Am. Proc. 31:550-555.

Allmaras, R. R., G. W. Langdale, P. W. Unger, R. H. Dowdy, and D. M. Van Doren. 1991. Adoption of conservation tillage and associated planting systems. p. 53-83. In R. Lal and F. J. Pierce (ed.) Soil management for sustainability. Soil and Water Conserv. Soc. Ankeny, IA.

Benoit, G. R., and M. J. Lindstrom. 1987. Interpreting tillage-residue management effects. J. Soil Water Conserv. 42:87-90.

Burwell, R.E., and W. E. Larson. 1969. Infiltration as influenced by tillage-induced random roughness and pore space. Soil Sci. Soc. Am. Proc. 33:449-452.

Cassel, D. K., L. F. Ratliff, and J. T. Ritchie. 1983. Models for estimating in-situ potential extractable water using soil physical and chemical properties. Soil Sci. Soc. Am. J. 47:764-769.

Coen, G. M., and C. Wang. 1989. Estimating vertical saturated hydraulic conductivity from soil morphology in Alberta. Can. J. Soil. Sci. 69:1-16.

Cosper, H. R. 1983. Soil suitability for conservation tillage. J. Soil Water Conserv. 38:152-155.

Duley, F. L., and J. C. Russell. 1941. Crop residue for protecting row crop land against runoff and erosion. Soil Sci. Soc. Am. Proc. 6:484-487.

Franzmeier, D. P. 1990. Soil landscapes and erosion processes. p. 81-104. In W. E. Larson et al. (ed.) Proceedings of Soil Erosion and Productivity Workshop. 13-15 March 1989. Bloomington, MN. University of Minnesota, St. Paul.

Johnson, R. R. 1988. Soil engaging tool effects on surface residue and roughness with chisel-type implements. Soil Sci. Soc. Am. J. 52:237-243.

Kassa, Z. 1992. Pore water pressure and some associated mechanical responses to uniaxial stress in structured agricultural soils. MS thesis. University of Minnesota, St. Paul.

Khakural, B. R., G. D. Lemme, T. E. Schumacher, and M. J. Lindstrom. 1992. Effects of tillage systems and landscape on soil. Soil Till. Res. 25:43-52.

Laflen, J. M., G. R. Foster, and C. A. Onstad. 1985. Simulation of individual storm soil loss for modeling the impact of soil erosion on soil productivity. p. 285-295. In S. A. El Swaify et al. (ed.) Soil Erosion and Conservation. Soil Water Conserv. Soc., Ankeny, IA.

Larson, W. E., and S. C. Gupta. 1980. Estimating critical stress in unsaturated soils from changes in pore water pressure during confined compression. Soil Sci. Soc. Am. J. 44:1127-1132.

Logsdon, S. D., R. R. Allmaras, L. Wu, J. B. Swan, and G. W. Randall. 1990. Macroporosity and its relation to saturated hydraulic conductivity under different tillage practices. Soil Sci. Soc. Am. J. 54:1096-1101.

Logsdon, S. D., R. R. Allmaras, W. W. Nelson, and W. B. Voorhees. 1992. Persistence of subsoil compaction from by heavy axle loads. Soil Tillage Res. 23:95-110.

McBride, R. A., and M. L. Bober. 1989. A re-examination of alternative test procedures for soil consistency limit determination: I. A compression-based procedure. Soil Sci. Soc. Am. J. 53:178-183.

McKeague, J. A., C. Wang, and G. C. Topp. 1982. Estimating saturated hydraulic conductivity from soil morphology. Soil Sci. Soc. Am. J. 46:1239-1244.

Onstad, C. A., M. L. Wolfe, and C. L. Larson. 1984. Tilled soil subsidenceduring repeated wetting. Trans. ASAE 27:733-736.

Pidgeon, J. D., and B. D. Soane. 1978. Soil structure and strength relations following tillage, zero-tillage and wheel traffic in Scotland. p. 371-378. In W. W. Emerson et al. (ed.) Modification of soil structure. John Wiley & Sons, New York.

Pierce, F. J., W. E. Larson, R. H. Dowdy, and W. A. P. Graham. 1983. Productivity of soils: assessing long-term changes due to erosion. J. Soil Water Conserv. 38:39-44.

Ratliff, L. F., J. T. Ritchie, and D. K. Cassel. 1983. Field-measured limits of soil water availability as related to laboratory-measured properties. Soil Sci. Soc. Am. J. 47:770-775.

Rawls, W. J., and D. L. Brakensiek. 1988. An infiltration model for evaluation of agricultural and range management systems. p. 166-175. In Modeling agricultural, forest, and rangeland hydrology. ASAE Publ. ASAE, St. Joseph, MI.

Rawls, W. J., T. J. Gish, and D. L. Brakensiek. 1991. Estimating soil water retention from soil physical properties and characteristics. Adv. Soil Sci. 16:213-234.

Staricka, J. A., R. R. Allmaras, and W. W. Nelson. 1991. Spatial variation of crop residues incorporated by tillage. Soil Sci. Soc. Am. J. 55:1668-1674.

Staricka, J. A., P. M. Burford, R. R. Allmaras, and W. W. Nelson. 1990. Tracing the vertical distribution of simulated shattered seeds as related to tillage. Agron. J. 82:1131-1134.

Van Doren, D. M. Jr., and D. C. Reicosky. 1987. Tillage and irrigation. p. 391-428. In J. R. Wilcox (ed.) Soybeans: Improvement, production and uses. Agron. Monogr. 16. ASA, Madison, WI.

Voorhees, W. B. 1983. Relative effectiveness of tillage and natural forces in alleviating wheel-induced soil compaction. Soil Sci. Soc. Am. J. 47:129-133.

Voorhees, W. B. 1989. Root activity related to shallow and deep compaction. In W. E. Larson et al. (ed.) Mechanics and related processes in structured agricultural soils. NATO ASI Series Vol. 172:173-186. Kluwer Academic Publishers.

Voorhees, W. B. 1991. Yield effects of compaction--are they significant? Trans. Am. Soc. Agric. Eng. 34:1667-1672.

Voorhees, W. B. 1992. Wheel-induced soil physical limitations to root growth. Adv. Soil Sci. 19:73-95.

Voorhees, W. B., W. W. Nelson, and G. W. Randall. 1986. Extent and persistence of subsoil compaction caused by heavy axle loads. Soil Sci. Soc. Am. J. 50:428-433.

Wang, C., J. A. McKeague, and K. D. Switzer-Howse. 1985. Saturated hydraulic conductivity as an indicator of structural degradation in clayey soils of Ottawa area, Canada. Soil Tillage Res. 5:19-31.

Wischmeier, W. H., and D. D. Smith. 1978. Predicting rainfall erosion losses: A guide to conservation planning. Agric. Handb. 537. Sci. and Educ. Admin., USDA, Washington, DC.

Wosten, J. H. M., and M. Th. van Genuchten. 1988. Using texture and other soil properties to predict the unsaturated soil hydraulic functions. Soil Sci. Soc. Am. J. 52:1762-1770.

Zobeck, T. M., and C. A. Onstad. 1987. Tillage and rainfall effects on random roughness: A review. Soil Tillage Res. 9:1-20.

9 Weed Distribution in Agricultural Fields

David A. Mortensen
Department of Agronomy
University of Nebraska
Lincoln, NE

Gregg A. Johnson
Department of Agronomy
University of Nebraska
Lincoln, NE

Linda J. Young
Biometry Department
University of Nebraska
Lincoln, NE

Weed seedling and seedbank populations as well as soil physical and chemical properties are spatially variable. Soil pH, organic matter content, structure, and depth to groundwater all influence herbicide type and rate. In addition, preliminary data indicate that weed distributions influencing the decision to apply a herbicide are spatially variable. Characterizing this variation makes it possible to develop sampling methodologies for agronomists to accurately assess weed populations in a field, and also provides the necessary data to test the feasibility of intermittent herbicide application. Weed distribution studies were conducted in 4.05-ha sections of soybean (*Glycine max* L.) fields in eastern Nebraska. Weed density counts were grid sampled and tested for agreement with the negative binomial distribution using a chi-square test; the negative binomial distribution best represented the data. The associated "K" values, an index of aggregation in the population, were very low, indicating a high degree of aggregation or clumping of weed species. In general, larger K values were associated with smaller seeded weed species.

INTRODUCTION

Traditional weed control systems have included preplant or preemergence herbicide applications as a standard practice. For the most part, this approach has been very effective; however, their use has been prophylactic in nature since the herbicide application is made before weeds germinate and emerge. Recent improvements in bioeconomic weed management models (Mortensen and Coble,

---

Copyright © 1993 ASA-CSSA-SSSA, 677 South Segoe Road, Madison, WI 53711, USA. *Soil Specific Crop Management.*

1991), and machine controlled intermittent herbicide application (Thompson et al., 1991) provide new tools for improving the efficiency of herbicide use. The use of weed seedbank or seedling estimates coupled with economic threshold based bioeconomic models has resulted in significant reductions in pre- and postemergence herbicide use (Schweizer and Zimdahl, 1984; Gerowitt and Heitefuss, 1990). Intermittent application of herbicide in response to weed density has been shown to be technically sound and appears to be economically feasible in the foreseeable future. Felton et al. (1991) recently described a sensing device capable of discriminating between green vegetation and other ground covers, while Shropshire and Von Bargen (1989), and Guyer et al. (1986) demonstrated the feasibility of discrimination between weed and crop.

The successful implementation of bioeconomic weed management models, and the successful design of intermittent herbicide application devices requires that weed distributions be carefully described. Sampling protocols for crop scouts are dependent on this information. In addition, geographic information systems (GIS) maps of weed distributions and design criteria for plant sensing devices rely on accurate descriptions of weed distributions.

Weed Distributions

The influence of population distribution on estimation of statistical parameters is widely known in the entomological field where the distribution of insects, is well understood. The quantification of pest population distributions and the use of this distribution information for developing intermittent application technologies, or scouting procedures for use in pest management models has reduced or eliminated insecticide use on apples (*Pyrus malus* L.) (Metcalf, 1980), cotton (*Gossypium hirsutum* L.) (Sterling and Pieters, 1979), corn (*Zea mays* L.), and soybean (*Glycine max.* L.) (Kogan and Herzog, 1980). Characterizing the distribution of weeds or soil variables can be useful in formulating decision criteria for spraying. Until recently, few weed population distribution studies have been conducted.

The void in population distribution research has become evident in light of recent interests in applying herbicides based on their distribution, whether weeds are observed by a field scout or by plant detection sensors. Much of the past threshold related research has focused on weed-crop interference, leaving patterns of weed dispersion and distribution largely unresearched (Doyle, 1991). Improper assessment of the weed infestation may result in overestimation of yield loss, especially at high weed densities (Wiles et al., 1991; Hughes, 1989). Most field studies assume that weed species occur in monospecific stands and that the population is homogeneously distributed (Navas, 1991). On the contrary, preliminary studies indicate that weeds occur in multispecies, nonhomogeneous populations within a field (Thornton et al., 1990; Navas, 1991; Van Groenendael, 1988). Thornton et al. (1990) demonstrated the importance of adequate weed distribution characterization when estimating weed populations for economic threshold decisions. In their research, aerial photographs were taken of several winter wheat fields. Photographs were digitized and classified to estimate the *Avena fatua* infestation in the field. From these digitized images of the field it was determined that 18% of the field was infested with the weed, of which 94% was classified as light and 6% as heavy infestations. The average density was 1.4 plants m^{-2}, well below the calculated economic threshold. Observers from the ground estimated the weed infestation was moderate to severe across the entire field. In another study (Johnson et al., 1991), weed seedling distributions were recorded in several farms in eastern Nebraska. These distributions were described by grid sampling 4.05-ha sections of fields, with 22 m spacings for X and Y coordinates. All species studied occurred in an aggregated or clumped distribution, with common cocklebur (*Xanthium pennsylvanicum* L.) and velvetleaf (*Albutilon theophrasti* L.) exhibiting the greatest degree of clumping. Spatial analysis also showed that weed seedlings occurred in distinct clumps throughout the field (Fig. 9-1). The weedy monocot population was well below the economic threshold in 50% of the field. Interestingly, from the field edge, farmer cooperators and researchers agreed that weed populations appeared uniformly distributed at moderate to high infestation levels.

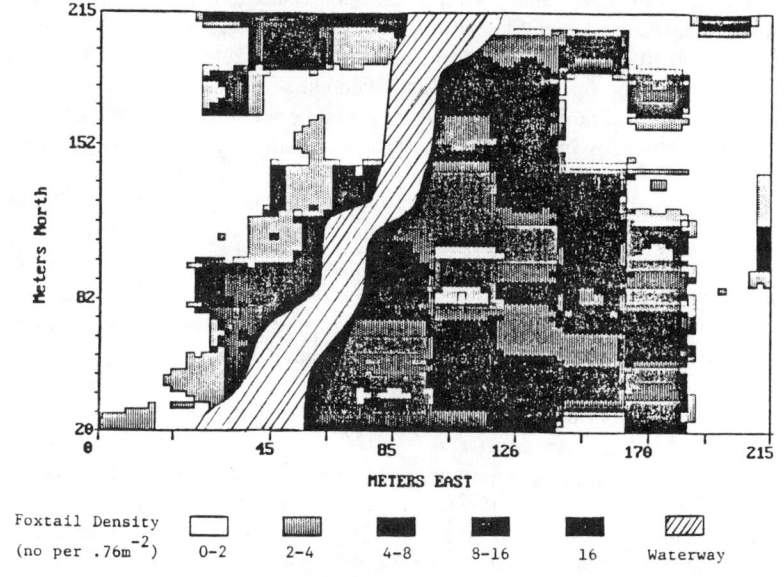

Fig. 9-1. Spatial distribution of *Setaria sp.* in soybean near Ceresco, NE. Counts were transformed using ln(x+.5k).

An aggregated population is one in which the individuals are clumped or clustered in a range of densities (Cottam et al., 1957). Clumping or aggregation tends to result in the presence of large areas free or nearly free of weeds in spaces between clumps. Researchers at North Carolina State University studied the distribution of broadleaf weeds in 14 soybean fields (Wiles et al., 1991). In most cases, weed distributions were best described by the negative binomial. Other studies have also demonstrated that the distribution of most weed seed and seedlings fit this distribution (Marshall, 1988; Zanin et al., 1989).

Cottam et al. (1957), suggests that large fields with a randomly dispersed population occupying only one portion of the area will appear spatially aggregated. Such cases of pseudo-aggregation often result from a microsite that will not support a weed population. This is especially true in fields where variability in soil type, fertility, and drainage must be taken into consideration when evaluating the distribution of a weed population. Agronomic variables including choice of crop, soil fertility, herbicide, and tillage practice as well as previous field history all have pronounced effects on weed species diversity and distribution. Andreasen et al. (1991) studied soil properties affecting the spatial distribution of 37 weedy species in Danish cereal grain fields. Differences in crop type and clay content explained most of the within and between field variation in species frequency. Wiese (1985) stated that weed population distributions are more variable under reduced tillage cropping systems. Under reduced tillage systems weed seed dispersal is more dependent on natural dispersal mechanisms. Patchy weed distribution characteristic of reduced tillage systems is a result of natural seed dispersal mechanisms and growth habits of species adapted to reduced soil disturbance. A greater proportion of the population is made up of perennial and biennial weeds. Brown et al. (1990) intensively sampled three no-tillage fields in southern Ontario to characterize the distribution of weeds in the cropping system. Distinctive patches of perennial and biennial species were observed in two fields and patches of annuals were observed in the third field. Marshall (1989) demonstrated that the distribution of plants and seeds are a result of many interacting factors, including dispersal, soil disturbance, and other agricultural operations. The remainder of the chapter will review the procedures and results of several intensive weed distribution studies. The goal of this research is to describe the distribution of weeds commonly found in a soybean-corn crop rotation.

Methods and Analysis of the 1991 Intensive Field Survey

Intensive sampling was conducted at two field sites in 1991, one in Ceresco, NE and the other in Wahoo, NE, both located in eastern Nebraska. Sampling dates were June 21 and June 28 at the Ceresco and Wahoo sites, respectively. At each field site, a representative 4.05-ha area was selected for intensive sampling. Sampling was accomplished using a grid overlay system with each intersection or point on the grid (22-m spacing) being a sampling point. At each sampling point in the field, weed density by species was determined in eight consecutive sampling quadrats. A sampling quadrat comprised the area between 76-cm spaced crop rows for 1 linear meter. This resulted in a total of approximately 800 sampling points at each of the two field sites.

Data from the intensive field surveys were used to determine which of five discrete distribution models would best fit the weed frequencies observed. Five 2-parameter distributions (negative binomial, Poisson with zeros, Neyman type A, logarithmic with zeros, and the Poisson-binomial) were considered since they covered the extremes with respect to kurtosis and skewness and have been successfully used to describe the distributions of other biological data (Katti, 1966). The Poisson, binomial, and Thomas double-Poisson were also studied since they have been used in numerous biological studies. The fit of these distributions were evaluated using DISCRETE (Gates, 1988), software that computes a chi-square test and the Kolmogorov-Smirnov criterion for each distribution.

Our hypothesis, based on previous data, was that the negative binomial distribution would fit the weed population data. This distribution is defined by 2 parameters, the mean (μ) and a positive exponent k. K is a mean to variance ($\sigma 2$) relationship where:

$$k = \frac{\mu^2}{(\sigma^2 - \mu)} \tag{1}$$

The parameter k can be considered an index of aggregation in the population (Southwood, 1966). As the population variance approaches the mean, the distribution of the population approaches the Poisson, a random population with large values of k (Southwood, 1966; Bliss, 1953). Conversely, low values of k suggest an aggregated or clumped distribution. However, small values of k do not necessarily imply spatial aggregation (Nicot et al., 1984) just as the Poisson distribution does not necessarily imply the lack of spatial heterogeneity (Patil and Stiteler, 1974).

The first step in determining whether a population follows a negative binomial distribution is to compare the observed frequencies of the sampled values to the expected frequencies of the negative binomial distribution at values of k and μ derived from the observed population. The probability of a sample quadrat containing 0, 1, 2, plants etc. based on the negative binomial distribution is:

$$\Pr(0) = (k/\mu+k)^k \tag{2}$$

$$\Pr(x) = ((k+x-1)/x)(\mu/\mu+k) \Pr(x-1) \tag{3}$$
$$\text{where } x = 1,2,3...$$

The expected frequencies are determined by multiplying the probability by the number of samples. Agreement between the expected and observed frequencies at each population density (sample quadrat) was evaluated using the chi-square test.

Geostatistical Analysis

Geostatistical analysis is becoming a powerful tool for descriptive analysis of spatial dependence. A statistical interpolation technique known as Kriging will be used. Kriging is a process based on spatial autocorrelation that permits precise estimation of variables between sampling points for use in mapping a population. Several geostatistical programs were used to provide estimates for graphical manipulation of the data and also to determine correlation between several environmental variables affecting the population.

Before Kriging can be performed, a quantitative assessment of spatial dependency using a functional relationship between the spatial pattern of the sampled points and their observed values is defined (Marx and Thompson, 1987). A semivariogram is defined and constructed as shown in Fig. 9-2.

C_0, termed the nugget effect, is the discontinuity at the origin or the microdistributional effect (measurement error). C_0+C_1, commonly called the sill, is the semivariogram value for very large distances. The range (r) is the distance beyond which the variogram value remains essentially constant or the distance at which samples become independent and are no longer correlated with each other. The distance between points is denoted by (h). $\gamma(h)$ is the semivariogram function and is half the averaged squared difference between the paired data values (Isaaks and Srivastrava, 1989):

$$\gamma(h) = (1/2m) \Sigma\{[z(x_i)-z(x_i+h)]^2\} \quad \text{where m = number of pairs} \quad (4)$$

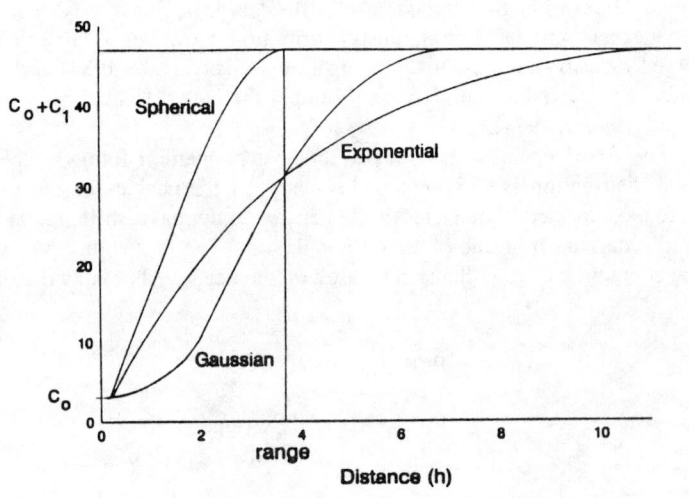

Fig. 9-2. Semivariogram model.

A limited number of functions can be used to calculate the semivariogram. These functions must be positive-definite to ensure that estimation variances will be nonnegative (Marx and Thompson, 1987). The most commonly used models for the semivariogram are spherical, exponential, Gaussian, linear, or logarithmic. Preliminary examination of the field data suggest that most can be fitted to the spherical model. The semivariogram function for a spherical model is calculated as:

$$\gamma(h) = 1.5(h/r) - 0.5(h/r)^3 \quad \text{where } h < r \qquad (5)$$

The spherical model reaches its sill at a finite distance equal to its range. If a spherical model is assumed, a straight line through the first several points would intersect the sill at about 2/3 of the range. Therefore, the range of influence is calculated to be 2r/3. Relationships between the Co, r, and Co+C_1 are used to estimate parameter values at unsampled locations. These kriged values are then used to produce a graphic map. GS+[1] was used to map the kriged values (Fig. 9-1). ARC-INFO[2] is also being used to analyze several spatial datasets from the Ceresco and Wahoo sites. Using ARC-INFO, correlation analysis between weed populations and field site physical and chemical properties is possible to further refine weed population maps. In addition, ARC-INFO can be used with other numerical information to draw correlations between sampled variables and other site characteristic variables such as elevation or soil type, to help further examine the effect of these secondary variables on weed population dynamics. ARC-INFO shows tremendous promise in evaluation of several spatially dependent variables simultaneously.

RESULTS FROM THE INTENSIVE FIELD SURVEYS

The Ceresco farm site was planted to soybean and was treated with a preemergence 35.6 cm wide band application of trifluralin and alachlor. Soybean was planted in 76-cm spaced rows and were at the cotyledon to V1 stage at the time of sampling. This field site was chosen because it had relatively high weed diversity and weed infestation level was moderate to low. A grassy waterway dissected the 4.05-ha section of the field and therefore was divided in two sections, the field area east of the waterway and the field area west of the waterway. Dividing the field in two sections was necessary to perform the kriging estimation. Because of the presence of the waterway, some sample quadrats fell within this natural border and were therefore not included

---

[1]GS+: Gamma Design Software, Box 201, 457 Bridge Street, Plainwell, MI 49080.

[2]ARC-INFO: Environmental Systems Research Institute, Inc., 380 New York Street, Redlands, CA 92373.

in the analysis. The result was a total of 764 sampling points of which 247 were west of the waterway and 517 were east of the waterway.

Preliminary statistical analysis indicated the data were highly skewed. Although some quadrats contained as many as 190 weeds/0.76m^{-2}, the mean density over the entire area was relatively low (Table 9-1). It is also interesting to note that a majority of the quadrats sampled contained no weeds and again shows the skewness of the density counts. This apparent skewness, where most data are confined to only a few cells, is indicative of a clumped or aggregated distribution.

Table 9-1. Distribution of weed species east and west of the waterway at Ceresco, NE in 1991.

Weed species	Site	0	1	2	3	4	5	>6
		---------- % of quadrats sampled ----------
Foxtail	East	20	10	7	6	4	3	50
	West	29	18	12	7	6	2	26
Pigweed	East	68	12	8	5	1	1	5
	West	34	17	15	11	6	5	12
Large crabgrass	East	96	1	1	0.8	0.8	0	0.4
	West	76	5	6	3	2	.4	7.6
Common lambsquarters								
	East	96	3	0.4	0.6	-	-	-
	West	96	2	0.6	0.4	-	-	-
Velvetleaf	East	94	3	1	0.2	0.6	0.4	0.8
Shattercane	East	94	3	1	0.5	0.7	0.3	0.5
Common cocklebur	East	93	5	0.7	0.3	0.7	-	-

Table 9-2. Results of the agreement with the negative binomial distribution of data for the Ceresco site in 1991.

Weed species	Area	Range[1]	Prob.	Chi-square k value
Foxtail	East	0-130	0.02 **	0.479
	West	0-114	0.06 NS	0.362
Pigweed	East	0-28	0.04 *	0.203
	West	0-86	0.15 NS	0.546
Large crabgrass	East	0-10	0.02 *	0.023
	West	0-50	0.15 NS	0.095
Velvetleaf	East	0-13	0.56 NS	0.038
	West	0-1	-	-
Common cocklebur	East	0-4	0.16 NS	0.094
	West	0	-	-
Shattercane	East	0-10	0.77 NS	0.039
	West	0-1	-	-
Common lambsquarters				
	East	0-3	-	-
	West	0-3	0.19 NS	0.058

* = Significant at the 0.05 probability level, ** = Significant at the 0.01 probability level, NS = Not significant at the 0.05 probability level, and - = Inadequate information to perform chi-square test.

[1] Minimum and maximum weed density per quadrat.

Field weed density counts were tested for agreement with eight discrete distributions using a chi-square test. On the west side of the waterway, no significant difference between the actual and the expected frequencies of the negative binomial distribution were observed for foxtail (*Setaria* sp.), pigweed (*Amaranthus* sp.), large crabgrass (*Digitaria sanguinalis*), and common lambsquarters (*Chenopodium album*) (Table 9-2). There were not enough data to perform a chi-square test for velvetleaf (*Abutilon theophrasti*) and common cocklebur (*Xanthium strumarium*). On the east side of the waterway, there was no significant difference between the actual and the expected frequencies for foxtail, velvetleaf, common cocklebur, and shattercane (*Sorghum bicolor*). Although there was a significant difference for pigweed and large crabgrass, once again the negative binomial distribution best represented the data.

The associated k values were also very low suggesting a high degree of aggregation or clumping. It is interesting to note that the k values for foxtail and pigweed, smaller seeded species, were larger than k values associated with common cocklebur and velvetleaf, large seeded species. The larger k values associated the smaller seeded species is likely related to mechanisms of weed seed dispersal. The magnitude of the k value is likely related to weed seed dispersal. The smaller seeded species are more likely to be wind blown and carried by water. In addition, dispersal by the combine is likely to play a more significant role for smaller seeded species as seed are produced up through grain harvest.

As mentioned earlier, weed density maps were constructed based on kriged values. These maps show the spatial aggregation present in selected weed species populations.

At the Wahoo site, pigweed was the predominant weed species complex. Field density counts were tested for agreement with the negative binomial distribution using a chi-square test. As was the case at the Ceresco site, the negative binomial fit the distribution as well as or better than other distributions tested. The associated k value of the negative binomial distribution was 1.4, indicating aggregation or clumping in the weed population.

SUMMARY

Numerical distribution must not be confused with spatial distribution. The negative binomial distribution is a numerical distribution and suggests the degree of aggregation but does not define the location of the arrangement of these clumps in a field. By mapping numerical data, we can get an idea of the configuration of the numerical distribution. One way to do this is through the use of geographic information systems or GIS techniques which use spatial dependence to estimate the density of unsampled locations. GIS techniques will be further employed as a tool to evaluate several spatially dependent variables with respect to the weed distribution.

The sampling technique outlined in the chapter provided the necessary data to characterize weed distributions occurring in two farmsteads in eastern Nebraska. The same grid system dimensions will be used in future sampling and will include a larger number of fields in the coming field season. Maintaining the same grid system from field to field and year to year makes it possible to estimate a common k value (provided the negative binomial distribution holds). The k value will then be used to construct a sequential sampling method intended for use by farmers and consultants. The distribution data will also be used to define the machine requirements for a sensor-controlled spot herbicide sprayer presently under development at the University of Nebraska.

Two discrete farm system types will be studied to describe the distribution and infestation level of several weed species. Therapeutic (postemergence herbicide), and prophylactic herbicide (preemergence) use practices will represent the two farm types. A minimum of six farms of each type will be intensively sampled. These farm types represent a range of practices that will enable us to isolate the influence of agronomic practices, events and field attributes that influence weed populations. In addition, we plan to sample at two times during the growing season, spring and fall, in order to further refine the spatial aggregation with respect to life stages.

REFERENCES

Andreasen, C., J. C. Streibig, and H. Haas. 1991. Soil properties affecting the distribution of 37 weed species in Danish fields. Weed Res. 31:181-187.

Bliss, C. I. 1953. Fitting the negative binomial distribution to biological data. Biometrics June 1953.

Brown, R. B., G. W. Andersen, B. Proud, and J. P. Steckler. 1990. Herbicide application control using GIS weed maps. ASAE Paper 90-1061. Am. Soc. Agric. Eng., St. Joseph, MI.

Cottam, G., J. T. Curtis, and A. J. Catana, Jr. 1957. Some sampling characteristics of a series of aggregated populations. Ecology 38:610-621.

Doyle, C. J. 1991. Mathematical models in weed management. Crop Prot. 10:432-444.

Felton, W. L., A. F. Doss, P. G. Nash, and K. R. McCloy. 1991. To selectively spot spray weeds. Am. Soc. Agric. Eng. Symp. 11-91:427-432.

Gates, C. E. 1988. Discrete: A computer program for fitting discrete frequency distribution. p. 458-466. In L. McDonald et al. (ed.) Lecture notes in statistics. Springer-Verlag, Berlin, Germany.

Gerowitt, B., and R. Heitefuss. 1990. Weed economic thresholds in cereals in the Federal Republic of Germany. Crop Prot. 9:323-331.

Guyer, D. E., G. E. Miles, M. M. Schreiber, O. R. Mitchell, and V. O. Vanderbilt. 1986. Machine vision and image processing for plant identification. Trans. Am. Soc. Agric. Eng. 29:1500-1507.

Hughes, G. 1989. Spatial heterogeneity in yield-weed relationships for crop-loss assessment. Crop Res. 29:87-94.

Isaaks, E. H., and R. M. Srivastava. 1989. In Introduction to applied geostatistics. Oxford Univ. Press, New York.

Katti, S. K. 1966. Interrelations among generalized distributions and their components. Biometrics 22(1):44-52.

Kogan, M., and D. C. Herzog. 1980. Sampling methods in soybean-Entomology. Springer-Verlag, New York.

Marshall, E. J. P. 1989. Distribution patterns of plants associated with arable field edges. J. Appl. Ecol. 26:247-257.

Marshall, E. J. P. 1988. Field-scale estimates of grass populations in arable land. Weed Res. 28:191-198.

Marx, D., and K. Thompson. 1987. Practical aspects of agricultural kriging. Arkansas Agric. Exp. Stn. Bull. 903.

Metcalf, R. L. 1980. Changing role of insecticides in crop protection. Annu. Rev. Ent. 25:219-256.

Mortensen, D. A., and H. D. Coble. 1991. Two approaches to weed control decision-aid software. Weed Technol. 5:445-452.

Navas, M. L. 1991. Using plant population biology in weed research: a strategy to improve weed management. Weed Res. 31:171-179.

Nicot, P. C., D. I. Rouse, and B. S. Yandell. 1984. Comparison of statistical methods for studying spatial patterns of soilborne plant pathogens in the field. Am. Phytopathol. Soc. 74:1399-1402.

Patil, G. P., and W. M. Stiteler. 1974. Concepts of aggregation and their quantification: A critical review with some new results and applications. Res. Pop. Ecol. 15:238-254.

Schweizer, E. E., and R. L. Zimdahl. 1984. Weed seed decline in irrigated soil after six years of continuous corn (*Zea mays*) and herbicides. Weed Sci. 32:76-83.

Shropshire, G. J., and K. Von Bargen. 1989. Fourier and Hadamard transforms for detecting weeds in video images. ASAE Paper 90-7522. Am. Soc. Agric. Eng., St. Joseph, MI.

Southwood, T. R. E. 1966. Ecological methods with particular reference to the study of insect populations. Methuen and Co. Ltd., London.

Sterling, W. L., and E. P. Pieters. 1979. Sequential decision sampling. In Economic thresholds and sampling of *Heliothis* species on cotton, corn, soybeans and other host plants. Southern Coop. Ser. Bull. 231:85-101.

Thompson, J. F., J. V. Stafford, and P. C. H. Miller. 1991. Potential for automatic weed detection and selective herbicide application. Crop Prot. 10:254-259.

Thornton, P. K., R. H. Fawcett, J. B. Dent, and T. J. Perkins. 1990. Spatial weed distribution and economic thresholds for weed control. Crop Prot. 9:337-342.

Van Groenendael, J.M. 1988. Patchy distribution of weeds and some implications for modelling population dynamics: a short literature review. Weed Res. 28:437-441.

Wiese, A. F. 1985. Weed control in limited-tillage systems. Monograph Ser. No. 2. Weed Sci. Soc. Am., Champaign, IL.

Wiles, L. J., H. J. Gold, G. G. Wilkerson, and H. D. Coble. 1991. Modeling weed patchiness for improved yield loss estimates and postemergence control decisions. Ph.D. Diss. North Carolina State Univ.

Zanin, G., A. Berti and M. C. Zuin. 1989. Estimation du stock semencier d'un sol laboure ou en semis direct. Weed Res. 27:407-417.

10 Value of Managing Within-Field Variability

Frank Forcella
Research Agronomist
USDA Agricultural Research Service
North Central Soil Conservation Research Laboratory
Morris, MN

The prospect of managing a field with spatially varying soil characteristics is gaining momentum. This is due to the emerging capability for real-time sensing of soil variability within fields, and the rapid advances being made in the development of inexpensive geographic positioning systems. Knowledge of within-field soil variability may permit agriculture to evolve into a far more precise and environmentally sound endeavor. Both the practitioner and the environment are likely to benefit from such prescription farming.

Prior to implementation of these new technologies, necessary prerequisites will be analyses of the gross value of information describing within-field variability. Because a cost is associated with acquiring data on within-field variability, the value of this information requires calculation. In nearly homogeneous fields the value may be slight from a manager's perspective. On the other hand, in highly heterogeneous fields the value of this information may be considerable. At what point along a scale of variability does such information become important? And, do the varying costs associated with differential management within fields affect this threshold? The objective of this chapter is to provide basic concepts involving the initial decision of whether management of within-field variability is a worthwhile endeavor.

SPATIALLY VARIABLE MODEL FIELDS

A series of 11 hypothetical fields, each 10 ha, were created (Fig. 10-1). Each field differs from all others in a spatially variable soil characteristic. For simplicity, this soil characteristic is binary, either 0 or 100, and each 1-ha unit of the field can be described as either 0 or 100. Standard deviations can be calculated for each field, these range from 0 to 50 (Fig. 10-1), and they subsequently will be referred to as an index of within-field variability (WFV). WFV remains constant for a field, regardless of the underlying values of the characteristic used in its calculation, as long as the standard deviation is divided by the range and multiplied by 100.

Copyright © 1993 ASA-CSSA-SSSA, 677 South Segoe Road, Madison, WI 53711, USA. *Soil Specific Crop Management.*

DIFFERENTIAL MANAGEMENT

The spatially-variable soils in these fields may permit differential management within each field. Such management might entail, for example, differential application of N fertilizer. Nitrogen fertilizer is a relatively inexpensive agrichemical, costing about $0.25/kg. As an example, assume the following: (i) corn yield response to N is linear until it reaches a plateau at 100 kg/ha. (ii) Soil A (from Fig. 10-1) has a residual inorganic N content of 60 kg/ha and will require a 40 kg/ha ($10/ha) application. Lastly, (iii) Soil B has a residual inorganic N content of 20 kg/ha and will require an 80 kg/ha ($20/ha) application. The 40 kg/ha difference between the two soils in inorganic N represents a $10/ha application differential. That is, a farm manager would save $10/ha by applying less fertilizer to soil A than B. Expressed differently, if a farm manager applied the same amount of fertilizer to both soils, three outcomes are possible: (i) Soil A would be overfertilized while Soil B received the correct amount; (ii) Soil B would be underfertilized while Soil A received the correct amount; or (iii) with an average amount applied, Soils A and B would be simultaneously over- and underfertilized. Whatever the case, with the absence of prescription fertilization in a spatially variable field, costs for misapplication would arise due to both over- and underapplications of fertilizer.

CALCULATION OF MISAPPLICATION COSTS

Overapplication costs, in their simplest form, represent waste; they arise because too much product was applied to the relevant soil and the crop simply could not make use of it. Overapplication of an agricultural product or practice also could decrease crop yields, but this is assumed to be negligible in the simple examples used here. In contrast, underapplication costs represent the loss in crop yield because too little product was applied to the relevant soil. A commodity/product ratio (C/P) is associated with underapplication costs. C/P ratios of 1:1, 3:1, and 5:1 assume that the crop commodity is valued equally, three times more, and five times more, respectively, than the product applied to produce it.

Over- and underapplication costs are calculated relative to average application costs for a field. Average application costs are calculated by averaging the needs of the crop across the field and disregarding soil specific management. Using the N fertilizer example described above (i.e., a $10/ha differential), the averaged N application rates for each of the eleven fields shown in Fig. 10-1 would be as follows: 40, 44, 48, 52, 56, 60, 64, 68, 72, 76, and 80 kg N/ha, respectively. Averaged costs in each field associated with such management would be: 10, 11, 12, 13, 14, 15, 16, 17, 18, 19, and 20 dollars/ha, respectively (Table 10-1). Because these fields were not managed by prescription, the misapplication costs were 0, 18, 32, 42, 48, 50, 48, 42, 32, 18, and 0 dollars for each 10-ha field, assuming a C/P ratio of 1:1 (Table 10-1). Misapplication costs are equivalent to money saved if soils were managed by prescription.

The extent of misapplication costs is not constant, it changes with the degree of WFV. If cumulated over an entire field, misapplication costs are low when WFV is low, and rise exponentially with increasing WFV (Fig. 10-2). In Fig. 10-2, total misapplication costs are the sum of costs or losses associated with over- and underapplication of an agricultural product when a C/P ratio of 1:1 is assumed.

Table 10-1. Example of calculation of misapplication costs using N fertilizer costs of $10/ha for Soil A and $20/ha for Soil B in 11 10-ha fields (see Fig. 10-1).

Field No.	1	2	3	4	5	6	7	8	9	10	11
WFV Index	0	30	40	46	49	50	49	46	40	30	0

------------------prescribed application cost---------------------

Hectare											
A	10	20	20	20	20	20	20	20	20	20	20
B	10	10	20	20	20	20	20	20	20	20	20
C	10	10	10	20	20	20	20	20	20	20	20
D	10	10	10	10	20	20	20	20	20	20	20
E	10	10	10	10	10	20	20	20	20	20	20
F	10	10	10	10	10	10	20	20	20	20	20
G	10	10	10	10	10	10	10	20	20	20	20
H	10	10	10	10	10	10	10	10	20	20	20
I	10	10	10	10	10	10	10	10	10	20	20
J	10	10	10	10	10	10	10	10	10	10	20
Average:	10	11	12	13	14	15	16	17	18	19	20

------------------------misapplication cost[1]------------------------

Hectare											
A	0	9	8	7	6	5	4	3	2	1	0
B	0	1	8	7	6	5	4	3	2	1	0
C	0	1	2	7	6	5	4	3	2	1	0
D	0	1	2	3	6	5	4	3	2	1	0
E	0	1	2	3	4	5	4	3	2	1	0
F	0	1	2	3	4	5	4	3	2	1	0
G	0	1	2	3	4	5	6	3	2	1	0
H	0	1	2	3	4	5	6	7	2	1	0
I	0	1	2	3	4	5	6	7	8	1	0
J	0	1	2	3	4	5	6	7	8	9	0
Summation:	0	18	32	42	48	50	48	42	32	18	0

[1] Misapplication cost for each field is calculated by summing the absolute difference between the average application cost per hectare and prescribed application cost per hectare.

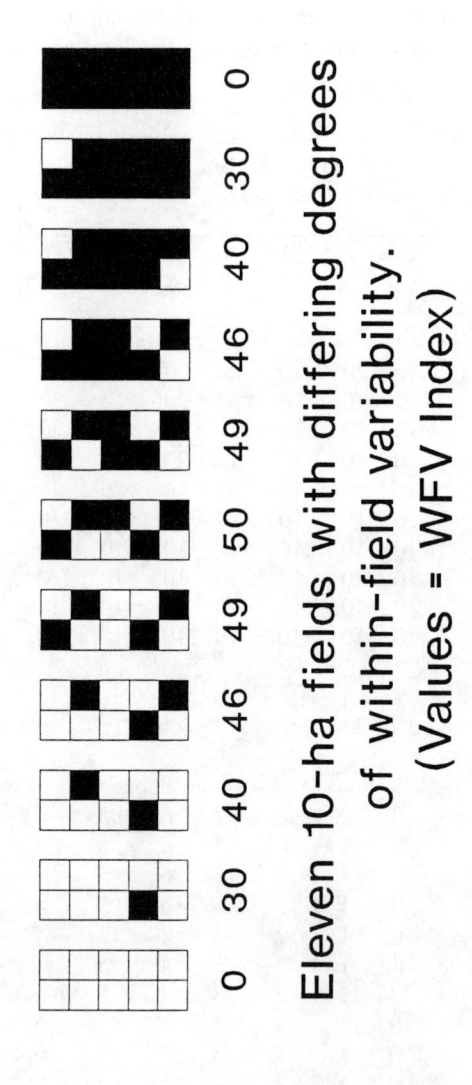

Fig. 10-1. Eleven 10-ha fields with differing degrees of within-field variability. Numerical values are WFV Indices. Soil A represented by white squares, soil B - black.

VARYING APPLICATION DIFFERENTIALS

Four application differentials are shown in Fig. 10-2. The lowest differential, $10/ha, may represent the N fertilizer example described above. In this case, with slight variability (WFV = 30), total misapplication cost is a mere $2/ha; even with maximum variability (WFV = 50), total cost of misapplication is only $5/ha. These amounts are so low that the value of differential management is nil. However, as the application differential increases to $20/ha (some fertilizer or herbicide applications), and $50/ha and $100/ha (some herbicide applications), the cost and/or financial gain due to misapplication and the value of prescription farming rises rapidly (Fig. 10-2).

Total cost of misapplication also rises quickly as the C/P ratio increases. For example, with maximum WFV (50) and at the $20/ha differential, misapplication costs with C/P of 1:1, 3:1, and 5:1 are $10/ha, $20/ha, and $30/ha, respectively (Table 10-2).

SUMMARY

These simple analyses lead to important conclusions regarding within-field variability. From a purely economic perspective, management of within-field variability is worthwhile, but only if the degree of soil variability is large enough to justify the costs of obtaining this information and managing it differentially. Little economic justification would exist for differentially managing a field with an inexpensive soil-applied herbicide such as trifluralin, whose efficacy is inversely related to soil organic matter. With this chemical, the cost of applying the maximum recommended rate may be as little as $5/ha. Even with high within-field variability, savings due to differential application would be minimal, and almost certainly less than the cost of detecting spatial patterns of soil organic matter.

In contrast, differential application of highly priced agrichemicals may be justified economically even in fields with relatively minor spatial variability. The primary conclusion, then, is that the value of managing within-field variability is highly dependent upon the degree of spatial variability and the cost of differential application.

Lastly, another important consideration regarding the value of managing within-field variability is environmental containment, integrity, and safety of agrichemicals in soils. These concepts are more difficult to calculate than that of economics. Nevertheless, should the management of within-field variability lessen the overall agrichemical load in both the agricultural and nonagricultural environments, then the value of such management increases appreciably.

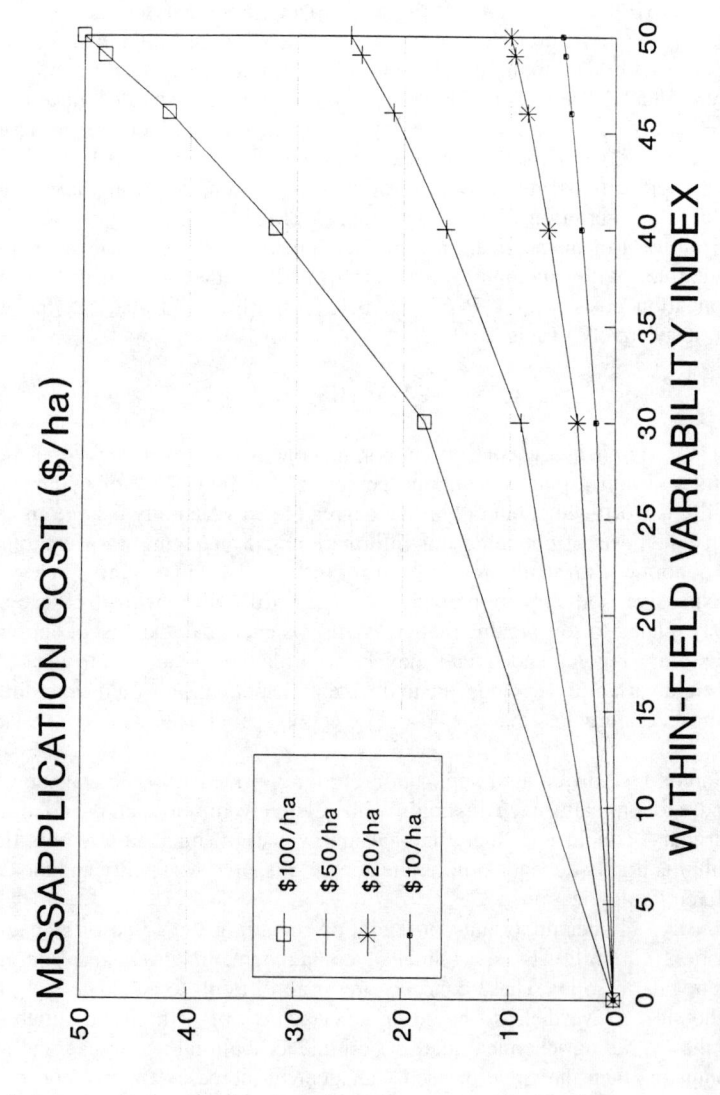

Fig. 10-2. Misapplication cost as a function of WFV index. Cost = (C$/P$ * Y$/2) + Y$/2, where C$ is the commodity price, P$ is the applied product price, and Y$ is the per-hectare application differential between soil types.

MANAGING WITHIN-FIELD VARIABILITY

Table 10-2. Estimated misapplication costs for the 11 fields shown in Fig. 10-1. WFV may result in differing management costs. Examples of $10, $20, $50, and $100 per ha cost differentials are provided.

Field no.	1	2	3	4	5	6	7	8	9	10	11
WFV Index	0	30	40	46	49	50	49	46	40	30	0
$10 Differential ($10 to $20/ha)											
Avg. Cost	100	110	120	130	140	150	160	170	180	190	200
Over	0	9	16	21	24	25	24	21	16	9	0
1:1 Under	0	9	16	21	24	25	24	21	16	9	0
Total	0	18	32	42	48	50	48	42	32	18	0
3:1 Under	0	27	48	63	72	75	72	63	48	27	0
Total	0	36	64	84	96	100	96	84	64	36	0
5:1 Under	0	45	80	105	120	125	120	105	80	45	0
Total	0	54	96	126	144	150	144	126	96	54	0
$20 Differential ($5 to $25/ha)											
Avg. Cost	50	70	90	110	130	150	170	190	210	230	250
Over	0	18	32	42	48	50	48	42	32	18	0
1:1 Under	0	18	32	42	48	50	48	42	32	18	0
Total	0	36	64	84	96	100	96	84	64	36	0
3:1 Under	0	54	96	126	144	150	144	126	96	54	0
Total	0	72	128	168	192	200	192	168	128	72	0
5:1 Under	0	90	160	210	240	250	240	210	160	90	0
Total	0	108	192	252	288	300	288	252	192	108	0
$50 Differential ($10 to $60/ha)											
Avg. Cost	100	150	200	250	300	350	400	450	500	550	600
Over	0	45	80	105	120	125	120	105	80	45	0
1:1 Under	0	45	80	105	120	125	120	105	80	45	0
Total	0	90	160	210	240	250	240	210	160	90	0
3:1 Under	0	135	240	315	360	375	360	315	240	135	0
Total	0	180	320	420	480	500	480	420	320	180	0
5:1 Under	0	225	400	525	600	625	600	525	400	225	0
Total	0	270	480	630	720	750	720	630	480	270	0
$100 Differential ($10 to $110/ha)											
Avg. Cost	100	200	300	400	500	600	700	800	900	1000	1100
Over	0	90	160	210	240	250	240	210	160	90	0
1:1 Under	0	90	160	210	240	250	240	210	160	90	0
Total	0	180	320	420	480	500	480	420	320	180	0
3:1 Under	0	270	480	630	720	750	720	630	480	270	0
Total	0	360	640	840	960	1000	960	840	640	360	0
5:1 Under	0	450	800	1050	1200	1250	1200	1050	800	450	0
Total	0	540	960	1260	1440	1500	1440	1260	960	540	0

Average cost of management (Avg. Cost) disregards WFV, and when this occurs there arise two misapplication penalties over- (Over) and underapplication (Under), as well as their sum (Total). "Over" represents waste, in that the crop can not make use of the product; whereas "Under" represents loss in crop production because too little of the product was applied. The ratios, 1:1, 3:1, and 5:1, represent the value of lost crop commodity compared to the product applied.

11 Working Group Report

Richard R. Johnson, Chair
W. Voorhees, Recorder
F. Forcella, Discussion Paper

Other Participants:	R. Allmaras	W. Nelson
	R. Ascheman	R. Pavelski
	H. Cheng	W. Pan
	D. Clay	F. Pierce
	J. Culley	E. Schnug
	D. Lytle	D. Tyler
	D. Mortensen	R. Wagenet
		R. Wolkowski

The objective of managing variability should be to manage each site in a field at an optimum level to produce products at the greatest return without damaging the environment. In many food or forage crops, both quality and yield are of large importance in characterizing value of the final product. Quality plays an important but lesser role in feed grains. The farmer and his advisors will determine if the objective of optimization of inputs and production has been attained. A measure of environmental optimization will probably come from regulatory groups or the perceptions of society.

FACTORS INFLUENCING VARIATION MANAGEMENT

Managing variability first requires identification of information needed. Data needed may change as new problems arise, as knowledge of a field evolves, or as management goals are altered. In some cases, information may already exist in a database, i.e., long-term climatic data or soil morphologic information. Where databases exist, a judgement must be made on adequacy for the proposed use. Where databases do not exist, or exist with insufficient detail, tests or assays must be conducted to define existing variation.

In the process of obtaining information on existing variation, consideration should be given to determine if variation:
1. Is predictable, i.e., does it occur in a random or nonrandom pattern,
2. Is natural or human-induced,
3. Is static or dynamic -- if dynamic, will it vary either in time, space, or both.

Copyright © 1993 ASA-CSSA-SSSA, 677 South Segoe Road, Madison, WI 53711, USA. *Soil Specific Crop Management.*

For example, a particular weed may be present only in a restricted field area. Further investigation indicates that this weed pest is associated with a former livestock containment area used several years ago. This information reveals that the pest is nonrandom, was induced by human activities, and has not readily spread either in time or space. Control strategies for this pest can be quite different from a highly mobile disease pathogen that is occurring with no apparent pattern within a field.

Another integral factor to managing variation is a thorough understanding of how the crop or cropping system responds to inputs designed to manage this variation. Perhaps a simple change in cultivar or minor change in crop rotation can control variation. If an added input is needed, a dose-response curve for that input is of prime importance. Several other factors that should be considered include possible interaction of the input with soil type, crop cultivar, fertility conditions, other applied inputs, and future cropping plans. When variable rate technology is used, consideration should also be given to impacts such as traffic patterns (Will excessive trips be required in areas not requiring treatment?) or response time of equipment being used (Can equipment alter rate rapidly enough to control the variation?).

Placement of management inputs can also significantly control variation. For example, soils high in pH may favor use of incorporated banded vs. surface broadcast P. Likewise, high pH favors use of postemergence vs. soil-applied pesticides. Tradeoffs in placement often exist such as reduced N loss of urea-containing fertilizer occurs with soil incorporation while less soil erosion occurs with tillage systems minimizing incorporation of protective surface plant residue.

Environmental impacts should also be considered when managing variability. On soils vulnerable to leaching, chemical choice and application timing can reduce impacts on the environment. Judicious use of tillage, crop rotation, and animal manure may also reduce chemical needs. Soil erosion can be reduced with inputs such as conservation tillage, crop rotation, and contour farming. Choosing an optimum cropping system for the environment often involves many tradeoffs making the management of variability across a field difficult. For example, a Corn Belt farmer often can reduce soil erosion most by using continuous no-tillage corn production. Yet, using a corn/soybean (*Zea mays* L.)/[*Glycine max* (L.) Merr.]) crop rotation with some tillage may have several other environmental benefits, i.e., less insecticide and herbicide, lower N fertilizer rates, and higher crop yields. The ability to choose among these tradeoffs is often difficult, and not obvious on a given field.

Profit and feasibility of implementing decision rules must be strongly considered when managing variability. Alternatives that increase profits will be adopted most rapidly. Many farmers will consider profit-neutral alternatives if other obvious advantages are apparent, i.e., less environmental damage or reduced risk. The scale of the operation will also impact adoption. In a highly variable field, large farm operators may be restricted by large machine sizes. Conversely, small operators may be restricted by management skills or capital, particularly if they are part-time farmers whose farm generates only a small fraction of their income.

WHAT WE KNOW

Adequate information exists to make many sound crop management decisions. There is room to refine some of the areas that will be discussed below, but future refinements of knowledge will not likely lead to major gains in productivity, profitability, or environmental protection. Rather, these refinements will improve precision or extend knowledge to specific situations. Other currently well-understood areas will require continual updating as new technology out-dates older technology. For example, optimum crop cultivar and pesticide choice are often known, but will change with advances in crop breeding and pesticide development.

Mineral nutritionists have developed good tests to measure soil pH, organic matter, and P. Potassium can be measured with reasonable accuracy. Nitrogen can be measured, but method and time of sampling is in need of considerable refinement. Some micronutrients can be tested, but in general, this area is much less advanced than for other elements. We are able to make recommendations on crop nutrient needs at a level paralleling our understanding of testing for availability of particular elements.

In the area of pest control, a majority of pests can be easily identified. Threshold populations for many insects have been identified to define when treatment is justified. Thresholds for weed species are being identified, but disease thresholds are understood to a lesser degree. Where pesticides are required for control, prescriptions for rate and time of application are available. How pesticides interact with each other, soil properties, crop rotations, and crop cultivars are also reasonably understood. General information on pesticide use is presented on pesticide labels, and considerable additional information is also available through local dealers and extension offices.

Data are also available on climatic probabilities affecting crop growth and development. Databases on soil physical properties are also reasonably developed. However, these databases are not always used, and they may not be in a format or scale that facilitates optimum use.

Considerable information is available on optimum variety choice, seeding rates, date of planting, row spacing, and seeding depth. The influence of crop rotation on crop yield and input requirements is also generally well understood, especially for widely used cropping systems.

Amount and timing of crop water needs also is known generally. Since 1980, our understanding of how alternative tillage systems affect crop productivity has expanded greatly. Tillage interactions with crop rotation and soil type are also available for most major cropping systems and areas. Poorly drained soils generally yield best where tillage is deeper and incorporates more residue below the soil surface. Conversely, well-drained soils generally yield best when residue is left on the soil surface, and no-, or shallow-tillage is used. Exceptions to these rules exist, but are often understood within a cropping region. Adequate planting and tillage equipment exists to create and operate in these different environments.

The good knowledge base described above has been tested and proven in static environments, i.e., environments where inputs are held at a constant level throughout the field. Much less is understood about applying such information in situations where the environment is dynamic and where it is desirable to change inputs throughout a field. The following section will describe how our knowledge base should be extended to manage variability.

RESEARCH NEEDS

Six areas of research needs were identified that should be addressed. These needs are as follows:
1. Data capture and analysis.
2. Design and protocol for experiments.
3. Integrated crop management.
4. Soil taxonomy and database utility.
5. Machinery, sensors, and field operations.
6. Cost/benefit/risk analysis.

The first two topics were identified as crucial. The remaining four topics were judged to be equally important among themselves, but less important than the first two topics. All six topics are explained in greater detail below.

1. Data Capture and Analysis

Outlines of soil mapping units are generally irregular in shape, and may or may not follow landscape contours. Many other forms of site variability also may be irregular. In contrast, the sensing of crop yield and quality normally is done in a grid pattern, where grid spacing may or may not be appropriate to underlying site variability. The difficulty of integration or overlay of these two schemes must be overcome.

Integration and analysis of site variability information is especially difficult. Use of new analytical tools, such as geographic information systems (GIS), and development of appropriate statistical methods are necessary before site variability can be managed reliably and profitably.

There often is not a direct relationship between the precision of an input attribute and a measured yield response. For example, a soil attribute may vary by 100% but the yield response to that variation may be only 1%. Similarly, the spatial distribution of a soil attribute may be very uniform but yields show a wide range of values.

Variability may be dynamic; climate variability has a scale of hours whereas some soil characteristics may be stable over decades. Changing economics will also change the relative importance of different attributes.

In order to respond to dynamic conditions of the situation, a farm operator must be provided a continuously updated knowledge base, and be able to use that update.

Much is known about how soil and climate affect a few crop species, and varieties within those species. Whether this knowledge can be applied to other species/varieties is not well known.

Any management scheme addressing soil variability should also have the capability to incorporate the animal component, especially from the standpoint of manure management.

2. Protocol for Experimental Design and Analysis

Certain technologies may not transfer equitably to all farms. Small operations likely will not have the same labor, capital or machinery complements as a larger farm. How should new technology be scaled?

The research community generally deals with small plots of land. How should resulting knowledge and experience be transposed to large fields? Public research institutions often do not have the infra-structure to support large field-scale research.

In designing experiments and analyzing data, how should variation be categorized? Variation may be (i) random, clumped, or regular; (ii) dynamic or static (stationary); and (iii) temporal or geographic. Creative new ways of statistical thinking involving a holistic approach, will be necessary. Analyses may be based more on regression analysis and multivariate analysis than on traditional analysis of variance.

3. Integrated Crop Management (ICM)

When attempting to relate crop yields to site variability, the extent of yield variability attributable to influences from pests, weather, etc. must be determined.

How should variability relationships from plant-soil monocultures be extended to crop rotation schemes?

Within integrated crop management, there are also integrated pest management concerns. While threshold values for certain pests are known, these may change over space and time. Control of a pest in one area may increase the pest pressure in another area. Changing patterns of control strategies (gradual shift from chemical to physical to biological) must be recognized.

Tillage or plant residue management to control soil erosion can increase pressure from certain pests. Research is needed to determine the proper balance for a given situation.

4. Soil Taxonomy and Databases

Are current methods of soil taxonomy description giving the needed information with respect to scale and parameters being measured? If the spatial scale of our data application is changed, will the current diagnostic tools be adequate?

When mapping soil variability, how do soil biota affect the interpretations?

5. Machinery, Sensors, and Field operations

Research is needed to determine the best ways to use real-time sensing of soil variability (as opposed to a static soil map). At what scale can inputs be managed? Practically, can a tillage implement be changed within a few meters or seconds? Should we? From a machinery management perspective, research is needed to determine effects of traffic patterns and field operation configurations for time efficiency and controlling soil compaction.

Accurate and economical on-the-go sensors of yield and quality, weed species, soil nutrients, and soil water content are needed.

A nondestructive sensor of yield is needed to make decisions regarding preharvest preparation, marketing, and machinery scheduling.

Soil variability requires ways to differentially apply irrigation water. Technology is needed to control a center pivot system to apply varying amounts of water depending on soil parameters and crop demand.

The control of all the variability-related data must be kept simple and small to accommodate limited space and operator flexibility in a truck or tractor cab.

Remotely sensed data must be in digital form, timely, and provide for sufficient reaction time.

6. Cost/Benefit/Risk management

The degree of variability within a field dictates the economic and environmental feasibility of managing variability. Similarly, the market price of the commodity grown on the field determines economic justification for managing variability. Consequently, in-depth analyses of the degree of practical variability within a field must be matched with ranges of potential commodity prices to determine whether field variability should be managed.

Are current tools good enough if the scale of management is changed? Can new tools be afforded by potential users?

Considerable effort must be made to educate and train people on this emerging technology, recognizing the diversity of managerial skills at all levels of the farming operation.

How should the benefits of farming by soil be assessed? Is profit the only criterion? Recognition of long-term potential stewardship aspects is necessary. Moreover, the possibility that governmental regulations may require changes that impact the profit margin must be considered.

An effective overall evaluation of farming by soil type needs to incorporate emerging GIS technology, along with computer-assisted decision aid technology.

SECTION III

ENGINEERING TECHNOLOGY

12 Sensing and Measurement Technologies for Site Specific Management

Steven C. Borgelt
Agricultural Engineering Department
University of Missouri
Columbia, MO

The development of sensors, data collection and processing systems, and control systems was ranked as the highest engineering research priority for agriculture in 1984 (ASAE, 1984). In 1991, this same research priority ranked second behind the development and evaluation of management practices for water quality (ASAE, 1991). The development of sensor technology and control strategies for managing variability in soils and cropping systems addresses both of these research priorities.

Customizing the rates of crop production inputs, such as herbicides, pesticides, fertilizers, and irrigation water at every location in a field, may help accomplish the goals of (i) environmental protection, by minimizing overapplication; and (ii) maximum profitability, by optimizing inputs and yields. The development of sensors and control systems may encourage widespread acceptance and adoption of site specific agriculture.

This chapter summarizes sensor development past and present. However, the summaries in these pages do not contain a complete worldwide survey of the research, development, and products available. A complete survey would be too lengthy. My presentation will address the summarized technologies and discuss others. The summaries briefly describe a product, technology, or research project, and then provide information for the reader to pursue an in-depth investigation of the sensor at will.

This chapter is an attempt to provide accurate, correct, and current information. However, some information may have changed and may be outdated. The summaries are not an endorsement of the products or technologies.

Copyright © 1993 ASA-CSSA-SSSA, 677 South Segoe Road, Madison, WI 53711, USA. *Soil Specific Crop Management.*

The outline of this chapter is as follows:
YIELD SENSORS
    European Production
    Foreign Developments
    Domestic Developments
ON-THE-GO GRAIN MOISTURE SENSORS
SOIL ORGANIC MATTER SENSORS
SOIL CHEMICAL SENSORS

YIELD SENSORS

European Production

Claydon Yieldometer

Description: This sensor measures the volume flow of grain leaving the combine grain bin auger (Fig. 12-1). Grain flows into the top of the Yieldometer and accumulates on a stationary paddle wheel. As the grain level approaches a level sensor, the sensor activates a relay which controls an electromechanical clutch. The clutch then connects a mechanical chain drive with the clean grain elevator drive system to rotate the paddle wheel and meter grain into the combine grain tank. Each half revolution of the paddle is sensed and an output signal is sent to the display unit mounted in the cab, which can provide a printout of yield or store the data on a removable memory card. The company claims an accuracy of +/- 1%. It is currently available as an option on European Claas combines, and may be available on North American machines in the near future. Price: $6,000 - $10,000.

A hydraulically driven Claydon Yieldometer was used at Texas A&M University for yield mapping research (Bae et al., 1987; Searcy et al., 1989) and the same one is being used, with modifications, at the University of Missouri.

Information contacts:
Mr. John T. Claydon	Mr. Jorge Bethege
Claydon Yieldometer LTD	CLAAS OHG
Gaines Hall, Wickhambrook	Postfach 1140
Newmarket, Suffolk	D-4834 Harsewinkel
England CB8 8YA	Germany
Phone: 440-820327	Phone: 49 52 47 12963
	FAX: 49 52 47 12925

SENSING AND MEASUREMENT TECHNOLOGIES 143

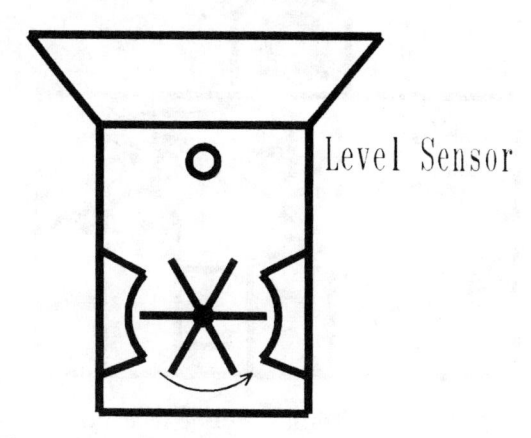

Figure 12-1. Schematic of the Claydon yieldometer.

Flowcontrol

Description: Flowcontrol monitors grain flow rate by means of gamma rays absorption (Fig. 12-2). The system consists of three units -- a gamma ray emitter, a detector, and a display unit. The emitter is mounted under the grain flow from the clean grain elevator with the detector mounted directly above the emitter. As the grain passes through a measuring gap between the emitter and detector, it reduces the intensity of the gamma radiation registered by the detector. This reduction is proportional to the grain mass flow rate. The company claims an accuracy of +/- 1% and states that the device is not affected by moisture content.

The price of the monitor is 18,033 DK (about U.S. $3,000).

Reports indicate the device is not available as a retrofit and may not be available for import into the United States.

Information contacts:

Dronningborg Maskinfabrik A/S
Udbyhojvej 113, P.O. Box 80
DK-8900 Randers
Denmark
Phone: 45 86 42 58 55
Sales FAX: 45 86 40 58 24

Overgaard/Thoustrap
Box 96
DK-8900 Randers
Denmark
Phone: 45 86 40 40 44
FAX: 45 86 40 70 10

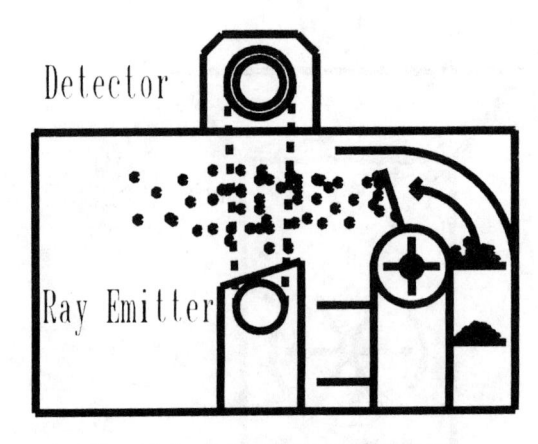

Fig. 12-2. Schematic of flowcontrol.

Belgium

Description: This sensor uses the change of momentum of moving grain impacting against a curved plate as an indicator of flow rate (Fig. 12-3). The curved plate is mounted at the outlet of the clean grain elevator, so that the grain coming from the elevator is forced to change direction. The changing momentum of the grain causes friction and impact forces on the curved plate. In the current sensor design, a leaf spring opposes motion of the plate and an inductive displacement sensor is used to measure the resulting deflection. This measured displacement is proportional to the force exerted on the plate, and therefore, to the grain mass flow rate. The sensor developers have shown accuracies of +/- 2-3%. (Vansichen and DeBaerdemaeker, 1991).

The development of this sensor is a cooperative effort between Ford-New Holland, Belgium and the Agricultural Engineering Department of the Katholieke Universiteit Leuven. An European patent application has been filed.

Information contacts:
>   Mr. Raf Vansichen
>   Agricultural Engineering
>   Kardinal Mercierlaan, 92
>   B-3030 Leuven Heverlee
>   Belgium
>   Phone: 16 22 09 31
>   FAX: 16 22 18 55

Mr. G. Strubbe
Ford-New Holland
Leon Claeysstraat 3a
B-8210 ZEDELGEM
Belgium

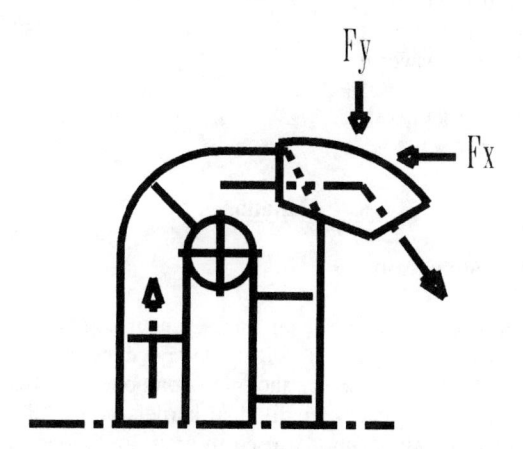

Fig. 12-3. Schematic of Belgian sensor.

Silsoe Research Institute, Great Britain

Description: A capacitive sensor is being developed and tested at the Silsoe Research Institute. The dielectric of the sensing capacitor changes as the grain flow through the sensor changes. Calibration of the sensor is material dependent and varies with the material distribution in the sensing volume. Accuracy of this sensor is moisture content dependent, although tests have shown that the effects of moisture content are small. The system is nonintrusive and insensitive to transmitted vibration. Semi-annular capacitance plates are mounted around the discharge auger tube and the device is enclosed in a brass tube that acts as a shield to reduce stray capacitance effects. (Stafford et al., 1991).

Information contact:
   Dr. John Stafford
   Silsoe Research Institute
   Wrest Park
   Silsoe, Bedford MK45 4HS
   England
   Phone: 44 525 860000
   FAX: 44 525 860156

RDS Technology Ltd., Great Britain

Description: A grain yield measurement system is being developed. Due to patent considerations and proprietary concerns, detailed information is not available. Details on the device are expected in the fall of 1992.

Information contact:
> Mr. Richard Danby
> RDS Technology Ltd.
> Stroud Road Nailsworth
> Gloucestershire GL6 OBE UK
> Phone: 44 453 834084
> FAX: 44 453 835521

Canada

University of Saskatchewan

Description: A grain flow sensor for use under combine sieves has been developed from piezo-film strips. A piezo-electric material, high polar Poly-Vinylidene Film (PVDF) was used as the main component of this experimental grain flow monitor. The impact of individual kernels on the film is monitored and recorded, a concept similar to that used in grain loss monitors. The sensor is mounted underneath the combine sieve and samples a portion of the grain from the sieve. In laboratory tests, grain flow determined by the sensor showed an average error of 4.5%, with a maximum error of 7.5%. Field tests showed relatively large measurement errors (Pang and Zoerb, 1990).

Information contact:
> Dr. Gerald C. Zoerb
> Agricultural Engineering Department
> University of Saskatchewan
> Saskatoon, Saskatchewan
> Canada   S7N 0W0
> Phone: (306) 966-5319
> FAX: (306) 966-5334

United States Developments

University of Idaho

Description: This grain-yield monitoring concept uses a permanent magnet motor to drive the clean grain system. A 12-volt DC motor, equipped with a shunt in the power line, powers the clean grain auger under the combine shoe. The current supplied to drive the motor is indicated by measuring the voltage across the shunt. Field tests have shown that the measured voltage is linearly related to quantity of grain in the clean grain system.

Research on this concept for yield measurement and mapping continues.

Information contacts:
    Dr. Charles Petersen
    Agricultural Engineering Department
    University of Idaho
    Moscow, ID 83843
    Phone: (208) 885-7906
    FAX: (208) 885-8923

    Dr. Geoffrey Shropshire
    Agricultural Eng. Department
    University of Idaho
    Moscow, ID 83843
    Phone: (208) 885-6182
    FAX: (208) 885-8923

Kansas State University - Pivoted Auger

Description: An auger is mounted with one end pivoted and the opposite end supported by a load cell (Fig. 12-4). Grain flows from the existing clean grain auger into the pivoted end and travels through the auger to be discharged to the grain tank. The signal from the load cell is filtered and used as measurement of grain flow. The biggest drawback to this system is the limited installation space available on a combine (Wagner and Schrock, 1989). Lab and field data indicate an accuracy of +/- 3%.

    Note: Another concept is being developed and considered for commercialization. Due to patent considerations and the proprietary nature of the sensor, details of the sensor cannot be disclosed.

Information contact:
    Dr. Mark Schrock
    Agricultural Engineering Department
    Seaton Hall
    Kansas State University
    Manhattan, KS
    Phone: (913) 532-5580
    FAX: (913) 532-6944

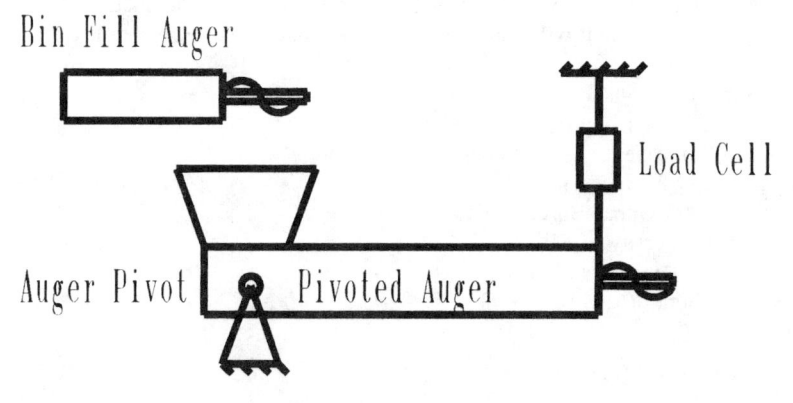

Fig. 12-4. Schematic of pivoted auger.

South Dakota State University - Bushel Buddy

Description: Yield is determined on a volume basis in the combine grain bin. An ultrasonic sensor mounted above the combine bin determines the depth or change in depth of the grain. This change in grain depth, along with the hopper dimensions, can be used to find a change in grain volume over a traveled distance. Two preliminary tests were performed in 1991, and all components functioned acceptably. A possible improvement considered for this system would be the use of multiple ultrasonic sensors to determine the total topography of the grain surface in the combine grain bin. Grain moisture content is determined by a capacitive sensor mounted in the grain tank fill auger.

No plans for commercialization have been made.

Information contacts:

Mr. Kent Klemme
Agricultural Engineering Department
Box 2120
South Dakota State University
Brookings, SD 57007
Phone: (605) 688-5141
FAX: (605) 688-4917

Dr. Don Froehlich
Mechanical Eng. Department
Box 2219
South Dakota State University
Brookings, SD 57007
Phone: (605) 688-5426

USDA Agricultural Research Service - University of Minnesota

Description: This sensor is a conical weighing device mounted in the clean grain system of a combine, between the clean grain elevator and the auger that delivers grain to the grain tank. Grain is dropped into a sheet metal funnel and flows through the opening in the bottom of the funnel onto a point-up metal cone supported by a load cell. The grain then falls into the auger that delivers it into the grain tank. Changes in the load cell signal are correlated to changes in the grain flow rate onto the point-up cone. The developers indicate that a considerable amount of noise has been encountered in the load cell signal. Further development will continue to improve performance of the device.

Information contact:
Dr. John Lamb
Soil Science Department
439 Borlaug Hall
1991 Upper Buford Circle
University of Minnesota
St. Paul, MN 55108
Phone: (612) 625-8743

USDA Agricultural Research Service - University of Illinois

Description: A method of measuring the depth of grain on individual flights of the clean grain elevator is being evaluated. A light source and photodiodes are mounted in the clean grain elevator. The light that is received at the photodiodes gives an indication of the quantity of grain on each flight. Limited laboratory tests have been conducted, verifying the validity of the sensing concept. A test stand to perform further tests is being constructed. Additional laboratory and field tests are planned.

Information contact:
    Dr. John Hummel
    USDA Agricultural Research Service
    Agricultural Engineering Department
    University of Illinois
    Urbana, IL 61801
    Phone: (217) 333-0808
    FAX: (217) 244-0323

Farmers Software Association

Description: A grain flow measurement system is being developed for measuring and mapping grain yield variations. The device uses a simple principle and is located in the clean grain system. Due to patent considerations and proprietary concerns, detailed information is not available. It is reported that existing measuring and monitoring techniques are being employed in a unique fashion to measure grain flow.

Information contact:
    Mr. Neil Havermale
    Farmers Software Association
    P.O. Box 660
    Fort Collins, CO 80522
    Phone: (800) 237-4182
    FAX: (303) 223-4093

J.P. Faivre Partnership

Description: A grain flowmeter has been designed and is being tested as add-on equipment for existing combines. The output signal is designed to be compatible with existing sprayer monitors and display consoles. Due to patent considerations and the proprietary nature of the sensor, details cannot be disclosed. The device will be marketed through Dawn Equipment. A target marketing date is set for August/September 1992. Target price: under $1000.

Information contact:
Mr. Steve Faivre
J.P. Faivre Partnership
1913 W. Fairview Drive
Dekalb, IL 60115
Phone: (815) 756-1812
FAX: (815) 756-1007

Rodvelt Agritronics - Bin Watch

Description: Light-emitting diodes (LEDs) in the combine cab indicate the amount of grain in the combine grain tank. A strip of electronic sensors mounts in the combine grain bin and a programmable control box mounts in the cab. As grain levels rise or fall, the sensor at each level sends signals to light the LEDs for that level. Models can be equipped with one, four, or eight sensors or can be custom made with 16 sensors. This device was not developed to precisely monitor grain flow, but could be used for that purpose. Price (1990): Single sensor ($75), four sensors ($125), eight sensors ($175).

Information contact:
Mr. Ron Rodvelt
Rodvelt Agritronics
Box 14543
Shawnee Mission, KS 66215
Phone: (913) 338-1629

ON-THE-GO GRAIN MOISTURE SENSORS

These grain moisture sensors use capacitance to measure moisture content and are temperature compensated for improved accuracy. All provide a digital readout of moisture content.

Moisture Trac

Model 5010 - A small hole is cut in the clean grain auger. As grain flows through the bin fill auger, a small amount of grain slips through the hole into a small sampling chamber. There, a sensor measures the moisture content of the grain. As the moisture content value is displayed, the sample chamber empties through a rotating release disk. Price: $1395

Model 3000 - A fin-type blade sensor mounts inside the combine grain bin fill auger. A short length of auger flighting is removed for installation. As grain passes by the sensor blade in the auger tube a continuous readout of moisture content is provided. Price: $695.

Information contact:
   Shivvers, Inc.
   614 W. English
   Corydon, IA 50060
   Phone: (515) 872-1005
   FAX: (515) 872-1593

David Manufacturing Co.

The Cal-U-Dri system uses a stainless steel fin-type blade sensor that mounts near the top of the combine bin fill auger and extends into the auger. A short length of auger flighting is removed for installation. As grain is augured past the sensor it continuously updates the moisture content reading. An alarm will sound when grain moisture content reaches a preset value. Price: $795.

Information contact:
   David Manufacturing Co.
   1600 12th Street N.E.
   Mason City, IA 50401
   Phone: (515) 423-6182
   FAX: (502) 685-6678

Moisture Control Systems

The Micro 4 system uses a fin-type blade sensor that resides in a test chamber such as a sampling bucket or bin. After the test chamber fills and the sensor is uniformly covered with grain, a moisture content reading is taken. This sensor does not provide a continuous sampling of grain moisture.

Information contact:
   MPD/MCS
   316 E. Ninth Street
   Owensboro, KY 42303
   Phone: (502) 685-6533
   FAX: (502) 685-6678

SOIL ORGANIC MATTER SENSORS

Soil organic matter can be estimated by measuring soil darkness at one or multiple wavelengths. Two types of soil organic matter sensors have been developed.

Single Wavelength, Landscape Dependent Sensor

This sensor projects a single wavelength of red light onto the soil. The reflected energy is measured to estimate the organic matter content for a specific landscape. The sensor is rugged and relatively inexpensive. The sensor requires recalibration for new landscapes and moisture levels. The sensor and related control equipment have the trade name - S.M.A.R.T. - Soil Monitoring & Application Regulator by Tyler. Tyler Limited Partnership holds the rights for sale of the system. Dr. Larry Gaultney, agricultural engineer at Purdue University was responsible for its initial development.

Price: $7500 for the base unit, all options add up to $2,000 additional.

Information contact:
  Mr. Don Mcgrath, President
  Tyler
  P.O. Box 249
  East Highway 12
  Benson, MN 56215
  Phone: (612) 843-3333

Multiple Wavelength, Landscape Independent Sensor

This sensor projects multiple wavelengths light onto the soil surface. The sensor then measures and records the reflected light energy. This sensor has been calibrated for soils that span a geographic range and a range of soil moisture. This technique uses more complex technology than the single wavelength sensor. More complex techniques can increase cost and require more maintenance. To date, limited results have been reported on field investigations. Date to commercialization is uncertain. This sensor was developed by Dr. Ken Sudduth and Dr. John W. Hummel, USDA Agricultural Research Service, and Mr. Bob Funk, AGMED,Inc. The sensor has been patented and licensed to AGMED, Inc., Springfield, IL for commercialization.

Information contact:
  Dr. Ken Sudduth
  USDA Agricultural Research Service
  Agricultural Engineering Bldg.
  University of Missouri
  Columbia, MO 65211
  Phone: (314) 882-4090
  FAX: (314) 882-1115

SOIL CHEMICAL SENSORS

Soil Doctor

Description: The Soil Doctor is an on-the-go soil properties analysis and prescription fertilizer application system. The complete package consists of a sensor, monitoring electronics and a control system for varying fertilizer

SENSING AND MEASUREMENT TECHNOLOGIES

application rates.

The system has been field tested since 1987 and was introduced for use in 1990. The following are brief descriptions of the products.

Model 90 (Fig. 12-5). This model has been replaced by the new systems. A sensor measures the chemical composition of the soil. A small sensing device is located in the bottom of an applicator knife. A water-based solution is injected through a thin slot in the knife, into the soil, just ahead of an insulated sensing electrode. The solution mixes with the soil and under electrical stimulation a reaction occurs that allows the electrode to accurately detect the amount of nitrate nitrogen in the top foot of soil. Price: $12,995.

For 1992 two new models of systems have been introduced. Both of the new systems use two coulter openers and the Soil Doctor AXT adds trailing sensing blades (Fig. 12-6).

RE - The Soil Doctor RE uses a rolling electrode sensing system. Fluid is not injected into the soil in this system. The rolling electrode sensor uses an electronic interrogation system to examine soil properties. The sensing depth and sampled soil volume are expanded from the Model 90. Price: $9,990.

AXT - The Soil Doctor AXT provides a variety of electrode-equipped sensor knives and analysis solutions for special needs. In this system, a fluid is injected into the soil by the first knife, the solution mixes with the soil and under electrical stimulation the trailing sensor knife examines the soil type, organic matter, soil moisture, and nitrate levels to prescribe and deliver fertilizer on-the-go. Price: $12,995.

An anhydrous ammonia control option adds $3,765 to the price of the systems.

Information contact:
    Mr. John W. Colburn, Jr.
    President and CEO
    Crop Technologies, Inc.
    1811 Upland Drive
    Houston, TX 77043
    Phone: 800-637-2767
           (713) 973-2767
    FAX: (713) 379-4393

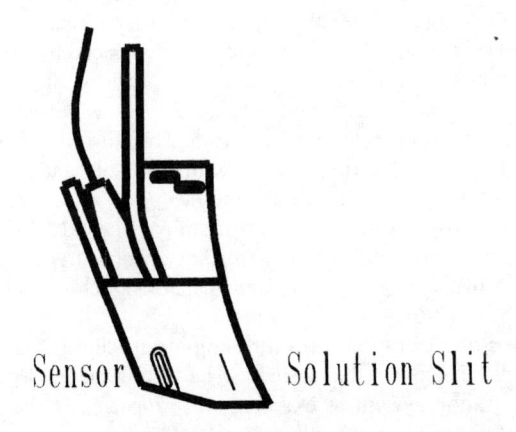

Fig. 12-5. Schematic of soil doctor model 90.

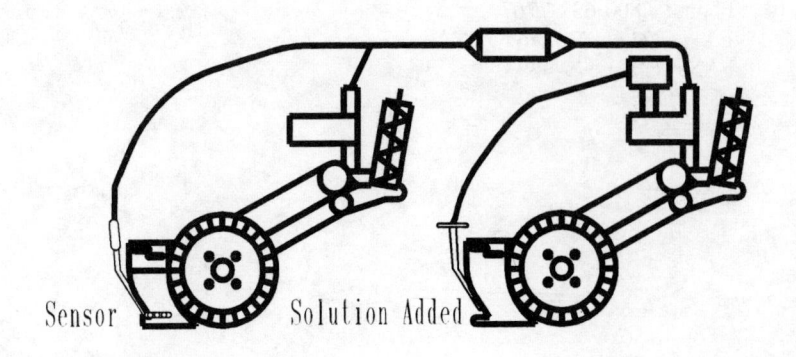

Fig. 12-6. Schematic of soil doctor model AXT.

Ion Selective Measurement Techniques

The following devices all operate by measuring specific ions that are present in a soil solution.

Ion Selective Field Effect Transistors

Description: Research using Ion Selective Field Effect Transistors (ISFETs) to determine the nitrates in a soil sample is underway. Ion selective sensors are mounted on computer chips and when used with specific membranes, ion concentrations can be measured. Laboratory tests with ISFETs are showing reasonable responses to nitrate ions. Developing a system using ISFETs to evaluate in-field soil nitrate is the goal of the work.

This research is a cooperative effort of Hitachi, Ltd. and the University of Illinois.

Information contacts:

Dr. John W. Hummel
USDA Agricultural Research Service
Agricultural Engineering Department
University of Illinois
Urbana, IL 61801
Phone: (217) 333-0808
FAX: (217) 244-0323

Dr. Stuart Birrell
Agricultural Eng. Department
University of Missouri
Columbia, MO 65211
Phone: (314) 882-2350
FAX: (314) 882-1115

Electrochemical Nitrate Measurements

Description: Research using a nitrate-selective electrode in an electrochemical cell has been reported. On-the-go sampling is followed by nitrate extraction. The extraction is then pumped through an electrochemical flow cell, where nitrates are sensed by a nitrate-selective electrode. Laboratory and field experiments have indicated that ion-selective electrode technology is adaptable to automated field monitoring of soil nitrate levels. Currently, the time and methodology of obtaining nitrate extraction appear to be limiting factors. Further development is aimed at the extraction process and interpretation of the measured nitrate signal. The system may be ready for commercial application as early as 1994.

Information contact:
Dr. John F. Adsett
Agricultural Engineering Department
Nova Scotia Agricultural College
P.O. Box 550
Truro, Nova Scotia, Canada, B2N 5E3
Phone: (902) 893-6711
FAX: (902) 893-4547

Hand-held Ion Meters

Description: Several shirt-pocket size hand-held meters with ion selective sensors and digital readouts are available. Individual meters will measure the following ions in solution: nitrate, potassium, pH, sodium, and salinity (electrical conductivity). These meters provide results in 1 to 10 min, depending on the ion measured. Prototype meters that measure ammonium and calcium are being tested. The technology was developed by Horiba, Ltd. The meters are distributed by Spectrum Technologies, Inc.

Prices: pH $174, nitrate $309, potassium $294, Sodium $294, Salinity $224

Information contact:
>Mr. K. Michael Thurow, President
>Spectrum Technologies
>1210 S. Aero. Dr.
>Plainfield, Il 60544
>Phone: 800-248-8873
>(815) 436-4440

Nitrogen Sensor for Manure

Description: Gamma-ray radiation is used to read the N levels in manure slurry and control the application of the slurry. Reports indicate a Danish equipment manufacturer is using a gamma-ray sensor to sense N in manure slurry. A control system has also been developed that controls the application of the liquid manure at operator selected rates, using the sensor input in the control signal. The sensor can be retrofitted to existing slurry wagons. Import of the technology will be difficult due to gamma-ray radiation restrictions (McMullin, 1991).

Information contact:
>Samson
>Tange A/S
>Bjerringbrovej 10, DK 8850
>Bjerringo, Denmark
>Phone: 45 86658533

REFERENCES

ASAE. 1984. National research priorities 1984. Am. Soc. Agric. Eng., St. Joseph, MI.

ASAE. 1991. Research priorities 1991. Am. Soc. Agric. Eng., St. Joseph, MI.

Bae, Y. H., S. C. Borgelt, S. W. Searcy, J. K. Schueller, and B. A. Stout. 1987. Determination of spatially variable yield maps. ASAE Paper 87-1533. ASAE, St. Joseph, MI.

McMullin, E. Mid-March 1991. New from Denmark's ag technology show. Farm Industry News, Mid-March, p. 22-23. Webb Division, Intertec Publ., St. Paul, MN.

Pang S. N., and G. C. Zoerb. 1990. A grainflow sensor for yield mapping. ASAE Paper 90-1633. ASAE, St. Joseph, MI.

Searcy, S. W., J. K. Schueller, Y. H. Bae, S. C. Borgelt, and B. A. Stout. 1989. Mapping of spatially variable yield during grain combining. Trans. ASAE 32(3):826-829.

Stafford, J. V., B. Ambler, and M. P. Smith. 1991. Sensing and mapping grain yield variations. p. 356-365. In Automated Agriculture for the 21st Century. Proceedings of the 1991 Symposium. ASAE Publ. 11-91. ASAE, St. Joseph, MI.

Sudduth, K. A., J. W. Hummel, and M. D. Cahn. 1991. Soil organic matter: A developing science. p. 307-316. In Automated Agriculture for the 21st Century. Symposium Proceedings. ASAE Publ. 11-91. ASAE, St. Joseph, MI.

Vansichen, R., and J. DeBaerdemaeker. 1991. Continuous wheat yield measurements on a combine. p. 346-355. In Automated Agriculture for the 21st Century. Proceedings of the 1991 Symposium. ASAE Publ. 11-91. ASAE, St. Joseph, MI.

Wagner, L. E., and M. D. Schrock. 1989. Yield determination using a pivoted auger flow sensor. Trans. ASAE 32(2):409-413.

13 Positioning Technology (GPS)

David A. Tyler
Department of Surveying Engineering
University of Maine
Associate Director
National Center for Geographic Information and Analysis
NCGIA
Orono, ME

The global positioning system (GPS) was designed primarily as a military navigation system intended to provide 24-h, all weather, real-time positioning to within tens of meters worldwide. With the launch of the first test satellites in the late 1970s, engineers began to develop alternative means of using the satellite signals to allow a wide variety of positioning techniques. This chapter will describe the basic properties of the system and explain the principle ways in which it is used, particularly in the context of agriculture.

GLOBAL POSITIONING SYSTEM

Satellites

The current GPS constellation consists of 17 satellites (5 block I and 12 block II) in five orbital planes at an altitude of 20,000 km with 12-h orbits. The five block I satellites are the remains of the experimental constellation established in the early 1980s to test the concepts of the system and new hardware configurations. The block II satellites are the production models which will continue to be launched for the next few years. Block III satellites are in the design and construction phase and will eventually replace the Block II satellites.

The full constellation of 24 satellites, with 20 operational and three spares is planned to be fully operational in 1993. This constellation will provide 24-h coverage of 5 to 10 visible satellites worldwide.

The former USSR has implemented a navigation system call GLONASS which is very similar to the GPS. GLONASS consists of 24 satellites (21 operational and three spares) in three orbital planes at an altitude of 19,100 km with an 11.25-h orbit. At least one manufacturer is currently developing a receiver which will use both GPS and GLONASS signals (Danaher and Gerlach, 1991).

Copyright © 1993 ASA-CSSA-SSSA, 677 South Segoe Road, Madison, WI 53711, USA. *Soil Specific Crop Management.*

Signals

The GPS satellites transmit on two frequencies. L1 has a frequency of 1575.42 MHz and a wavelength, l = 19 cm. L2 has a frequency of 1227.60 MHz and a wavelength, l = 24 cm. On both frequencies there is a standard code (C/A code), a precise code (P code), and there will be an encrypted precise code (Y code). Although there is still a significant amount of debate over access to the codes, the present position of the Department of Defense is that the P code will only be available to the military and selected civilian users once the full constellation is in place in 1993. A message which includes orbital data, timing data, and system status is also imposed on the signal.

POSITIONING (NAVIGATION)

When the GPS was originally designed for military navigation in the 1970s, the concept was relatively simple. A receiver would simultaneously receive signals from three or more satellites and measure the transit time of the signals. If the satellite positions are known as a function of time, and the time of the signal observation is known, the position of the receiver can be calculated by a simple three-dimensional geometric computation. If the altitude of the receiver is assumed to be known, a reasonable assumption for a ship at sea level, three satellites are required to calculate a horizontal position. If the altitude of the receiver is unknown, as in an aircraft, or a land vehicle at an unknown position, and all three coordinates of the position are required, at least four satellites are required. The accuracy of the system was not intended to exceed a few meters which would be quite adequate for most navigation applications. No one during the early stages of system design, ever conceived of other ways in which the satellite signals would be used.

The first commercial GPS receiver designed for civilian use in the early 1980s, the Macrometer, was designed to ignore the coded signal completely and only use the carrier phase. Since that time, system designers have developed many navigation and position solutions based on different combinations of the C/A code, the P code, and the carrier phase from one or both frequencies. Solutions proposed and implemented today frequently are based on a combination of all of the data which are available. Several of these solutions are discussed below.

Pseudo Range

The pseudo range solution for the position of the receiver is perhaps the easiest to understand. A coded signal, the C/A code or P code, is transmitted from the satellites and an exact replica of the code is generated within the receiver. The coded signal arriving from the satellite is compared to the internally generated code, and a time delay, is observed. If the satellite and receiver clock are in perfect synchronization, the observed time delay is multiplied by the known velocity of propagation of the signal, and the distance from the satellite to the receiver is determined. This would require a very expensive clock in each receiver and is not practical. An alternative is to treat the receiver clock offset (error) as an unknown parameter to be solved with the position coordinates of the receiver. This is where the name "pseudo" range comes from. Pseudo ranges from four satellite positions will allow the computation of the receiver position, X,Y, and Z in an earth-centered Cartesian coordinate system and the clock offset. More than four ranges will provide redundancy.

Single Point (absolute)

When the procedure just described is employed with a single receiver and the absolute position of the receiver is computed, this is known as single point pseudo-range positioning. The position obtained will be corrupted by errors in the satellite orbit and computed clock offset, incorrectly modeled ionosphere and troposphere delays in the signal and possibly weak geometry. When the full constellation is available this kind of positioning is expected to yield an accuracy of 5 m with the P code and 50 m with the C/A code. The computations are simple and can be done in real time.

Differential

Accuracy of pseudo-range positioning can be increased substantially by placing one receiver in a fixed position, and computing the position of a roving receiver with respect to the fixed receiver. If the fixed receiver indicates a shift in position during the time of satellite observation, this is known to be caused by errors. If the roving receiver is relatively close (100 km) to the stationary base station so that the satellite signals are passing through the same window in the ionosphere and troposphere, the errors in position can be assumed to be the same at both stations. The positions of both stations are then corrected by either correcting the satellite ranges or by directly correcting the X,Y,Z coordinates. If the position computation and differential correction is to be carried out in real time, the base station and roving receiver must be linked with a radio frequency connection; otherwise the differential corrections are done in a post-processing procedure. When the full constellation is available, differential GPS is expected to yield an accuracy of 3 m with the P code and 5 m with the C/A code. This level of accuracy may be intentionally degraded by the Department of Defense by selective availability (SA).

A single base station can provide differential corrections for many roving receivers, and it is possible to establish a base station that will satisfy the needs of all users in a region. The accuracy of differential GPS will diminish with distance and for meter level positioning, a 100-km separation between the base and roving receivers is considered to be maximum. At the University of Maine, the Forestry School is experimenting with a base station set on the roof of a campus building. At this point, the receiver is only turned on when someone has scheduled a period of observation in advance, but in the future, as demand increases, it will be possible to keep the base station on continuously and provide differential correction data for blocks of time over a telephone - modem connection. This data could also be distributed over side bands on commercial FM radio stations or via a satellite link from a base station network to a user to allow real-time differential correction.

Establishing base stations at 200-km spacing to provide differential GPS data may be rather costly and inefficient. Loomis et al. (1991) have proposed a networked solution to meet local, regional, or worldwide needs for differential data. The user of such a system would calculate a differential correction based on a spatial averaging of the data from several base station in an area or region.

Carrier Phase

An alternative to using the coded data imposed on the satellite signal is to directly observe the carrier wave. The carrier wave on L1 has a wavelength of 19 cm. The phase observable is the difference between the doppler shifted phase of the signal from a satellite and the phase of an internally generated signal (Leick, 1990). If we ignore certain sources of error for the moment, this observation can be viewed as a measurement of a partial wavelength which is included in the range to the satellite. The total range is the partial wavelength plus an unknown number (the integer ambiguity) of complete waves. The integer ambiguity for a single receiver and the satellite and receiver clock offset errors are resolved by differencing the phase observations of single receivers on several satellites and then several receivers on several satellites. Errors caused by signal propagation through the ionosphere are eliminated by simultaneously observing both L1 and L2. All submeter positions with GPS will involve an observation of the carrier phase on one or both frequencies. The accuracy achievable when the full constellation is operational is expected to be 0.01 part per million (ppm) of the separation between receivers. Although carrier phase measurements could be used to determine a single absolute position, this is usually not done, and only relative or differential positioning will be discussed here.

Static

Static GPS positioning refers to the technique used in surveying where two or more receivers remain in fixed positions on survey stations for from 15 min to 3 h. The phase observations are collected after the session and post processed to obtain differential position vectors and station positions. This approach is applicable in agricultural application only to establish control positions and will not be discussed fully here.

Kinematic

Kinematic positioning refers to positioning where one or more receivers are moving and at least one receiver is on a fixed station. At least four satellites are continuously observed from all receivers. The integer ambiguity must be initially resolved for each receiver, and then the phase count is tracked continuously at each receiver. The roving receivers may be moved from station to station and left stationary for a few seconds or minutes at each station, or they may be tracked continuously with repeated relative positions computed at regular intervals. Position computations may be accomplished in a post-processing mode, or they can be done in near real time if radio links are established between fixed and roving receivers. Where differential positioning using the C/A or P codes and pseudo ranges will result in real time or post-processed positions accurate to a few meters, kinematic positioning with phase observations can produce positions accurate to a few centimeters. The problem with this approach is that the satellite signal must be continuously tracked. An instantaneous loss of the signal caused by moving behind a tree or other obstacle will cause a cycle slip, or loss of the solution for the integer number of wavelengths in a range to a satellite. Continuous kinematic positioning is practical now in wide open field such as one encounters in Montana, but the cycle slip problem must be resolved in order for this technique to be used universally.

Rapid Static

Rapid static is a position technique where receivers are stationed on control points for only as long as is required to resolve the integer ambiguities. This is not an approach with much use in agricultural applications and it will not be discussed here.

Integer Resolution "On The Fly"

The fundamental problem with continuous kinematic GPS positioning using the phase observable is cycle slips. Whenever the line of sight to one of the satellites is momentarily obscured by moving behind or under an obstacle, the whole cycle or integer ambiguity is lost, and a new solution is required. Traditionally, if one can use that term in GPS positioning, the problem was solved by returning to a known position and re-initializing the receiver or by remaining stationary for a long enough period to allow a static solution for the whole cycles. Both of these solutions are impractical for many application of GPS and research in recent years has been directed towards several alternatives that will allow a very rapid solution for the integer ambiguities (Hatch, 1990). Abidin (1991) has proposed and tested with simulated data, a combination of previously suggested methods.

At the risk of oversimplifying a complex computational task, the proposed solutions can be summarized as follows: If a moving receiver is tracking five satellites, and phase lock is lost on one satellite, a position solution using the other four satellites can be used to limit the number of possibilities for the whole cycle count to the fifth satellite, once the fifth signal is recovered. In like manner, if only four satellites are being tracked, and one signal is momentarily lost, a coded solution with P code or C/A code can be used to limit the search for possible integer ambiguities when the lost satellite signal is recovered. In both of these situations, when a cycle slip is recognized, alternative solutions for position are immediately attempted and a search is started to resolve the ambiguity as soon as the lost signal is recovered. The alternative solution is used to limit the possible family of solutions for the integer ambiguities. The best alternative solution will depend on: i) the type of data that is observed, i.e., single frequency, dual frequency, C/A code or P code, ii) the number of satellites that are observed and tracked, and iii) the geometric configuration of the satellites observed. In general, the more satellites that are observed and the more data that is collected, the more rapidly the integer ambiguity will be resolved.

What all this means is that a roving receiver after a cycle slip would stop and wait until the ambiguity is resolved. In the best of all situations, the resolution would be accomplished in a few seconds and the stop would barely exist. In a worst case, the stop could be tens of minutes. Given the current stage of both theoretical study and algorithm development, it appears that at least a dual frequency receiver will be required for a reasonably quick solution to this problem (Hatch, 1991).

AGRICULTURE APPLICATIONS

For the next few decades, GPS will be the method most of us will use to determine horizontal and vertical positions. Some enthusiasts are referring to GPS as a new public utility, a positioning utility, and when one considers the possibility of base station networks with differential corrections available over commercial radio, the public utility analogy is reasonable. GPS technology is still developing, and has come along way since the first commercial receiver just 10 years ago. It is now possible to purchase a 12 channel code and phase receiver about the size of a box of 10 3.5-in. computer disks for around $2,000. One manufacturer is producing a receiver on a single electronic circuit board not much larger than a playing card. The satellite constellation is still not complete and the data reduction software needs work, but determining a position, to within a few centimeters if necessary, is going to be very straight forward and easy in the future. We are not used to thinking of spatial position as an attribute that can be collected effortlessly and it may take time to develop substantial applications.

It will become very simple to find out where things are, and to put things where we want them. We will be able to track the position of a machine and log whatever data on soil condition, topography, crop yield, or weed density we need to collect and relate to position. We will also be able to place fertilizer, herbicide, pesticide, and seed wherever we want and relate these placements to all other data and information we have on a field. At a recent conference on automated agriculture, (ASAE, 1991) seven papers described applications of GPS to soil mapping, weed location, farm geographic information system (GIS) construction, fertilizer application, and yield monitoring. Several of these papers described research, but there were a few which described production operations.

REFERENCES

Abidin, H. 1991. New strategy for 'on the fly' ambiguity resolution. Proceedings of The Institute of Navigation ION GPS-91: 875-886.
ASAE. 1991. Automated Agriculture for the 21st Century. Am. Soc. Agric. Eng., St. Joseph, MI.
Danaher, J., and R. Gerlach. 1991. Digital Hopping GPS/GLONASS Receiver. Proceedings of The Institute of Navigation ION GPS-91: 69-76.
Hatch, R. 1990. Instantaneous ambiguity resolution. KIS symposium, Banff, Alberta, Canada.
Hatch, R. 1991. Ambiguity resolution while moving - Experimental results. Proceedings of The Institute of Navigation ION GPS-91: 707-713.
Leick, A. 1990. GPS Satellite Surveying. John Wiley and Sons, New York.
Loomis, P., L. Sheynblatt, and T. Mueller. 1991. Differential GPS Network Design. Proceedings of The Institute of Navigation ION GPS-91: 511-520.

14 Importance of Spatial Variability in Agricultural Decision Support Systems

G. W. Petersen
R. L. Day
C.T. Anthony
J. Pollack
Land Analysis Laboratory
Department of Agronomy
The Pennsylvania State University
University Park, PA

J. M. Russo
ZedX, Inc.
Boalsburg, PA

Farmers and other managers in the field must continually make production decisions using a variety of information sources that vary in quality and scale. As the opportunity for farm-level monitoring increases with advances in instrumentation, computers, and database management programs, strategies must be developed for collecting, integrating, and interpreting the many types of information that are necessary for decision making.

One evolving strategy for information management that is being fueled by an expanding computer technology industry with its steadily lowering prices for new products, is a decision support system (DSS). A DSS integrates and organizes all types of information required for a decision. In some professional circles, DSSs are synonymous with expert systems, decision-making models, and geographic information systems (GIS). No matter which definition is preferred, the underlying concept is that a DSS is designed to bring together all the information tools necessary to make a timely and confident decision.

OBJECTIVES

This chapter is written with two objectives. The first is to give a brief overview of the current state of DSSs. The second objective is to focus on how a user can judge whether a database meets the information requirements for a decision. The criteria to make this judgement could be part of a general control strategy for managing information for on-farm decision making. We will discuss how data necessary for a decision vary both in true value and quality as one changes spatial scale. A user can quantify this sensitivity to scale with some

---

Copyright © 1993 ASA-CSSA-SSSA, 677 South Segoe Road, Madison, WI 53711, USA. *Soil Specific Crop Management.*

simple analysis and, subsequently, construct criteria for judging whether the data are acceptable for a decision.

OVERVIEW OF DECISION SUPPORT SYSTEMS

A simple conceptual diagram of a DSS is shown in Fig. 14-1. Tracing the steps in the figure, information can be viewed as flowing from the environment via instrumented or human sensors as data to a database. The information as data is analyzed and manipulated for storage or transmission to a user as part of a decision process. The information processed for a decision results in an action to be executed within the environment. After the action is carried out, the environment is again monitored to begin a new cycle of information flow. Thus, information flows to and from the environment in an endless loop that begins with sensing and ends with action. A DSS integrates expert knowledge, management models and timely data to assist producers with daily operational and long-range strategic decisions.

The DSS in Fig. 14-1 depicts generically the basic components of real-world systems such as the Agriculture and Environmental Geographic Information System (AEGIS). AEGIS, shown as a schematic in Fig. 14-2, is a pilot DSS developed jointly by the Universities of Florida and Puerto Rico (IBSNAT, 1992). It is a good example of a modern DSS which is complete with databases, models, expert systems, and GISs. Since databases, models, expert systems, and GISs are emerging to be dominant components in modern DSSs, they will be described in more detail.

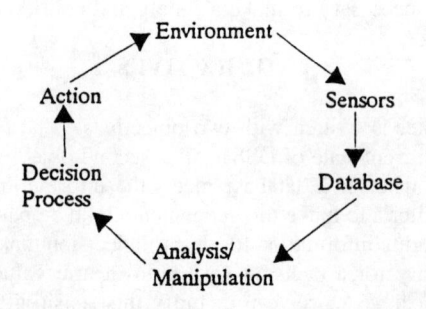

Fig. 14-1. Conceptual diagram of a decision support system.

AGRICULTURAL DECISION SUPPORT SYSTEMS

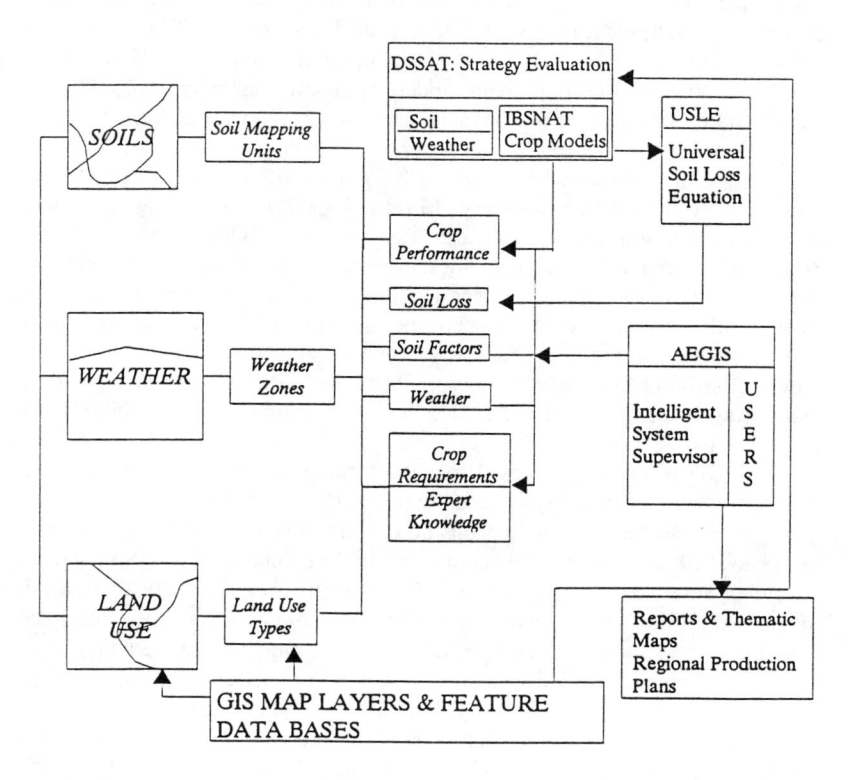

Fig. 14-2. AEGIS decision support system (modified after IBSNAT, 1992).

A Database Management System (DBMS) logically organizes and relates data for efficient retrieval and analysis (Atre, 1980). Geographic data, when stored as computer files, are typically described as being in "vector" or "raster" form. A database in vector form stores information as points, lines, or polygons. Map features are stored explicitly with x,y coordinates. A database in raster form stores information as a geometric array of rectangular cells, each with an assigned value (FAO, 1988). A major advantage of databases stored in vector form is their superior cartographic output from specialized data processing software, such as a GIS. Whereas, databases stored in raster form have the advantage of being easily input into computer models. An example of a database in vector form is TIGER data (Sobel, 1990). A well-known example of a database in raster form is the Digital Elevation Model (USGS, 1990).

Models are simplified abstractions or artificial representations of real-world phenomena (Lassiter and Hayne, 1971; Hall and Day, 1977). They are best typified by mathematical algorithms of biological and physical states and processes. Examples in agricultural science include simple models to simulate soil temperatures (McCann et al., 1991); more complicated models to estimate crop growth (Jones and Kiniry, 1986), erosion (Laflen et al., 1991), agricultural nonpoint pollution (Young et al., 1989), agricultural sustainability (Jones et al., 1991); and sophisticated field level models to quantify agrichemical dynamics, runoff, and erosion as part of a management system (Knisel, 1980).

Expert Systems are computer programs that simulate the problem-solving skills of one or more experts in a given field and provide recommendations to a problem (Coulson and Saunders, 1987). Expert systems draw and store inferences from information and are often called knowledge-based systems. Because they frequently use heuristics to generate recommendations they are known as rule-based systems. Expert systems track their chain of reasoning, thereby allowing a user to query the system as to why a particular recommendation was given. They have been developed to assist apple (*Pyrus malus* L.) growers (Rajotte and Bowser, 1991), to produce interpretive geologic engineering maps (Usery et al., 1988), and to determine crop suitability for a given plot (Nevo and Amir, 1991).

A GIS is a computer-based spatial information system designed for data capture, management, analysis, and mapping (Berry, 1987). It is a DBMS specifically designed for simultaneous processing of spatial and related attribute data (FAO, 1988). The basic components and functionality of a GIS include: a data input subsystem that can capture data directly from a sensor or from an analog map; a data storage and management subsystem that handles locational information and attributes associated with a location independently; a data analysis and processing subsystem that performs a variety of tasks such as information aggregation, buffer generation, map generation, and statistical summaries; and a data output subsystem that presents information in either tabular or map form (Marble, 1984; Dangermond, 1990). The potential of GISs to manage soils information has been recognized for some time (Burrough, 1982; Cunningham et al., 1984; Rogoff, 1982). Soil scientists have used GISs to capture and create interpretive maps (Reybold and TeSelle, 1989), to develop soil-landscape models (Bell, 1990), to model agricultural nonpoint pollution potential (Petersen et al., 1991a), and to combine with remote sensing for land resources analysis and management (Petersen et al., 1991b).

EVALUATING DATABASE SUITABILITY FOR USE IN DECISION SUPPORT SYSTEMS

There are an increasing number of databases available for on-farm decision making. Many of these databases may not be appropriate because the values are out of range for a decision, were collected at too crude or too fine a scale, or are too variable to be used confidently. To judge the appropriateness of a database for a decision, a user must compare the values in a data set to the data requirements for a decision. This judgement of data needs from the perspective of a decision has been addressed in detail in a previous paper (Russo and Bass, 1990). For the present discussion, it is only important to note that the user must develop objective criteria to judge if data are appropriate for decision making.

The development of criteria to judge a database requires knowledge of the temporal and spatial scales represented by the data (Petersen et al., 1990). It also requires some measure of how confidently one can use a single datum as a true value for a particular scale. The sensitivity of a given model to its input data must be considered to determine the suitability of the data for the model.

Error analysis techniques quantify errors in model predictions associated with various levels of reliability in input data (Fritton, 1974) and have been applied to quantitative predictive models such as CREAMS (Knisel, 1980). The sensitivity of a model to each input variable must be determined separately since sensitivity may be different for each variable. Also the sensitivity of a single input variable may vary depending on the values of all other input variables. For example, assume a crop growth model requires available P and pH as input. The model may be quite sensitive to available P at neutral pH values but less sensitive to P under more acidic conditions. The error in model predictions caused by input data uncertainties may be so large that the input data is not adequate to place any confidence in the model predictions. If this occurs, the user must either find another model that is less sensitive to the input data or collect more accurate input data. Sensitivity limits can be established that define the reliability of input data that are needed to produce meaningful model predictions. The sensitivity of a given model to input data quality must be determined to assess the suitability of the data for the model.

In an agricultural management application, the reliability of input data for a field or management unit can be determined by collecting multiple observations within the management unit and determining the variability of the values. If the variability if too high (i.e., outside the model sensitivity limits), then the mean value of that input variable cannot be reliably used in the model since the predictions would contain unacceptable error.

HYPOTHETICAL SCENARIO

A hypothetical scenario is used in this chapter to demonstrate how one would assess a database for its suitability of scale and quality for a decision-making model. In this scenario, an on-farm user wishes to use a model as a decision-making tool to determine how much P should be applied on fields at three farm sites. The farms are at three locations within a county; and are situated in three different micro-environments. The user considers each farm site to be divided into "cells." These cells are part of a rectangular array of equally sized cells covering the county. The user would employ the same management action within a 3x3 or nine cell area. Therefore, when a decision is made to apply an amendment, it would be on a management unit of at least nine cells.

The user treats the decision-making model as a "black box." That is, the user does not understand the workings of the model but only knows the data input requirements for it to run correctly. The user has chosen a soil fertility survey that was conducted early in the season as a source of input data for the model. As part of the survey, the amounts of the major soil nutrients important to crop growth were measured for every field cell. The user has two questions about the field data. First, is the field data appropriate for the model at the scale of a nine cell area? For example, the variability of residual P within a nine cell area may be so large that the mean value for the nine cell area falls outside the sensitivity limits of the model. Second, does the data have to be collected for every cell each time the model is used? The second question is particularly important to the user because a survey may have to be conducted at the start of each season to determine the amount of P resident across the entire crop land. Annual surveys, especially if they require intense sampling, could be costly.

In our scenario, an existing database was selected to represent hypothetical soil fertility survey data collected across the three farms; it is the U.S.G.S 1:24,000 scale Digital Elevation Model (DEM) data set. Although this data set contains geographically addressed digitized elevations, these values can be imagined as representing the P amounts measured in the survey. The chosen DEM data form a rectangular array within a single county in Pennsylvania (Fig. 14-3). As can be seen from the shaded-relief image in the inset of Fig. 14-3, the chosen data array captures the varied relief of the area. Using a near-neighbor, cell-centered averaging scheme and beginning with a single cell, the means and standard deviations of the P amounts were generated for four spatial scales: 1, 9 (3x3), 81 (9x9), and 729 (27x27) cells. The means were calculated to derive an average value for a given scale, and the standard deviation was computed to derive a measure of variability. Means and standard deviations were computed for the three farm sites.

As part of our imagined scenario, a set of numerical limits were arbitrarily specified as the input requirements for the hypothetical model. The hypothetical decision-making model for P applications was assumed to be validated by a sensitivity analysis over a wide range of input values regardless of spatial scale. The sensitivity analysis revealed that the model would function as designed for input P values ranging from 200 to 300 units. The analysis also found that the mean value could be used as a true value within an acceptable level of risk if the standard error at a 95% confidence level was no greater than +/- 15 units.

The computed means of the DEM data for the three farm sites in the county are presented for the four chosen spatial scales in Fig. 14-4. The four means have been connected with a straight line with the assumption that other means could be interpolated for scales lying between the depicted ones. In the same figure, the acceptable range of input values that were determined through sensitivity analyses for the hypothetical model are shown as a shaded area. Wherever a mean value at a given scale falls outside the shaded area, then the mean for that site cannot be used as input into the model. In other words, the model would not be valid for the data at that particular site and spatial scale. As can be seen in Fig. 14-4, the means for site 1 at all scales fall outside the shaded area, and consequently, should not be used as input into the model. Conversely, the means at the other two sites satisfy the data input requirements

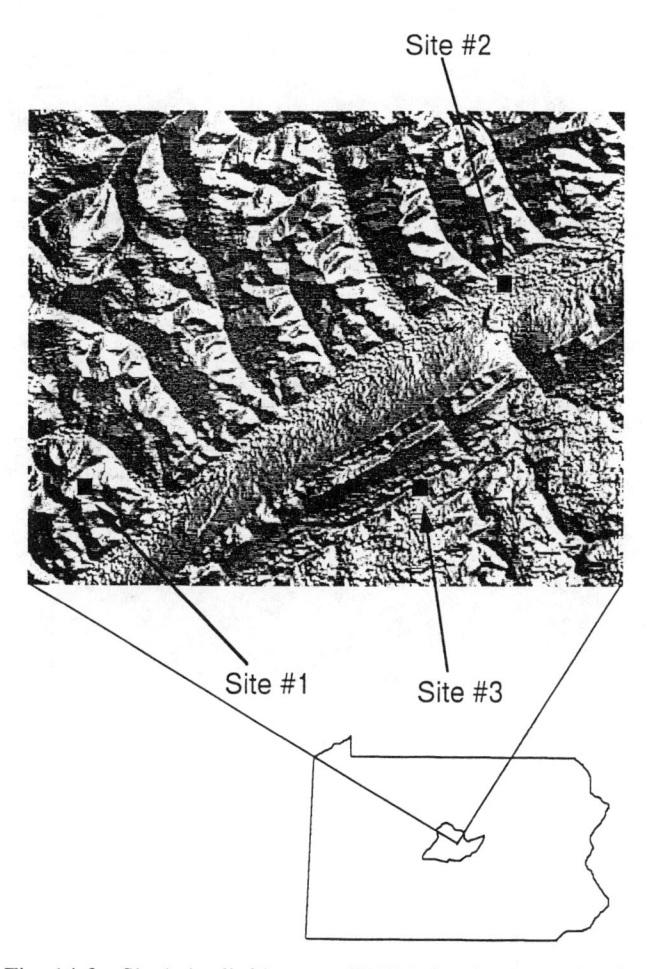

Fig. 14-3. Shaded relief image of DEM date in Pennsylvania.

of the model.

The calculated standard errors at a 95% confidence level for the same three farm sites are presented for the same four spatial scales in Fig. 14-5. In a manner similar to the means, the three standard errors have been connected with a straight line with the assumption that other errors can be interpolated for scales lying between the depicted ones. The confidence interval is shown in the figure as a shaded area to demarcate the error range about the true mean that would be acceptable to the model. If the variability, as computed by the standard

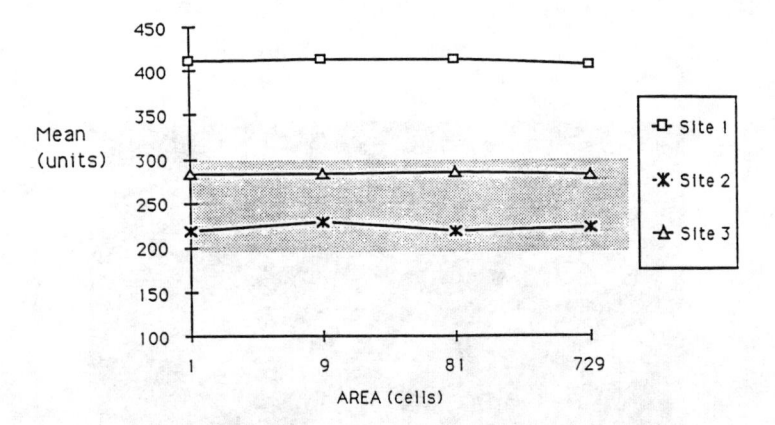

Fig. 14-4. Means of P units by spatial scale for three farm sites. Means in shaded area are acceptable as input into model.

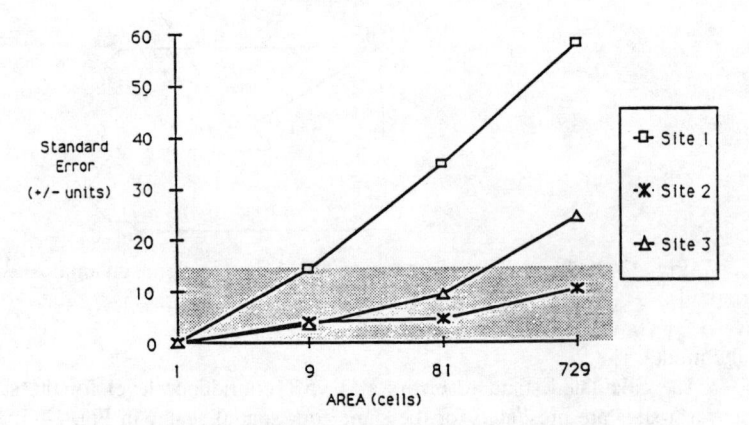

Fig. 14-5. Standard errors of P units by spatial scale for three farm sites. Standard errors in shaded area indicate means are acceptable as input into model.

error, is found to lie above this shaded area for any particular scale of a given site, then the mean for that scale may have an unacceptable amount of variability when used as input into the hypothetical model. As can be seen in Fig. 14-5, the variability for each farm site increases with increasing spatial scales. Only site 1 at the spatial scales of 81 and 729 cells, and site 3 at the scale of 729 cells, have standard errors that would result in the means at those scales being unacceptable as input into the model. The user, who is now aware that the means at all spatial scales at site 1, and the mean at the 729 cells scale at site 2 cannot be used as input into the model, has two choices. The user can find another model that is validated for the data found at those scales, or can employ the existing model with a higher level of risk. How high a level of risk the user is willing to accept depends on how important the decision is to the overall management of the farm.

Having the results of an analysis such as shown in Fig. 14-4 and 14-5, the user can draw some definitive conclusions about conducting future surveys for P amounts at each of the farm sites. The first conclusion is that the original survey data would allow the user to use the model at the management resolution of nine cells at sites 2 and 3. The second conclusion is that P amounts at site 2 could have been collected at the crudest spatial resolution of 729 cells and still have satisfied the data input requirements of the hypothetical model. The third is that a future survey at site 3 would have to be done at a spatial scale of at least 81 cells to have field values that would satisfy the model input requirements. The fourth and last conclusion is that the current model was unacceptable at any scale for the survey data taken at site 1. The user would have to find a new model with input requirements that were compatible with the field data collected at site 1.

The increase in data variability with spatial scale can be graphically depicted with landscape maps. From a map perspective, the behavior of the input data can be judged across a region rather than just for a few specific locations. The values for each cell in the rectangular array of DEM data are shown as greyscale landscape maps for the three higher spatial scales in Fig. 14-6. As one moves from a lower to higher scale, the number of cells used to compute the mean and standard error for a particular location (center cell) increases from 9 to 729. The relative values of the standard error are shown as shades of grey in the maps. The lighter the shade of grey, the greater the variability associated with the mean value for a particular site. Concentrating on one map for any spatial scale, one can see a mosaic of low and high variability areas as depicted by the variation in grey across the region. The most variable areas (brighter shades) are along mountain slopes and across other terrain having variable landforms. Viewing the three maps in succession from lower to higher scales, one can see a brightening of many sites as the amount of variability increases for most locations. By scanning these landscape variability maps, a user could designate suitable sites for model implementation. Conversely, for unsuitable sites, the user must find alternative models or additional data sources.

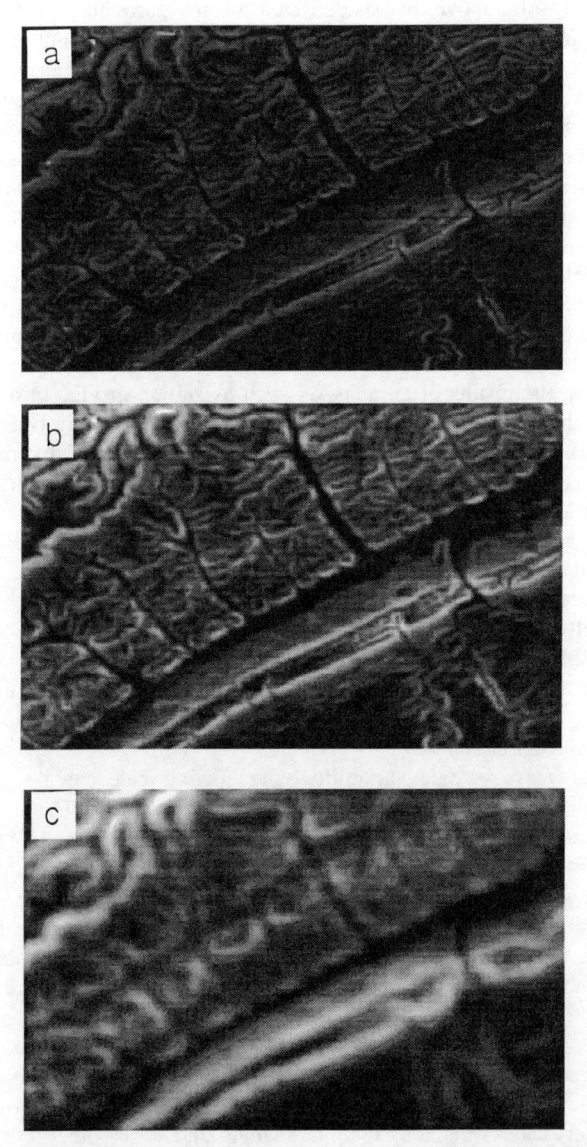

Fig. 14-6. Landscape images of variability as a function of three spatial scales: (a) 9, (b) 81, and (c) 729 cells. The lighter the shade of grey the greater the variability associated with the area surrounding a particular cell.

CONCLUSION

It is clear from our hypothetical scenario that a user must know about the scale and quality limitations of data before it is used in a DSS. Otherwise, data types of varying degrees of reliability will be integrated together in GIS and in expert system subsystems. The result of this integration will be predictions having little value for the decision on hand.

Criteria for judging information types should be incorporated into DSSs. This can be conceptualized by first viewing information flow in a DSS as moving in a direction opposite to that shown in Fig. 14-1. Instead of viewing the information flow as beginning with the assessing or sensing of the environment and ending with decision making, we could view the flow as starting with the decision and ending with the environment. By first considering the data requirements of the decision-making process, databases or other information types could be judged for their acceptability for a decision. Statistical measures must be used to judge whether a database satisfies the scale and data requirements for a decision. In most cases, simple statistical measures, such as employed in our hypothetical scenario, could be used as criteria to judge whether a database satisfies the scale and data behavior requirements for a decision. Having judged a database for its acceptability, the user can move confidently ahead and include it with other compatible information sources that are necessary for a particular set of decisions.

REFERENCES

Atre, S. 1980. Database: Structured techniques for design, performance, and management. John Wiley and Sons, New York.

Bell, J. C. 1990. A GIS-based soil-landscape modeling approach to predict soil drainage class. Ph.D. thesis. The Pennsylvania State University, University Park.

Berry, J. K. 1987. Computer-assisted map analysis: Potential and pitfalls. Photogr. Eng. Remote Sens. 53:1405-1410.

Burrough, P. A. 1982. Computer assistance for soil survey and land evaluation. Soil Surv. Land Eval. 2:25-36.

Coulson, R. N., and M. C. Saunders. 1987. Computer-assisted decision-making as applied to entomology. Annu. Rev. Entomol. 32:415-437.

Cunningham, R. L., G. W. Petersen, and C. J. Sacksteder. 1984. Microcomputer delivery of soil survey information. J. Soil Water Conserv. 39:241-243.

Dangermond, J. 1990. A classification of components commonly used in geographic information systems. p. 30-51. In D. J. Peuquet and D. F. Marble (ed.) Introductory readings in geographic information systems. Taylor and Francis, London.

FAO. 1988. Geographic Systems in FAO. Food and Agriculture Organization of the United Nations, Rome.

Fritton, D. D. 1974. Evaluation of experimental procedures by error analysis. J. Agron. Educ. 3:43-48.

Hall, C. A. S., and J. W. Day, Jr. 1977. Systems and models: terms and basic principles. p. 5-36. In C. A. S. Hall and J. W. Day, Jr. (ed.) Ecosystem modeling in theory and practice: An introduction with case histories. John Wiley and Sons, New York.

IBSNAT. 1992. Linking DSSAT to a geographic information system. International Benchmark Sites Network for Agrotechnology Transfer. Agrotechnol. Transfer. 15:1-6.

Jones, C. A., P. T. Dyke, J. R. Williams, J. R. Kiniry, V. W. Benson, and R. H. Griggs. 1991. EPIC: An operational model for evaluation of agricultural sustainability. Agric. Systems 37:341-350.

Jones, C. A., and J. R. Kiniry. (ed.) 1986. CERES-Maize: A simulation model for maize growth and development. Texas A&M University Press, College Station.

Knisel, W. G. (ed.) 1980. CREAMS: a field-scale model for chemical, runoff, and erosion from agricultural management systems. Conserv. Rep. 26. USDA, Washington, DC.

Laflen, J. M., W. J. Elliot, L. J. Lane, and G. R. Foster. 1991. WEPP: a new generation of erosion prediction technology. J. Soil Water Conserv. 46:34-38.

Lassiter, R. R., and D. W. Hayne. 1971. A finite difference model for simulation of dynamic processes in ecosystems. p. 367-440. In B. C. Patton (ed.) Systems analysis and simulation in ecology. Academic Press, New York.

Marble, D. F. 1984. Geographic information systems: an overview. p. 18-24. In Proceedings, Pecora 9 Conference, Sioux Falls, SD.

McCann, I. R., M. J. McFarland, and J. A. Witz. 1991. Near-surface bare soil temperature model for biophysical models. Trans. ASAE. 34:748-755.

Nevo, A., and I. Amir. 1991. CROPLOT: An expert system for determining the suitability of crops to plots. Agric. Systems 37:225-241.

Petersen G. W., J. M. Hamlett, G. M. Buamer, D. A. Miller, R. L. Day, and J. M. Russo. 1991a. Evaluation of agricultural nonpoint pollution potential in Pennsylvania using a geographic information system. Final report for grant no. ME89279. Pennsylvania Department of Environmental Resources. Harrisburg, PA.

Petersen G. W., G. A. Nielsen, and L. P. Wilding. 1991b. Geographic information systems and remote sensing in land resources analysis and management. Suelo Planta 1:531-543.

Petersen G. W., J. M. Russo, R. L. Day, and J. C. Bell. 1990. Generating high resolution resource data bases for global applications. p. 312-319. In Proceedings of the International Conference and Workshop on Global Natural Resource Monitoring and Assessments: Preparing for the 21st century. Venice, Italy.

Rajotte, E. G., and T. Bowser. 1991. Expert systems: An aid to the adoption of sustainable agricultural systems. Proceedings: Sustainable Agriculture

Research and Education in the Field. National Academy Press. Washington, DC.

Reybold, W. U., and G. W. TeSelle. 1989. Soil geographic databases. J. Soil Water Conserv. 44:28-29.

Rogoff, M. J. 1982. Computer display of soil survey interpretations using a GIS. Soil Surv. Land Evaluation. 2:37-41.

Russo, J. M., and B. L. Bass. 1990. Decision-based meteorological data. Presentation Paper 904051. 1990 International Summer Meeting, American Society of Agricultural Engineers, Columbus, OH.

Sobel, J. 1990. Principle components of the Census Bureaus' TIGER file. p. 112-119. In D. J. Peuquet and D. F. Marble (ed.) Introductory readings in geographic information systems. Taylor and Francis, London.

Usery, E. L., P. Altheide, R. R. P. Deister, and D. J. Barr. 1988. Knowledge-based GIS techniques applied to geological engineering. Photogrammetic Eng. Remote Sens. 54:1623-1628.

USGS. 1990. Digital elevation models - data user's guide. U.S. Geol. Surv., U.S. Dep. of Interior, Reston, VA.

Young, R. A., C. A. Onstad, D. D. Bosch, and W. P. Anderson. 1989. AGNPS: A nonpoint-source pollution model for evaluating agricultural watersheds. J. Soil Water Conserv. 44:168-172.

15 Working Group Report

J. Schueller, Chair
M. Mailander, Recorder
S. W. Searcy and D. S. Motz, Discussion Paper

Other Participants:	R. Alcock	G. Petersen
	S. Borgelt	E. Sadler
	J. Chaplin	M. Schrock
	D. Gandrud	J. Schumacher
	C. Goering	W. Smith
	L. Hanson	K. Sudduth
	K. Klemme	D. Waits
	R. Knutson	L. Ziegler
	D. Luhn	

ENGINEERING TECHNOLOGY FOR SITE SPECIFIC CROP MANAGEMENT

Soil specific crop management is an old idea that is coming back into vogue. When humans learned to cultivate crops for their food source, they were able to manage only small plots. The reliance on human power as the primary energy source limited agriculture to small areas, As the animal, and later mechanical, power became available, a farmer was capable of managing larger areas. As production areas increased, the ability to respond to variations within the land areas was diminished. The development of agronomic production tools has historically concentrated on greater productivity, based on the assumption of homogeneity within field units. Farmers have long acknowledged that fields are not homogeneous, but have not previously had the tools to manage for that variability. Today, computers provide the ability to manage the large quantities of data required to describe agricultural fields. This ability to obtain and use data has generated interest in the soil specific management of crops. However, today this is more of a concept than an established practice.

Statement of the problem

For soil specific crop management, the primary objective is to manage agricultural fields as a series of smaller areas with differing characteristics. These different characteristics may occur both between and within soil types. In this document, the term "site specific" will be used as a more general term.

Copyright © 1993 ASA-CSSA-SSSA, 677 South Segoe Road, Madison, WI 53711, USA. *Soil Specific Crop Management.*

To use this management strategy, differences within the field must be measured and located geographically. Knowledge of field variability will allow the production inputs to be customized for a particular geographic location. This management approach will ideally maximize the crop production, while minimizing potential negative effects from excessive amounts of chemicals and fertilizers.

This concept of site specific production is analogous to the industrial concept of "just-in-time" manufacturing. The just-in-time concept involves intensive management that controls costs by minimizing the inventory on hand at any one time. This concept of minimizing inventory and the associated costs is also appropriate in site specific management. The desire is to provide just enough seed, chemicals, and fertilizer to support the production potential of a given site without providing excess. While conceptually simple, this management approach is technically difficult.

Site specific management requires a systems engineering approach. Tools are needed that can incorporate position location technologies, a geographical database, computer modeling, and automatic control of machine functions. Developing scientific and engineering knowledge to support site specific management is a great challenge for agricultural researchers. Agricultural producers will require a set of integrated tools, if this management technology is to be widely adopted. The purpose of this chapter is to describe the status of the engineered systems available, and to identify the areas where further research and development is needed.

Impact of Engineering Technologies

The ultimate success of any site specific management techniques will hinge on the availability of engineered systems to implement the management decisions. The production machines of today must be improved to implement this increased management intensity. In addition, new tools must be developed to obtain and handle the vast amounts of data that are required before management decisions can be made.

The need for engineering technologies can be grouped into five functional areas. These different functions are:

site specific data acquisition,
spatial data handling and processing,
chemical, biological and hydrologic process modeling,
positioning systems, and
control of field machines.

For each of these functions, the human interface must be considered. Ease of use and safety are important issues throughout the concept of site specific management. The current status and needs in each of these five areas will be addressed individually. However, the reader must keep in mind the overriding need for the integration of these functions into an operational system.

To date, these functions have been, for the most part, developed independently, with minimal ties to the others.

SITE SPECIFIC DATA ACQUISITION

The entire concept of site specific crop management is predicated on the assumption that data is available to describe in-field variability. The description of spatial variability depends upon two constraints. First, the parameter of interest must vary, and second, there must exist a means of measuring that variability. While the significance of soil variability is a more appropriate concern for soil scientists, the measurement of that variability is an engineering task. The sensing of variability is probably the most important step in site specific management. Without accurate maps, varying application rates are no more appropriate than an average, uniform rate. Obtaining this descriptive information about a field is expensive using today's techniques.

The field information can be measured in three distinct manners, continuously, discretely, and remotely. Continuous measurement is done by a sensor operating as a machine is driven through a field. This sensor will give actual measurements along a path through the field. The second approach is the use of sampling at discrete points within the field. This is the method most widely used today. It is highly effective, but labor-intensive. The third method is the use of remote sensing techniques to obtain field images. Analysis of these images can describe the conditions of the field at the time that the images were formed.

Continuous Sensors

The development of continuous sensors has occurred primarily during the last decade. Continuous sensing devices have been developed for organic matter, soil nutrients, crop pests, and crop yield. Of these, the organic matter and yield sensors are the most fully developed. Two competing designs of organic matter sensors are available (Sudduth et al., 1991), each with its own advantages and disadvantages. Both of these devices have been licensed to commercial firms for further development. In-field testing with organic matter sensors by industry has begun, and incorporation of these devices into commercially available machines is occuring.

The measurement of yield on a spatial basis has been performed by several researchers in recent years (Searcy et al., 1989; Vansichen and De Baerdemaeker, 1991; Auernhammer and Muhr, 1991). In each of these cases, combines have been instrumented to record the grain flow rate while harvesting. Yield meters are commercially available (primarily from European manufacturers). However, those units are limited in function to measuring grain flow. To obtain a record of the yield variation within a field, two tasks are necessary. The flow of harvested material must first be measured at some point within the harvester. Secondly, the flow rate must be related back to the field location where the material was standing. All harvesters exhibit some lag between the time and position in which the biomass is actually harvested and the position where the flow is sensed. The reverse transformation from flow rate at a point in time to standing yield in field coordinates has proven to be difficult. Further development of tools to generate yield maps for combines and other harvesters is needed. The development of small grain yield sensors is being done now by industrial concerns, but little attention has been paid to other crops (forages, fiber crops, vegetables, and root crops.) A significant investment in research on yield measurements for other important crops will be necessary. The measurement of yield is the ultimate feedback loop for the site specific management technique, and is an important part of the overall system.

The development of a continuous sensor that can determine the soil nutrient content is both of great difficulty and great interest. The ability to measure nutrient content while driving through the field would eliminate the necessity of manually sampling. This would be a significant labor and cost savings.

The methods used in the laboratory for measuring soil nutrients are both slow and based on ion-specific electrodes, which must be emersed in a solution for relatively long time periods. The transfer of these techniques to a field machine is difficult. Research in academia has concentrated on three different approaches. The use of ion specific field effect transistors has been attempted with some success. Attempts have also been made to transfer the use of electrodes and solutions to a field machine. Adsett and Zoerb (1991) used a continuous soil sampling device, a liquid nitrate extraction cell and nitrate-selective electrodes to analyze the nitrate content of soils while driving through the field. One commercial nitrate sensor is available (Crop Technologies, Inc.), but its effectiveness is not yet widely accepted. All of the above mentioned techniques have attempted to measure soil N. Attempts to measure K, P or micronutrients have not been published. The lack of a continuous, mobile device for sensing soil chemical properties is and will be a factor limiting the wide spread adoption of site specific management techniques. Increased research emphasis is needed in this critical area.

Sensors are also needed to measure soil physical parameters such as tilth, texture, and coverage with crop residue. Soil physical parameters are often critical to seed germination and viability. Measuring and controlling residue on the soil surface has become a issue due to legislative mandate in the 1990 Food Security Act. The requirement that farmers maintain a specified percentage of soil cover has outstripped the ability of farmers or regulators to measure the soil cover. Sensors are needed immediately.

The automated detection of insects, weeds, or diseases on plants would be an extremely valuable capability and would allow reduced rates of chemicals in production agriculture. The current method for detecting plant stresses due to pests, weeds, or disease is human observation. To date, research on sensing techniques has been conducted primarily on the identification and location of weeds. This is taken two distinct approaches. The first is to record the location of the weeds for later application of herbicides (Ollila et al., 1990). Two approaches used are the detection of weed locations during harvest, with subsequent treatment in the following spring, and the detection of weeds in remotely sensed images. Both require postprocessing of the information, the generation of a map of weed locations and a field operation to apply the herbicides. While this approach can and is being done commercially, it is less desirable because of the delay between sensing and application.

The alternative and preferable method is to detect the weeds while traveling through the field and vary herbicide application rates in response to weed presence. The accurate detection of weeds is the technically difficult aspect of this approach. A commercial device (Felton et al., 1991) is available to detect and spray weeds while driving through a fallow field. This device is based on the reflectance differences between the green weeds and the soil/crop residue background. The detection of weeds in the midst of a green crop is a much more complex task. Researchers investigating the reflectance characteristics of weeds and crop have reported insufficient differences to distinguish the two classes of plants (Nitcsh et al., 1991; Thompson et al., 1990). Efforts have been made to identify weeds based on shape (Guyer et al., 1986) using machine vision. Success has been limited and no commercial devices are available. Successful development in this area would provide great savings in chemical cost and environmental impact. However, the ability to sense weeds within a growing crop is many years from commercial development. For the immediate future, the development of weed maps will be the only viable option.

Discrete Sampling

The most common method used today for obtaining field information is the discrete sampling within a field based on a grid. This method can provide the greatest information about a field, since the samples can be analyzed for many chemical properties. Soil physical properties are also possible, if the samples are taken correctly. While this technique seems to do an adequate job of describing a field, it is expensive due to labor and sample analysis costs, and requires a lengthy period to obtain the data.

An orthogonal grid is most commonly used to distribute the samples uniformly over the field. The resolution of that grid (distance between sample points) is normally determined by rule of thumb or the economic constraints of the maximum number of samples which can be afforded. While the optimum resolution for a specific field may be best determined by soil scientists, the resolution used does have an impact on the engineering system. The resolution required will impact the positioning accuracy and data processing operations. An additional uncertainty in the current method is the means used to interpolate between sample points. Kriging has become the accepted practice of researchers. The development and use of semivariograms for kriging is quite computationally intensive. The literature has not established kriging as more appropriate for crop management than other simpler techniques such as block averaging.

The grid sampling technique unfortunately takes no advantage of a priori knowledge. While the relative variability within and between soil types may be unknown, it is reasonable to expect that previous knowledge of soil group location and conditions could aid in reducing the number of samples required. In order to minimize cost, techniques are needed to reduce the need for annual sampling, better use a priori knowledge, and to automate the sampling process. Greater automation is needed in both the sampling process and the analysis of the samples in order to provide a more timely response. The time available for field operations is often quite limited, and those operations cannot wait for data to return from a lab.

Remote Sensing

The use of remote sensing for describing field variation is probably the most developed method. Government sponsored research on satellite sensing has been conducted since the 1960s. However, this information is not widely used today for several reasons. The images are rather expensive for individual farmers to obtain and use. The information obtained from either aerial or satellite images is limited to the soil surface or plants that are growing on that surface. Either method for obtaining images must rely on the use of ground truth to determine what is actually represented in the image. If satellites are used, there are some time constraints dictated by the orbit of that satellite. The current global registration accuracy of +/- 1 km per image presents a significant problem in the use of satellite images. However, despite these disadvantages, remote sensing does provide several important advantages. In a single image, an entire field, or several fields, can be characterized. The periodic orbit of the satellites provides an excellent time series of information for the monitoring of changes in a crop. The obtaining and analyzing of remotely sensed images is in commercial operation.

Aerial images provide a more cost effective and timely source of remotely sensed data. This technique is currently used in production agriculture, with a reliance on a human expert to interpret the image. As remote imaging is incorporated into a site specific management system, there will be a need for image processing techniques to extract the pertinent information.

Both aerial and satellite based sensors require ground truth information. Improved techniques are needed to obtain that ground truth and combine it with the remotely sensed data in near real time. Techniques, such as the fusion of ground and remote sensor, should be investigated.

Research and Development Needs

Site specific crop management requires the ability to obtain spatially variable data about a given field. The continued development of sensing methods which can provide that information is critical to the ultimate success of this management technique. Several technologies need to be developed through the support of industrial and academic research.

1. Continuous sensors, capable of field operation, are needed for a variety of soil and crop parameters. These include yield, soil physical and chemical

properties, plant stress (insect, disease, water, nutrient) and weed identification.

2. Improved techniques for determining the pattern and density of discretely sampled data are needed. Automated collection systems which use a priori knowledge should be used to maximize the information content and minimize the cost. Improved chemical analysis methods are needed to reduce the time and expense required to characterize a field.

3. Continued development of remote sensing techniques to describe field conditions is necessary. Rapid and accurate ground truth information is critical to the widespread use of remotely sensed data.

SPATIAL DATA HANDLING AND PROCESSING

Site specific crop management techniques, will by necessity, generate large amounts of data. This amount of data will quickly exceed the ability of agricultural producers to manipulate and analyze it using their conventional techniques. Therefore, computer software is essential to this management technique. Such software must incorporate the characteristics of a geographical information system, a statistical analysis package, an image processing library, and a graphic display package. Currently, no package is available which includes all of these characteristics.

Current practitioners of site specific management use commercially available programs and transfer data between those manually. This situation is less than ideal. The development of an integrated package with the above characteristics is underway both in academia and industry. While the development on several fronts is desirable for a rapid development, it also generates the possible proliferation of different formats for handling data. Data for a given field will likely be available in several different forms, depending on the source of that information.

Data from a continuous sensor will be in the form of a time series of data points with a sensor value and the current field position. This information must be combined and processed into a form that is a function of geographical location, rather than time. For the same field, discrete sample data may be available. In this form, each of the data points will be geographically located, but may consist of multiple values. This data must be handled differently from continuous data. Data may be available for the field of interest from governmental databases such as the Soil Conservation Service, the Agricultural Stabilization and Conservation Service and the U.S. Geologic Survey. Remotely sensed images will represent a fourth and different form of data.

A geographical database system is needed that can accept any of the above data forms and convert them into attributes of a field. With several attributes available, this system must be capable of recommending application rates and forming these rates into a map. This application rate map must be recorded in a format that can be taken to the field and used in application machinery. The form and operation of such a database is properly the choice of the developer. However, standardized input and output data formats, based on the Open Systems Interconnection model, should be identified. This standardization will be necessary to establish a market sufficiently large enough to justify the development of associated tools. As a standard format for floppy disk drives aided the expansion of the personal computer market, standard data formats will facilitate the adoption of site specific management techniques.

Research and Development Needs

The development of integrated database software for site specific management is needed. Efforts are currently underway, and will likely continue. There are needs that are not being addressed, however. Engineers, soil scientists, and agronomists should cooperate to identify standard formats for data handling. The following data types should be standardized using the Open Systems Interconnection model.

1. A standardized data structure and physical format is needed for parameters measured as a machine moves through a field.

2. Several standard image formats are now available for data from a digitized image. One or more of these should be supported.

3. Standards are needed to allow the transfer of site specific information from large governmental databases.

4. Standard application rate map formats and physical devices are needed to provide interchangeability of tractor and implements.

CHEMICAL, BIOLOGICAL, AND HYDROLOGIC PROCESS MODELING

To achieve fully the promise of site specific crop management, accurate predictions of the chemical, biological, and hydrologic processes occurring are necessary. To the farmer that promise is the opportunity to maximize yields while minimizing the production input costs. Current management techniques are based on economic and biological principles. Methods are needed for prediction of crop response to nutrient and water availability, weed competition, and weather. These methods have been developed and are currently used by the better producers. However, these conventional management techniques may not be optimum when smaller areas of production are managed independently.

Site specific management also holds a promise for the nonagricultural parts of society that is of equal, if not greater, importance. The urban and suburban dwellers are now longer satisfied with a cheap, plentiful food supply. They are demanding that food be produced with techniques that have a less negative impact on the environment. To do that, production methods must incorporate knowledge of the chemical and hydrologic processes that occur, and the local conditions that affect those processes. Although agricultural producers may wish otherwise, the nonagricultural majority of society will have increasing control of agricultural policy and regulation. The development and adoption of management practices that minimize the environmental impacts is expected by our society.

Much is known about the yield response of most crops to soil nutrients and fertilizer applications. This information is routinely used by extension personnel and crop consultants to make fertilizer rate recommendations. Many models have been developed as well to describe the growth of many important crop species. These range from simple empirical models to complex, process oriented models that attempt to include the major biological functions of the plant. However, most (if not all) of these models assume homogeneous conditions. Very little attempt has been made to predict the variability in the field using crop models. McCauley (1992) applied the GOSSYM cotton model for a range of soil conditions in a Texas field. The process was quite intensive, because the cotton was modeled independently at each distinct soil condition. Techniques are needed to simplify the crop modeling under variable conditions. With currently available software, only researchers will use these models for site specific management.

The modeling of chemical and hydrologic processes suffers from the same limitations identified above. Water and solute movement models can be classified into two groups, research and management. These differ greatly in the comprehensiveness of the data required and the quantitativeness of the predictions. Relatively few of these models have been adequately validated with field data. Little effort has been made to use these models to manage fertilizer or pesticide applications. Onken (1990) developed a regression based management model that attempted to minimize the residual nitrate remaining in the soil after a corn crop. Nitrate Leaching and Economic Analysis Package (NLEAP) (Shaffer et al., 1991) was developed to provide soil-based estimates of nitrate leaching potential, and to use that information in the recommendation of fertilizer rates. NLEAP is based on empirical relationships for various soil classifications. The authors suggest that a better method would be to use local data. These efforts represent a start toward chemical recommendation methods that will include environmental and economic optimums.

Computer models could also be useful in identifying those fields which may not benefit from site specific management. Although most authors indicate that nonuniform rates of fertilizers do provide an economic return, that is not true in all cases (Carr et al., 1991). It is difficult to determine the exact reason for a lack of response, but it is reasonable to assume that some fields will be so uniform that variable applications are not justified. Determining the variability required for economic justification of site specific management could be done with computer models and some preliminary soil sampling.

Research and Development Needs

Although modeling is a tool that is growing in its use for agriculture, little attempt has been made to combine models with the site specific crop management concept. The following are areas that should be addressed.

1. Modeling techniques that incorporate three-dimensional soil variability are needed. More complex, process oriented models are needed for the research community, and simpler, management oriented models are needed by producers and consultants.

2. Models incorporating biological, chemical, hydrologic, and economic processes are needed for management of chemical applications. Rate recommendations methods that specifically include environmental constraints will be necessary in the near future.

3. The identification of fields where site specific management would be beneficial could be simplified by the use of computer models. The means for identifying those areas should be developed.

4. Models are needed which can be integrated with a site specific database and can be used to support farm mangement plans. The ability to update the status of the crop as the growing season progresses will be critical.

POSITIONING SYSTEMS

The site specific management concept is based on the ability to repeatedly locate a position within a field. While on-the-go sensors may allow the response to individual soil characteristics, the real-time determination of all pertinent soil characteristics will be neither economically or technically feasible. Prerecorded maps will serve that purpose. The ability to identify position is implicit in the development of field maps. One common approach is to map the soil and field data into a grid. Each grid cell represents a management unit. The optimum size of that unit depends on several factors; the variability of the soil, the accuracy of the positioning system, and the amount of money the farm manager is willing to spend to obtain data for the map. The grid size cannot be so small that positioning system cannot accurately determine in which cell the machine is operating. This suggests that the minimum grid dimension is limited by the accuracy of the positioning system. It is suggested that the smallest grid dimension be a minimum of twice the position accuracy. If a grid of 0.5 ha per cell were used, this would require approximately 30 m accuracy.

Unfortunately, the optimum grid size (and corresponding required positional accuracy) is unknown. There is no currently available method for determining this optimum size. In current practice, "rules of thumb" predominate. This optimum grid size must be determined by soil and plant scientists, before engineers can determine the most appropriate positioning system.

The ideal positioning system would provide highly accurate locations, require little or no user input, and operate autonomously under all environmental conditions. Today, such a system does not exist, but work is progressing towards a suitable solution.

A common positioning method is dead reckoning. It has been used in several instances (Stafford et al., 1991; Ollila et al., 1990; Vansichen and De Baerdemacker, 1991; Robert et al., 1991) because of low cost and relative ease of implementation. Although inexpensive, the accuracy is limited, especially in situations where the machine is unable to return periodically to a known position.

Triangulation between multiple beacons is the basic principle for most positioning systems, although the signal used may vary. Gordon and Holmes (1988) modified a laser system for position determination. Palmer (1991) describes a radio-based commercial prototype for positioning agricultural machines. The LORAN navigation system covers North America and the adjacent coastal water, and is used today for positioning by at least one commercially available fertilizer application system. Searcy et al. (1989) used a commercial microwave system to record a combine's position during yield mapping. These systems have reported relative positional accuracies of 25-200 m for LORAN to less than 1 m for some optical systems. They also have the capability of providing real-time positioning. While post-processing (calculating travel path from previously recorded signals) may be acceptable for data recording operations such as yield mapping, real-time positioning is required for site specific application rate control.

These systems all have various faults which make them unsuitable for broad application in site specific agriculture. Dead reckoning relies on the operator to update periodically the known position. Night operations may make it impossible for the operator to determine location accurately. Also, fatigue and distractions may affect the accuracy and timeliness of the location. Laser, radio, and microwave systems all rely on an electromagnetic signal being reflected back to the vehicle to determine location. This requires reflectors or repeaters to be set up around the perimeter of the field at known locations. Additionally, a minimum number must be always visible to determine position. Line-of-sight problems become significant when hills, trees, or other obstructions block the signal path. Multipath problems also degrade accuracy when buildings or other objects reflect the signal. Due to their limitations, these systems are currently not sufficiently robust for widespread adoption.

In recent years, the global positioning system (GPS) has become the positioning system of choice for both research and commercial development. GPS has been used by Alsip and Ellingson (1991) for soil sensing; Colvin et al. (1991) for yield mapping; and by Larson and his colleagues (Larson and Robert, 1990) for variable fertilizer application. Recently, the accuracy of GPS has been degraded since the reimplementation of selective availability (SA) after Desert Storm. Alsip and Ellington were fortunate enough to conduct their work while

SA was off and reported accuracies of less than 5 m operating in single receiver mode. With SA on a single receiver can only provide accuracies from 30 to 100 m. This has a profound effect on GPS's utility for site specific agriculture. With 100 m accuracy, the smallest grid that could be used would measure 200 x 200 m, 4 ha in size. This is generally considered to be not much better than LORAN, and unacceptable accuracy.

Methods of improving upon the accuracy of GPS have been developed. Differential GPS can provide 3 to 5 m accuracy in real time. The drawback is that differential equipment is quite expensive. One reference station and related radio equipment can cost anywhere from $10,000 to $90,000. This cost is prohibitive for single users. Establishing shared reference stations may offer a solution. In cooperation with local governments or agribusinesses, interested users could share the cost of establishing a differential network. Following the introduction of low cost GPS boards for Original Equipment Manufacturers (OEMs) in 1991, differential GPS is becoming more affordable. Remote differential receivers can be embedded in the application hardware. Establishment of local area differential GPS networks is not a new concept. The FAA and several airports are investigating the use of differential GPS for tracking all vehicles around the airport, planes as well as service vehicles (Klass, 1991). Other modes of operation are possible with centimeter and millimeter level accuracies. However, these are quite expensive and require post-processing. This level of accuracy is not needed for site specific management, although some researchers interested in vehicle guidance are investigating these modes.

The GPS is not immune to the problems that plague other positioning systems. Atmospheric conditions can have significant effects on GPS accuracy. Twenty-four hour GPS coverage is not scheduled until 1993. GPS also suffers from multipath errors. Trees and buildings can interfere with both satellite and differential correction signals. However, the greatest problem with GPS is the cost of achieving acceptable accuracy. Due to the SA degraded signal, the expense of two receivers and a radio link is required for acceptable accuracy. Despite strongly voiced desires of civilian users, it is unlikely that SA will be turned off. For this reason, a civilian satellite-based positioning system is probable after the turn of the century. The introduction of OEM boards has brought the price of GPS down significantly. The technology for an accurate, inexpensive positioning system is available. However, the cost/benefit ratio is not yet at a level that will allow widespread agricultural use. Price/performance improvements will continue, given the rapidly growing market. The rapid pace of commercial development will likely provide a positioning system of acceptable accuracy and cost well ahead of other components of the site specific crop management system.

Research and Development Needs

Although competitors are attempting to provide positioning systems that will improve on the deficiencies of the GPS, it is likely to be widely accepted for site specific management systems. While GPS receivers do not now provide the performance and price desired, the rapid pace of commercial development will provide those improved units. No efforts by agricultural researchers are necessary in this area.

CONTROL OF FIELD MACHINES

Site specific management will not affect the basic function of most agricultural machinery. For most machines, the design changes required would be those necessary to accommodate the addition of sensors or controllers. Many modifications will be primarily electronic in nature, but some machines may require mechanical improvements to allow adequate control. Unfortunately, some basic design parameters, such as required response time, desired accuracy and precision, are dependent on unknown system variables properties (optimum grid size, positional accuracy, and crop).

Machines must be able to implement the variable rates that are associated with site specific crop management. They will both use data in the control of application rates and generate data during machine operations. This means that information must pass both to and from a stationary computer with the site specific database. In addition, information must pass back and forth between control devices on mobile equipment. For example, in a planting operation, it may be desirable to control the plant population and the application rate of pre-emergence fertilizer. This situation would require the transmittal of data between the tractor and the two control functions.

The large amounts of data to be transmitted and the flexibility required for the wide variety of field operations will require the use of standardization. The International Standards Organization (ISO) has developed an Open Systems Interconnect (OSI) methodology. Standardization can come at any of seven different levels. Currently, negotiations are underway within committees of the American Society of Agricultural Engineers, the Society of Automotive Engineers, and the Equipment Manufacturers Institute for the development of communication standards at the lowest levels of OSI. These standards for connector type and communications protocol will be necessary to foster the wide development of equipment which will meet the needs of the site specific management. In addition to the lower level standards, an agreement upon a common format for application maps, and the physical device used to transfer the data from management machines to mobile equipment must also be developed. As floppy disk drive standards provided a larger market for all personal computer manufacturers, a standardized method for transferring field data will assist in the marketing of site specific management equipment.

The use of these new management techniques will place new design demands on the equipment involved. As a prime power unit, the tractor will be used with planting, fertilizing, and harvesting. The tractor must provide communication to the controllers and sensors for each of those operations. A flexible display system with easy operator input will be necessary. Electronic displays will be designed to provide information on an "as need to know" basis. Much of the information passing back and forth from the tractor controller and various implement nodes will not be required by the operator. Therefore, it should only be available when requested. The tractor must be designed to accommodate the positioning system receiver so machine position can be determined at any point in time. Finally, the standard communication systems mentioned above must accommodate all of the implements which might be attached.

Implements will have to be redesigned or retrofitted to accommodate controllers and sensors. This will mean less reliance on ground drive for power and more use of electrical or hydraulic motors, which can be easily controlled by computer. Implements must be designed to accommodate multiple controlled operations. It is entirely conceivable that seeding rate and multiple chemical or fertilizer application rates will be controlled simultaneously. Implements will be redesigned to provide more accurate control of the functions involved. Precise management will require precise applications. Many of these design changes could be accommodated with third party add-on devices.

Research and Development Needs

Most of the field machinery needs will be met by the modification of current designs or the addition of third party add-on devices. Most changes will involve the addition of sensors and control systems to improve the flexibility of the machine. Standardization of the physical and data layers of the communication link between tractor and implement is currently underway. These efforts need to be reinforced and encouraged. In addition, further standardization at the application level will be required for the rate maps and measured data.

RECOMMENDATIONS

The site specific crop management techniques will require a coordinated system of engineered tools. While exceptional producers may be able to piece together a system from commercially available components, broad adoption of this technology will require a system of easy-to-use, reliable machines and software. The human interaction with these machines and software will be a critial aspect. The entire concept of site specific crop management is based on massive amounts of information. The tools used to implement this managment scheme must provide the user a means of using the data, without being overwhelmed by it. This systems approach must be considered by both academic researchers and commercial developers. Action on the following recommendations will speed the adoption of site specific management.

1. A wide array of sensing techniques must be developed to support the

inexpensive and timely characterization of soil and plant properties.

2. Software must be available for the analysis of geographical data. This software must incorporate all the functions of the site specific management concept.

3. Computer models of crop production must include biological, chemical, hydrologic, and economic constraints. Chemical and fertilizer application recommendations should be made using all these constraints.

4. Standards should be adopted at those points in the system where data interchange is required. Manufacturers should compete on the basis of form, function and cost, not through the use of proprietary data formats.

REFERENCES

Adsett, J. F., and G. C. Zoerb. 1991. Automated field monitoring of soil nitrate levels. p. 326-335. In Proceedings of the 1991 Symposium on Automated Agriculture for the 21st Century. ASAE Publ. 11-91. ASAE, St. Joseph, MI.

Alsip, C., and J. Ellingson. 1991. Computer correlation of soil color sensing with positioning for application of fertilizer and chemicals. p. 317-325. In Proceedings of the 1991 Symposium on Automated Agriculture for the 21st Century. ASAE Publ. 11-91. ASAE, St. Joseph, MI.

Auernhammer, H., and T. Muhr. 1991. GPS in a basic rule for environment protection in agriculture. p. 394-402. In Proceedings of the 1991 Symposium on Automated Agriculture for the 21st Century. ASAE Publ. 11-91. ASAE, St. Joseph, MI.

Carr, P. M., G. R. Carlson, J. S. Jacobsen, G. A. Nielsen, and E. O. Skogley. 1991. Farming soils, not fields: a strategy for increasing fertilizer profitability. J. Prod. Agric. 4(1):57-61.

Colvin, T. S., K. L. Karlen, and N. Tischer. 1991. Yield variability within fields in central Iowa. p. 366-372. In Proceedings of the 1991 Symposium on Automated Agriculture for the 21st Century. ASAE Publ. 11-91. ASAE, St. Joseph, MI.

Felton, W. L., A. F. Doss, P. F. Nash, and K. R. McCloy. 1991. A microprocessor controlled technology to selectively spot spray weeds. p. 427-432. In Proceedings of the 1991 Symposium on Automated Agriculture for the 21st Century. ASAE Publ. 11-91. ASAE, St. Joseph, MI.

Gordon, G. P., and R. G. Holmes. 1988. Laser positioning system for off-road vehicles. ASAE Paper 88-1603.

Guyer, D. E., G. E. Miles, M. M. Schreiber, O. R. Mitchell, and V. C. Vanerbilt. 1986. Machine vision and image processing for plant identification. Trans. ASAE 29(6):1500-1507.

Klass, P. J. 1991. GPS demonstration results push system into forefront for airport traffic plan. Aviation Week & Space Technology. Dec. 16-30. p. 43-45.

Larson, W. E., and P. C. Robert. 1991. Farming by soil. p. 103-112. In R. Lal and T. Pierce (ed.) Soil Management for Sustainability. Soil Water Conserv. Soc., Ankeny, IA.

McCauley, J. D. 1992. Modeling the effects of spatial agronomic inputs on crop yield. Unpublished M.S. thesis. Department of Agricultural Engineering, Texas A&M University, College Station.

Nitsch, B. B., K. Von Bargen, G. E. Meyer, and D. A Mortessen. 1991. Visible and near infrared plant, soil and crop residue reflectivity for weed sensor design. ASAE Paper 913006. Am. Soc. Agric. Eng., St. Joseph, MI.

Ollila, D. G., J. A. Schumacher, and D. P. Froehlich. 1990. Integrating field grid sense system with direct injection technology. ASAE Paper 901628. Am. Soc. Agric. Eng., St. Joseph, MI.

Onken, A. B. 1990. Crop nutrient budgets for optimizing fertilizer rates and minimizing environmental impact. In R. Jenson (ed.) Solutions to Non-Point Source Pollution. Proceedings of the 23rd Water for Texas Conference, Lubbock.

Palmer, R. J. 1991. Progress report of a local positioning system. p. 403-408. In Proceedings of the 1991 Symposium on Automated Agriculture for the 21st Century. ASAE Publ. 11-91. ASAE, St. Joseph, MI.

Robert, P. C., W. H. Thompson, and D. Fairchild. 1991. Soil specific anhydrous ammonia management system. p. 418-426. In Proceedings of the 1991 Symposium on Automated Agriculture for the 21st Century. ASAE Publ. 11-91. ASAE, St. Joseph, MI.

Searcy, S. W., J. K. Schueller, Y. H. Bae, S. C. Borgelt, and B. A. Stout. 1989. Mapping of spatially variable yield during grain combining. Trans. ASAE 32(3):826-829.

Shaffer, M. J., A. D. Halvorson, and F. J. Pierce. 1991. Nitrate leaching and economic analysis program (NLEAP): model description and application. p. 285-322. In R. F. Follett et al. (ed.) Managing nitrogen for groundwater quality and farm profitability. SSSA, Madison, WI.

Stafford, J. V., B. Ambler, and M. P. Smith. 1991. Sensing and mapping grain yield variation. p. 356-365. In Proceedings of the 1991 Symposium on Automated Agriculture for the 21st Century. ASAE Publ. 11-91. ASAE, St. Joseph, MI.

Sudduth, K. A., J. W. Hummel, and M. D. Cahn. 1991. Soil organic matter sensing: a developing science. p. 307-316. In Proceedings of the 1991 Symposium on Automated Agriculture for the 21st Century. ASAE Publ. 11-91. ASAE, St. Joseph, MI.

Thompson, J. F., J. V. Stafford, and B. Ambler. 1990. Weed detection in cereal crops. ASAE Paper 907516. Am. Soc. Agric. Eng., St. Joseph, MI.

Vansichen, R., and J. De Baerdemaeker. 1991. Continuous wheat yield measurement on a combine. p. 346-355. In Proceedings of the 1991 Symposium on Automated Agriculture for the 21st Century. ASAE Publ. 11-91. ASAE, St. Joseph, MI.

SECTION IV

PROFITABILITY

16 Profitability of Farming by Soils

N. C. Wollenhaupt
Department of Soil Science
University of Wisconsin-Madison
Madison, WI

D. D. Buchholz
Agronomy Department
University of Missouri
Columbia, MO

For new practices to be widely adopted in production agriculture, the practices must yield a financial profit. Perhaps of equal or greater importance to society is that the practice maintain or improve the "quality" of our natural resources such as soil and water. The objective of this chapter is to identify the variables that contribute to the costs of farming by soils practices, and to summarize the costs and returns of existing farming by soils studies.

FARMING BY SOILS PRACTICES

The concept of farming by soils (FBS) is being implemented by two somewhat different strategies that will referred to as the soil potential and nutrient grid methods. With the soil potential approach, nutrients are varied within field boundaries based on soil survey map units or in combination with aerial photography such as false color infrared photographs. The underlying assumptions with this approach are: (i) that soil test nutrient levels can be adequately assessed by compositing soil cores collected within a soil type or grouping of similar soils, and (ii) that soils have differing yield potentials whereby significant nutrient management gains can be achieved by fertilizing for different yield goals.

The nutrient grid approach subdivides fields into small cells (2 to 3 acres) based on a grid. Soil cores are collected within the cell and composited for analysis. A more recent development has been to collect and composite soil cores within a small radius of the grid intersection point. The soil test data are summarized and nutrient management maps are created. With this approach, it is recognized that the nutrients such as N, P, and K vary independently of soil map units and of each other. If nutrients are to be managed within fields, then the soil test levels must first be determined and summarized to create a nutrient

---

Copyright © 1993 ASA-CSSA-SSSA, 677 South Segoe Road, Madison, WI 53711, USA. *Soil Specific Crop Management.*

Table 16-1. Costs associated with a conventional fertilization practice and farming by soils practices based on a 100-acre (40-ha) field.

Conventional practice		Soil potential		Nutrient grid	
($/acre)		($/acre)		($/acre)	
Soil sampling		Soil sampling		Soil sampling	
Labor		Labor		Labor	
1 h @ $25/h	0.25	2 h @ $25/h	0.50	5 h @ $25/h	1.25
Sample analysis		Sample analysis		Sample analysis	
1 @ $6	0.06	5 @ $6	0.30	40 @ $6	2.40
Fertilizer application	3.00	Fertilizer application	4.50	Fertilizer application	4.50
		Photography, data management, digitalization	1.50	Data management and map making	1.50
		EPROM (0-200 acres)	0.50	EPROM (0-200 acres)	0.50
	3.31		7.30		10.15

management map. Fertilizer rates are then varied based on the nutrient management maps.

Both the soil potential and the nutrient grid approaches ultimately lead to the creation of a nutrient management map that is transferred to an EPROM. The EPROM becomes the base map used by the automated application equipment to change rates or blends "on-the-go" while traveling across a field. Note, the computer does not determine the fertilizer application rate. The rates are predetermined for the map polygons and entered into the computer at the field entrance. Fertilizer rates may be varied within fields without the computerized equipment; however, most producers opt for the time and labor savings associated with the computerized application equipment.

The fertilizer rates may further be modified by taking into account soil yield potential differences. This has not been widely practiced due to some computer software limitations; however, the authors are aware that the software is being modified to accept multiple layers of information. For example, a yield potential map might be used to vary N, while a nutrient grid map could be used to vary potash and phosphate, and at the same time an organic matter map could be used to vary the amount of herbicide being impregnated on the fertilizer.

VARIABLES AND COSTS

Variables and costs associated with FBS are compared to a conventional practice for a hypothetical 100-acre field in Table 16-1. Costs of soil sampling, sample analysis, and fertilizer spreading for a conventional practice (one soil sample, one rate of fertilizer applied) are about $3.30 per acre. The FBS costs include a higher charge for spreading, mapping and data management, and additional soil sampling and analysis. The soil potential FBS program costs are about $7.30 while the nutrient grid method costs are $10.15 in this example. This compares to reported costs ranging from $10 to $15 per acre.

The FBS costs, therefore, average about $4 to $7 per acre more than conventional practice. These costs will need to be offset with increased yield or lower fertilizer expense due to less fertilizer being applied.

FARMING BY SOILS CASE STUDIES

We have summarized the results of four FBS studies conducted in Montana, Minnesota, North Dakota, and Missouri. The studies represent both soil potential and grid sampling FBS approaches. Emphasis will be placed on understanding how the FBS methods affect costs and returns. Readers are encouraged to consult the original papers for clarification on methodology.

Montana Study

Carr et al. (1991) conducted a FBS study at several locations in Montana during 1987 and 1988. They used the soil potential approach where soil unit boundaries were identified from soil surveys and modified by using false color infrared photography or satellite imagery. Some soil units consisted of multiple soil series.

Twenty soil cores were collected and mixed to obtain a composite sample for each soil unit. Fertilizer recommendations were made for each soil unit based on soil test results, yield goals, and Montana State University fertilizer guidelines. Crops grown were wheat (*Triticum aestivum* L.) or barley (*Hordeum vulgare* L.). Fertilizer treatments included a rate and formulation for each soil unit in the field as well as a whole field rate.

Grain yields and returns were shown to differ by soil units (Carr et al., 1991); however, when averaged for a field, yields and net returns were not significantly different between field and soil unit management strategies (Table 16-2). In two out of five fields, a fertilizer rate (optimum) other than the soil unit based rate produced higher returns than the field average rate. The additional returns were $22 to $24 per acre. Part of the difference was attributed to drought, where fertilizer recommendations were based on yield goals for a "normal" precipitation year. Optimum fertilizer treatments generally included less N, P, and K than soil unit treatments in dry 1988. At Havre, in 1987, actual yield was in excess of the soil unit yield goal. A higher fertilizer rate than the soil unit rate resulted in the highest yield and net return.

Table 16-2. Yield and net returns over fertilization and application costs resulting from field, soil unit, and optimum fertilizer treatments in five Montana fields.

Year	Location	No. soil units	Yield Check	Yield Field	Yield Soil unit	Yield Optimum	Net returns Check	Net returns Field	Net returns Soil unit	Net returns Optimum
			---------- bu/acre ----------				---------- $/acre ----------			
1987	Havre	3	32.8a*	37.1a	37.0ab	43.0b	83.50a	78.32a	80.38a	100.00b
1987	Power	2	31.7a	42.8b	40.5b	42.6b	88.47a	107.59b	103.65b	107.95b
1987	Whitehall	4	13.8a	33.4b	33.0b	40.5b	34.90a	62.87ab	65.41ab	82.33b
1988	Havre	4	18.9a	20.4a	20.8a	22.7a	67.62ab	52.48a	57.62ab	75.99b
1988	Townsend	2	34.5a	34.7a	34.0a	34.7a	57.89b	50.95ab	49.50a	57.89b

* Values followed by the same letter in rows under yield and net returns do not differ significantly at the 5% level of probability according to the student t-test.

Table 16-3. Fertilizer application rates based on soil tests and yield goals, corn yields, and net returns for a Minnesota field.[1]

Year	Treatment	\multicolumn{3}{c}{Application rates}	Net yield	Net return
		N	P_2O_5	K_2O		
		--------- (lb/acre) ----------			(bu/acre)	($/acre)
1989	Check	0	0	0	90	184
	Conventional	130	40	30	164	315
	Variable - A	140	40	30		
	B	75	55	30	168[2]	326
	C	50	35	30		
1990	Check	0	0	0	69	118
	Conventional	130	40	30	121	197
	Variable - A	130	55	0		
	B	110	40	0	122	204
	C	30	25	20		

[1] P. Robert (1992, personal communication).
[2] Weighted mean by acres by soil map units.

The results of this study are difficult to interpret; however, we can conclude that FBS did not reduce yields or profit. Based on soil unit responses, the authors concluded that the FBS strategy has potential but noted that the main limitations were determining appropriate yield goals and fertilizer recommendations. Soil Conservation Service soil survey yield goals were not appropriate.

Minnesota Study

Robert et al. (1990) conducted a FBS study at Lamberton, Minnesota during 1989 and continued the study in 1990 (P. Robert, 1992 personal communication).

Treatments included a no-fertilizer check, conventional application (uniform fertilizer and herbicide), variable rates of fertilizer-uniform herbicide, and variable rate fertilizer and herbicide. Soil series were grouped into three soil management groups. Soil cores were collected and composited for each soil group. Fertilizer rates were based on soil test results and yield potential for each soil condition (group).

Fertilizer rates, corn yields, and net returns are presented in Table 16-3, as reported by Schmitt and Fairchild (1991). The FBS did not reduce yields or net returns. However, in spite of placing fertilizer where it was most likely to match crop response potential, yields were also not significantly increased.

We summarized corn (*Zea mays* L.) yields by soil map units to obtain some insight as to the lack of yield increase from FBS in this study (Table 16-4). Ten map units were managed as three groupings. If we assume that the no fertilizer treatment yields reflect soil yield potential, it would appear that map units 654, 446, and 999B2 might be in the wrong groupings. Furthermore, if variable rate (VR) yields are subtracted from conventional yields, we find that VR increased yields on six map units and decreased yield on four map units. The large yield increases for map units 954C2 and 999B2, which received substantially less fertilizer with the VR application, are difficult to explain.

The last column in Table 16-4 compares VR fertilizer and herbicide with the conventional practice. In management group A, the yields are higher than from only VR fertilizer, possibly suggesting a herbicide-fertility interaction.

As with the Montana study, there appears to be a problem associated with definition of the soil map units and correlation with yield potential. Also, these data may indicate that the crop yield fertilizer responses based on soil testing vary by soil map unit. There are substantial yield variations associated with soil map units and fertilizer management.

Table 16-4. Corn yield as a function of fertilizer and herbicide application method, Lamberton, 1989. [1]

Soil grouping	Map unit[2]	No fertilizer	Conventional application	VR fertilizer-[3] conventional	(VR fertilizer &[4] herbicide)- conventional
			(bu/acre)		
A	8114	105	158	+17	+24
A	86	91	157	+12	+19
A	884	86	180	-11	+ 3
A	654	75	135	+ 6	+44
B	446	97	169	+ 5	+ 7
B	421B	81	189	-18	- 6
B	423	86	192	- 2	-24
C	421B2	63	177	-38	-29
C	954C2	56	125	+24	--
C	999B2	90	96	+67	+71

[1] P. Robert (1992, personal communication).
[2] Soil survey soil map units (1:20,000).
[3] Variable application yield-conventional application yield.
[4] Variable application fertilizer and herbicide yield-conventional applied fertilizer and herbicide yield.

North Dakota Study

A FBS study with barley (1990) and wheat (1991) was completed by Wibawa (1991), under the direction of Dr. Bill Dahnke. This study compared the nutrient grid and soil potential methods for managing fertilizer applications. The treatments for the wheat experiment (1991) are listed in Table 16-5 along with wheat yields.

Fertilizer applications based on grid sampling produced the highest wheat yields when a 49-ft grid spacing was used for soil sampling. With a 246-ft grid, however, fertilization based on soil test by soil series produced the same yield.

Although grid sampling (49 ft) produced significantly higher yields, the extra costs of soil sampling and analysis caused it to have the lowest net returns (Table 16-6). The soil test by series produced the highest return, but probably not statistically different than treatment 2 (246-ft grid).

In this study, fertilizing for a field yield goal produced higher yields than yield goal (potential) based on soil series. Due to timely summer rains, soils with low yield potential (shallow and droughty) produced above-average yields (Dahnke, 1992 personal communication).

Table 16-5. Effect of variable rate fertilization on wheat yield based on yield goal, soil fertility, and soil series in a field near Oriska, ND.[1]

No.	Treatment	Yield*
		(bu/acre)
1	No fertilizer (control)	26.8 d
2	Grid sample (246 ft), 60 bu/acre yield goal	30.5 bc
3	Grid sample (148 ft), 60 bu/acre yield goal	32.0 abc
4	Grid sample (49 ft), 60 bu/acre yield goal	32.9 a
5	Grid sample (49 ft), yield goal based on soil series	32.6 ab
6	Avg. soil test by series, yield goal based on soil series	30.1 c
7	Avg. soil test by series, 60 bu/acre yield goal	31.7 abc

* Values followed by the same letter do not differ significantly at $P < 0.01$.
[1] From Wibawa (1991).

Table 16-6. Cost and return for variable study near Oriska, ND.[1]

Treatment	Yield goal	Crop value[2]	Fertilizer applied	Soil sampling	Soil analysis	Added cost[3]	Net return
			($/acre)				
Control	None	80.46	0.00	0.00	0.00	0.00	80.46
246-ft grid	60 bu/acre	91.63	3.53	1.04	1.74	1.18	84.14
148-ft grid	60 bu/acre	96.11	6.75	2.90	4.84	1.18	80.44
49-ft grid	60 bu/acre	98.79	6.75	26.14	43.56	1.18	21.16
49-ft grid	Soil series	97.89	6.85	26.14	43.56	1.18	20.16
Test by series	Soil series	90.30	6.00	1.80	0.15	1.18	81.17
Test by series	60 bu/acre	95.21	5.79	1.80	0.15	1.18	86.30

[1] From Wibawa (1991).
[2] Wheat @ $3.00/bu.
[3] Digitized map and fertilizer spreading by computerized application.

This study shows that there are significant variations in soil test nutrient levels within a soil map unit. However, the cost of defining the nutrient levels is prohibitive in this field. It should be noted that the soil map units were on average only 145 ft wide. Any nutrient variation associated with the soil series could not be identified with a grid greater than one-half the width of the map unit.

Missouri Study

FBS studies in southeast Missouri have focused on managing P and K based on the nutrient grid approach. Fields are soil sampled on a 330-ft grid. Soil test data are summarized into five fertilizer management groups and a map is created. The number of groupings was limited to five due to limitation of the software used to drive the application equipment.

Soil test levels and fertilizer rates are shown in Table 16-7 for two fields, Fields 1 and 3. Field 1 has a wide range of soil test P levels and is a prime example where a single rate fertilizer program results in some areas of the field receiving extra phosphate, while other areas are consistently underfertilized. Buildup is occurring where the fertilizer application exceeds crop removal, whereas the crop is mining P in others areas in the field, further lowering soil test levels. Field 3 (Table 16-7) is an example of a field that has relatively good P and K soil test levels.

Buchholz (1991) outlined a procedure whereby it is possible to establish yield response to variable rate spreading by using Missouri's soil fertility index equation developed by Dr. T.R. Fischer (Fischer, 1974). Assuming a nonfertilizer (P and K) limiting yield of 150 bu/acre, it is possible to calculate the yield difference expected between a conventional and VR fertilizer application.

A breakdown of yields, income, and costs including fertilizer, fertilizer application, and soil testing/data analysis is presented in Table 16-8 for Fields 1 and 3. Variable rate application returned $25.16 per acre over a single rate application in Field 1. Variable rate returns were $1.21 per acre lower than a single rate approach in Field 3.

These examples demonstrate the importance of increasing yields with a VR program in order to offset costs of grid sampling, soil analysis, and fertilizer application with computerized equipment. Varying the fertilizer application rate within a field alone will not ensure a profit for VR applications. Note, these fields were only managed for P or K differences. Additional yield gains (profit) may have been achieved by varying N applications; however, the data needed to construct a yield potential map were not available.

Table 16-7. Soil test phosphorus and potassium and fertilizer rate for two southeast Missouri fields.[1]

	Field 1					Field 3			
	Soil test		Fertilizer rate			Soil test		Fertilizer rate	
Acres	P	K	P_2O_5	K_2O	Acres	P	K	P_2O_5	K_2O
	- (ppm) -		--(lb/acre)--			- (ppm) -		-- (lb/acre)--	
Single rate					Single rate				
80	33	--	20	--	173.6	18	201	40	30
Variable rate					Variable rate				
11.4	9	--	100	--	16.9	20	314	75	50
26.7	17	--	80	--	56.8	21	386	70	35
7.6	26	--	60	--	7.9	32	251	20	55
34.3	56	--	0	--	71.3	29	424	25	25
80			47^2		20.7	42	498	0	20
					173.6			41^2	31^2

[1] Data from Bill Holmes, Oran, MO. (After Buchholz 1991.)
[2] Weighted mean--mean of values weighted by acres.

Table 16-8. Gross return, fertilizer costs, and application/data management costs for two southeast Missouri fields.

	Crop yield	Crop[1] value	Fertilizer[2] cost	Other[3] costs	Value- costs
	(bu/acre)	---------------------- ($/acre) ------------------			
Field 1					
Single rate	134.5	336.25	5.00	3.31	327.94
Variable rate	150.0	375.00	11.75	10.15	353.10
Field 3					
Single rate	147.6	369.00	13.60	3.31	352.09
Variable rate	150.0	375.00	13.97	10.15	350.88

[1] Corn price = $2.50/bu.
[2] Phosphorus = $0.25/lb P_2O_5; potassium = $0.12/lb K_2O.
[3] Soil sampling, soil analysis, and fertilizer spreading.

SUMMARY

The costs of a FBS variable rate, fertilizer management program will run about $4 to $7 higher than a conventional, single rate per field program. The higher costs of the FBS program include special application equipment, additional soil sampling and analysis, data management, and map making.

The Montana and Minnesota soil potential FBS studies show that crop yields and returns can vary substantially by soil map units or groupings of map units. However, when VR yields and returns were averaged for a field, the results were similar to the conventional, whole field single fertilizer rate approach. Variable rate fertilizer application did not decrease yields when compared to a whole field single rate fertilizer application.

Two limitations of the soil potential approach seem to be related to definition of soil map units and determination of appropriate yield goals. Present SCS soil survey map units are not particularly sensitive to yield potential. If we can manage nutrients for 10 bushels of corn and 3 bushels of small grain yield increments, then we need a soils map that groups soils accordingly. With the development of the ability to measure real time yields with a combine, a crop yield map could be made that would serve as a basis for looking at soil physical properties influencing yield. Soil property ranges could be defined or grouped, and new map units developed that would lead to a useful yield potential classification based on soils. In addition, the mapping needs to be conducted at a scale appropriate for the use, perhaps 1:8000. Also, yield maps may eliminate the use of soil survey information all together.

The North Dakota study shows that fertilizer application based on a nutrient grid map can result in crop yield increases over managing nutrients averaged by soil series. The study raises a question as to what the appropriate sample spacing should be such that yields and profits are optimized.

The Missouri study showed a high return for VR fertilizer application when the grid nutrient FBS approach was used to manage P fertilization in a field that contained soils with low soil test P levels. When soil test P and K levels were in the medium to high range, then there was a slight income reduction with FBS when only P and K were managed.

Schmitt and Fairchild (1991) state that the economic viability of FBS must be made with the conviction that soil testing and recommendations are a proven practice. Buchholz (1991) has provided an example of how VR and conventional fertilizer practices can be evaluated for costs and returns prior to initiating field applications. We conclude that instead of attempting field comparisons of the practices, researchers should focus on better defining and predicting spatial changes of soil physical and chemical properties. This is supported by the experiences of Webster and Oliver (1990), who suggest that, "of the total variance of many soil properties in areas of 10 to 1000 hectares, as much as a quarter, or even half can occur within a few square meters." It is presently nearly impossible to make economic evaluations based on field-collected yields because of the variance of the chemical and physical properties that affect yield--not to mention variation due to climatic conditions.

Although we have emphasized some of the limitations associated with FBS, we strongly believe variable rate fertilization shows promise as a practice that will be profitable. Nitrogen management based on a "useful" yield potential map combined with P, K, and lime applications based on nutrient grid sampling appears to offer the most potential for yield improvement and efficient fertilizer use. We believe profitability will be greatest for fields that contain soils of contrasting texture (yield potential) and low soil test P and K levels.

Even if the FBS practice only breaks even, there are potential environmental benefits to variable fertilizer application that cannot be easily calculated. In addition, new or greatly expanded segments of the agriculture industry will evolve supporting crop consultants, equipment manufacturers, fertilizer dealers, and soil testing laboratories, while at the same time maintaining profit margins for farmers and low food prices for consumers. Computers, computer software, fertilizer application equipment, automatic soil analysis equipment, yield monitors, and global positioning equipment are but a few of the equipment areas that will benefit from the FBS practice.

Buchholz (1991) also noted that public perception and acceptance of fertilizing specific field areas with tailored application rates will be positive. We conclude that FBS fertilizer management will make "cents" and is good "sense."

ACKNOWLEDGMENTS

We want to thank Dr. Pierre Robert, University of Minnesota; Dr. Bill Dahnke, North Dakota State University; Pat Carr, Carrington Experiment Station, Carrington, North Dakota; and Bill Holmes, Oran, Missouri for sharing results of their FBS studies.

REFERENCES

Buchholz, D. D. 1991. Missouri grid soil sampling project. p. 6-13. In Proc. 21st North Central Extension-Industry Soil Fertility Conference, St. Louis, MO. Potash & Phosphate Institute, Manhattan, KS.

Carr, P. M., G. R. Carlson, J. S. Jacobsen, G. A. Nielsen, and E. O. Skogley. 1991. Farming soils, not fields: a strategy for increasing fertilizer profitability. J. Prod. Agric. 4:57-61.

Fischer, T. R. 1974. Some considerations for interpretation of soil tests for phosphorus and potassium. Missouri Agric. Exp. Stn. Bull. 1007.

Robert, P., S. Smith, W. Thompson, W. Nelson, D. Fuchs, and D. Fairchild. 1990. Soil specific management. p. 54-59. In Minnesota Agric. Exp. Stn. Misc. Publ. 62-1990.

Schmitt, M., and D. Fairchild. 1991. Economic perspective on variable rate fertilization. p. 28-34. In 21st North Central Extension-Industry Soil Fertility Conference, St. Louis, MO. Potash & Phosphate Institute, Manhattan, KS.

Webster, R., and M. A. Oliver. 1990. Statistical methods in soil and land resource survey. Oxford University Press, New York.

Wibawa, W. D. 1991. Variable rate of fertilization based on yield goal, soil fertility, and soil series. Unpublished M.S. thesis, North Dakota State University Library, Fargo.

17 Cost Analysis of Variable Fertility Management of Phosphorus and Potassium for Potato Production in Central Washington

Max Ward Hammond
Manager, Cenex/Land O'Lakes Agronomy
Research and Training Center
Ephrata, WA

Research using intensive soil sampling has shown that spatial variations in P and K soil test levels are very common in production fields of central Washington (Dow et al., 1973; James and Dow, 1972; Mulla and Hammond, 1988). These variations result from natural soil development patterns as well as human-made causes. The latter develop due to land leveling, cropping patterns, past fertility management practices and incorporation of smaller fields into larger fields for automated sprinkler irrigation purposes.

The spatial variability of P and K levels has been identified to be a cause for both yield and quality reductions in potato (*Solanum tuberosum* L.) production (Kunkel et al., 1971, 1978). Underfertilization of low testing areas will result in localized yield and quality reduction as well the reduction of overall uniform quality of the crop. Where P and K deficiencies contribute to early senescence of the crop, tuber size is reduced and specific gravities will be lowered due to reduction in photosynthesis and carbohydrate storage. As irrigation for the remainder of the season proceeds, excessive moisture levels develop due to low evapotranspiration in the senesced areas. This condition then contributes to excessive enlargement of lenticels, the development of "elephant hide," and an increase in soft rot development. Predisposition to blackspot is greater in areas that have senesced early.

Attempts to compensate for low fertility areas by increasing fertilizer application rates will result in overfertilization of the rest of the field. This usually does not cause problems with yield and quality with the exception that high K levels have been implicated in lowering of specific gravities (Gray and Hughes, 1978; Kunkel and Thornton, 1986). Overfertilization is not economically efficient or environmentally acceptable. Past programs to manage for fertility variability have been suggested (Dow et al., 1980; Dow, 1981; Dow and James, 1973; Kunkel et al., 1971, 1978). These have been difficult to implement due to difficulties in accurate soil mapping and lack of technology for making practical variable fertilizer applications in the field to compensate for

---

Copyright © 1993 ASA-CSSA-SSSA, 677 South Segoe Road, Madison, WI 53711, USA. *Soil Specific Crop Management.*

fertility variations.

With the advent of intensive statistical analysis programs made feasible by computer technology, accurate soil fertility maps and variable fertility management plans can now be developed (Hammond et al., 1988; Mulla and Hammond, 1988). The technology for variable fertilizer application, in both ratio of nutrients and application rates, as dictated by computerized management maps, is also available to production agriculture (Fairchild and Hammond, 1988).

With development of variable fertility management technology, the question of economic feasibility is of immediate importance to growers involved in potato production. A complete economic analysis must examine both input costs and crop production returns. The intent of this study is to examine fertilizer application efficiencies and variable fertility management costs. Data on yield and quality returns is limited due to the lack of adequate technology for mapping crop yield and quality in potato production. However, some basic assumptions will be made in the discussion at the conclusion of this report.

METHODS

A cost analysis was made of variable fertility management programs for eight production fields under center pivot irrigation. This analysis examined application efficiencies and input fertilizer and variable management costs.

The variable fertility management program for each field was based on the P and K soil test levels of soil samples collected on a 200-ft grid. The soil test data were then statistically analyzed via computer programs (Surfer[TM], Golden Software, Golden, CO) to show projected soil test P and K levels. As an example, the results for study field #1 are depicted by contour-type maps showing the areas of various soil test levels in Fig. 17-1 and 17-2. The numbers on the contour lines indicate the parts per million for the soil test levels. Data from 100, 200, and 400 ft grids indicate that results from a 200 ft grid are adequate but significant detail is lost at 400 ft (Mulla and Hammond, 1988). To assist in visualizing the various soil test levels, the contour-type maps were converted into three-dimensional images via computer as shown in Fig. 17-3 and 17-4.

The soil test data were further analyzed by a computer program to combine P and K data. These results were then organized into fertility management zones. Up to five fertility management zones may be developed by this program. For the purposes of this study three fertility management zones were developed and are depicted on a map as shown in Fig. 17-5. The computer program also gives the average soil test values for the management zones as well as the acreage for each zone (Mapmaker, Soil Teq, Inc., Waconia, MN).

Based on the soil test levels for the fertility management zones, preplant fertilizer recommendations were made for P and K using mono-ammonium phosphate (MAP) and muriate of potash along with an application for N. Nitrogen rate was equalized across treatments by varying the rate of urea applied.

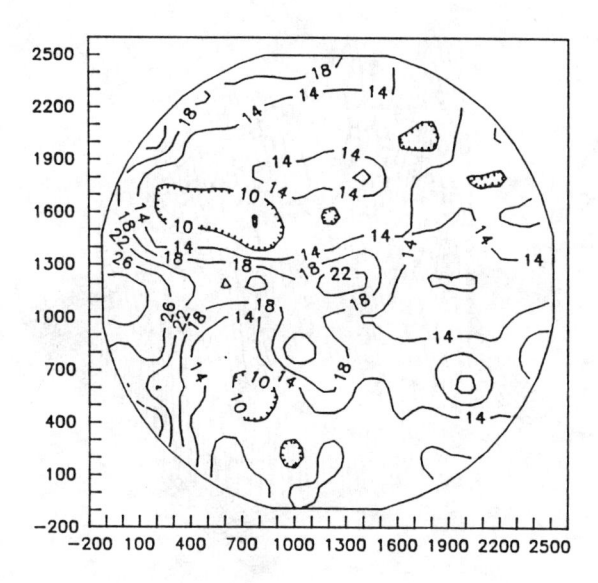

Fig. 17-1. Field no. 1 ppm soil test P.

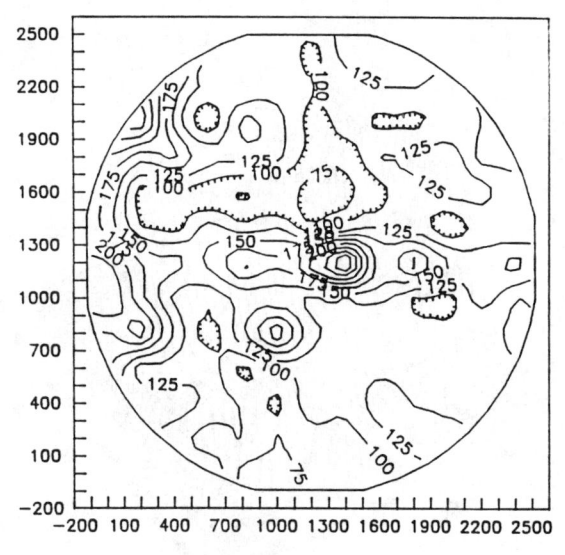

Fig. 17-2. Field no. 1 ppm soil test K.

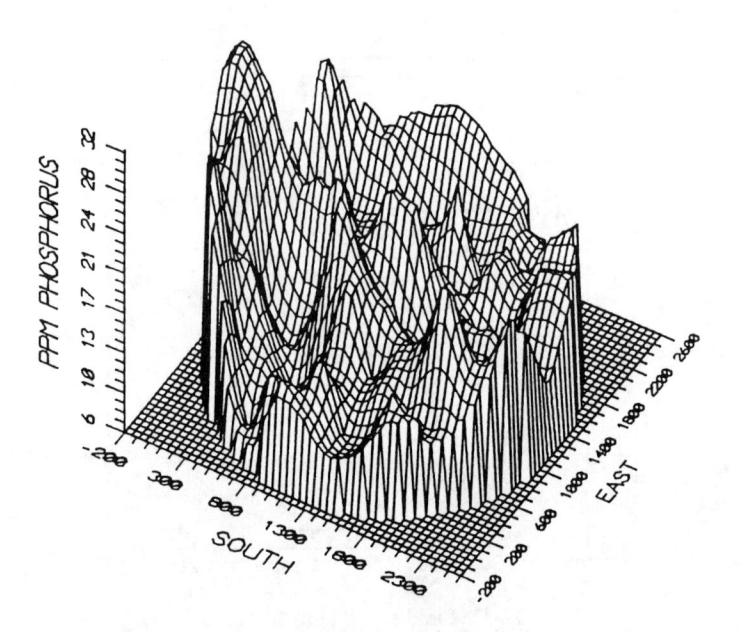

Fig. 17-3. Field no. 1 ppm soil test P.

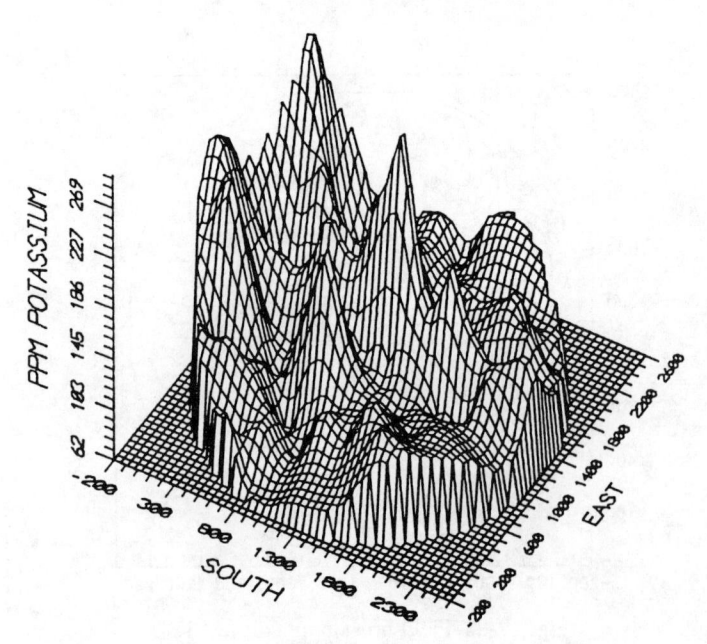

Fig. 17-4. Field no. 1 ppm soil test K.

COST ANALYSIS OF PHOSPHORUS AND POTASSIUM 217

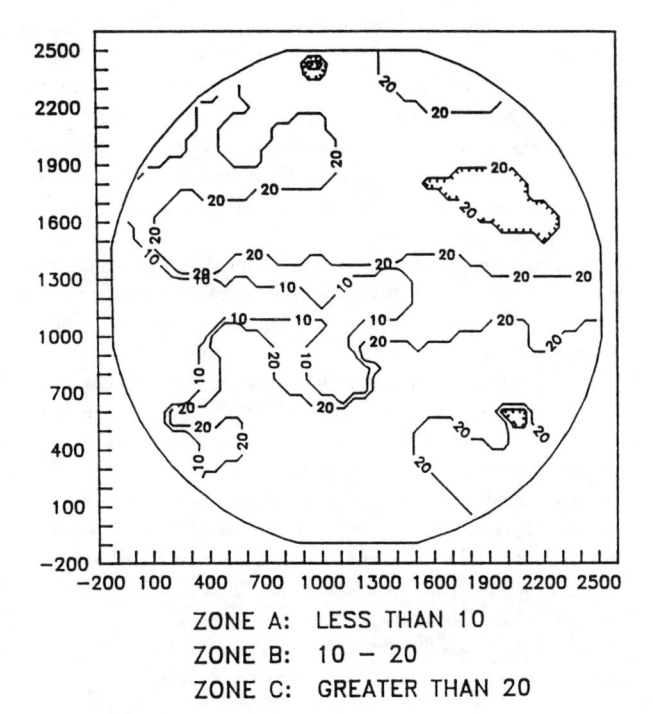

ZONE A: LESS THAN 10
ZONE B: 10 – 20
ZONE C: GREATER THAN 20

Fig. 17-5. Field no. 1 fertility management zones.

To develop a conventional fertilizer application program for each field, all soil test values were averaged to give overall composite soil test levels for P and K. The conventional fertilizer recommendations were based on these composite soil test values. Whole-field and management zone acreages and soil test levels for each study field are given in Tables 17-1A to 17-8A. Fertilizer recommendations in lb/acre for whole fields and management zones are also given in Table 17-1A to 17-8A.

To examine fertilizer application efficiencies, the variable application recommendations for each management zone were subtracted from the conventional whole-field recommendations and the net difference listed as "application rate error."

To determine fertilizer input costs, the acreages and fertilizer recommendations for each management zone were processed through a computerized fertilizer formulation program as was the conventional fertilizer recommendations for the whole field (AgriSource[TM], Cenex/Land O'Lakes, St. Paul, MN).

Various values used for calculation of fertilizer and management input costs are given in Table 17-9.

RESULTS AND DISCUSSION

The results of comparing variable management zone with conventional whole-field recommendations for application efficiencies for each study field are shown in Tables 17-1A to 17-8A. Results of the comparison between variable and conventional fertility management for fertilizer and management input costs are shown in Tables 17-1B to 17-8B. (Tables 17-1A to 17-10 are on p. 220-228.)

A summary of costs incurred for variable fertility management for P and K in comparison to conventional whole-field management for each of the eight study fields is given in Table 17-10.

With respect to fertilizer application efficiency, examination of the soil test data indicates that P and K levels in the variable management zones can differ significantly from the average for the whole fields. This difference contributes to application rate errors of both excess and insufficient applications. Fields with high soil test levels may not have as much error in excess application but there is always error due to insufficient application in low fertility level management zones.

Fertilizer input costs on a per acre basis are generally the same. Average increase in fertilizer costs per field over the eight study fields was 3%. Three of the eight fields had decreases in fertilizer input costs. Variable fertility management increased total costs of field services by 54%. Blending costs were essentially the same with the increase being attributed primarily to grid sampling, analysis of samples, and data management.

Across all study fields the average increase in fertilizer costs was $2.42/A with an average increase in management services of $9.49/A. This totals to an overall average increase of $11.41/A for variable fertility management of P and K.

Total costs for production of processing potatoes in central Washington averages approximately $2000/A. The increase in production costs due to variable fertility management is minor but the effects on crop return can be significant. Variable management is unlikely to gain increases in crop yield or quality in the management zones of high or intermediate fertility. Such increases are more likely to be gained in the zones of low fertility that would receive insufficient fertilizer application with conventional whole-field management. Since the acreage is low, any return due to increase in yield across the whole field will probably be minor. An overall increase of 1 to 2 ton/A at a contract price of $75/Ton will result in a $75-$150 return per acre. However, the low fertility zones can contribute to significant decreases in crop grade for the whole field. If by improving overall grade with an increase in contract price of $10 to $15 per ton and a 30 ton/A yield the return for variable fertility management can be $300 to $450/A. With minor increases in yield and significant increases in uniformity to improve crop grade, the returns on investment in variable fertility management of P and K should far exceed costs.

Though not directly related to potato crop production, preliminary data indicates an extended benefit of increased soil fertility uniformity for rotation crops following potato production (M. Hammond, 1992 unpublished data).

Finally, at a time of considerable environmental concern in regard to production agriculture, variable fertility management is a best management practice to ensure environmental conservation and sustainability of resources for production agriculture.

REFERENCES

Dow, A. I. (ed.). 1981. Principles of soil sampling for northwest agriculture. West. Reg. Ext. Publ. 9. Washington State Univ., Pullman.

Dow, A. I., and D. W. James. 1973. Intensive soil sampling - a principle of fertility management in intensive irrigation agriculture. Washington Agric. Exp. Stn. Bull. 781.

Dow, A. I., D. W. James, and T. S. Russell. 1973. Soil variability in central Washington and sampling for soil fertility tests. Washington Agric. Exp. Stn. Bull. 788.

Dow, A. I., T. S. Russell, and T. A. Cline. 1980. Soil test variability in sandy soils under two center-pivot irrigation systems. Washington Agric. Exp. Stn. Bull. 886.

Fairchild, D., and M. Hammond. 1988. Using computerized fertilizer application equipment for efficient soil fertility management. Proc. Far West Reg. Fert. Conf., Bozeman, MT. 11-13 July 1988. Far West Fert. AgChem Assoc. Pasco, WA.

Gray, D., and J. C. Hughes. 1978. Tuber quality. p. 504-544. In P. M. Harris (ed.) The potato crop: The scientific basis for improvement. Chapman & Hall, London.

Hammond, M., D. Mulla, and D. Fairchild. 1988. Development of management maps for soil variability. Proc. Far West Reg. Fert. Conf. Bozeman, MT. 11-13 July 1988. Far West Fert. AgChem Assoc., Pasco, WA.

James, D. W., and A. I. Dow. 1972. Source and degree of soil variation in the field: The problem of sampling for soil tests and estimating soil fertility status. Washington Agric. Exp. Stn. Bull. 749.

Kunkel, R., C. D. Moodie, T. S. Russell, and N. Holstad. 1971. Soil heterogeneity and potato fertilizer recommendations. Am. Potato J. 48:153-173.

Kunkel, R., and R. E. Thornton. 1986. Understanding the potato. Scientific paper no. 7267. Washington State Univ., Pullman.

Kunkel, R., R. E. Thornton, N. M. Holstad, D. C. Mitchell, and T. S. Russell. 1978. Sampling soils for potatoes. Proc. 1978 Wash. State Potato Conf., Washington State Potato Comm., Moses Lake, WA.

Mulla, D., and M. Hammond. 1988. Mapping of soil test results from large irrigation circles. Proc. Far West Regional Fert. Conf., Bozeman, MT. 11-13 July 1988. Far West Fert. AgChem Assoc. Pasco, WA.

Table 17-1A - Soil analysis and fertilizer recommendation - - Field #1.

Management area	Acres	Soil test P	Soil test K	Recommendations N	Recommendations P_2O5	Recommendations K_2O	Rate error P_2O5	Rate error K_2O	Rate error % Field
		ppm		lb/A					
Whole field	127	16	127	120	145	345	N/A		
Zone A	23	24	186	120	20	230	+125	+115	18
Zone B	34	14	140	120	180	320	- 35	+ 25	27
Zone C	70	14	102	120	180	400	- 35	- 55	55

17-1B - Fertilizer and management costs - Field #1.

	Fertilizer Cost/A	Fertilizer Total	Fertilizer Net Change/A	Service Sampling/Analysis[1]	Service Map	Service Blend	Service Spread	Service Net Change/A
Whole field	$112	14,244	N/A	$124	$0	$452	$635	N/A
Variable:		14,466	+1.75	$913	$127	$458	$921	$9.51
Zone A	$ 68							
Zone B	$116							
Zone C	$128							

[1]Based on one sample/0.92 acres for grid samples and one sample/approximately 30 to 50 acres for composite samples.

COST ANALYSIS OF PHOSPHORUS AND POTASSIUM 221

Table 17-2A - Soil analysis and fertilizer recommendation - - Field #2.

Management area	Acres	Soil test P	Soil test K	N	P_2O_5	K_2O	P_2O_5 Rate error	K_2O Rate error	% Field
		ppm		lb/A					
Whole field	137	27	238	120	0	125	N/A		
Zone A	81	29	262	120	0	80	0	+45	59
Zone B	42	27	203	120	0	200	0	-75	31
Zone C	14	15	203	120	155	200	-155	-75	10

17-2B - Fertilizer and management costs - Field #2.

	Fertilizer Cost/A	Fertilizer Total	Fertilizer Net Change/A	Sampling/ Analysis[1]	Map	Blend	Spread	Net Change/A
Whole field	$ 48	6,618	N/A	$124	$0	$222	$685	N/A
Variable:		7,180	+4.10	$976	$137	$235	$933	$9.12
Zone A	$ 42							
Zone B	$ 59							
Zone C	$ 94							

[1] Based on one sample/0.92 acres for grid samples and one sample/approximately 30 to 50 acres for composite samples.

Table 17-3A - Soil analysis and fertilizer recommendation - - Field #3.

Management area	Acres	Soil test P	Soil test K	Recommendations N	Recommendations P_2O_5	Recommendations K_2O	Rate error P_2O_5	Rate error K_2O	Rate error % Field
		ppm		lb/A					
Whole field	110	20	157	120	90	290	N/A		
Zone A	4	24	182	120	25	240	+ 65	+ 50	49
Zone B	29	17	169	120	130	265	- 40	+ 25	26
Zone C	27	14	92	120	180	420	- 90	-130	25

17-3B - Fertilizer and management costs - Field #3.

	Fertilizer Cost/A	Fertilizer Total	Fertilizer Net Change/A	Service Sampling/ Analysis[1]	Service Map	Service Blend	Service Spread	Service Net Change/A
Whole field	$ 92	10,125	N/A	$ 62	$0	$328	$550	N/A
Variable:		9,799	-2.96	$786	$110	$315	$798	$9.71
Zone A	$ 70							
Zone B	$ 85							
Zone C	$131							

[1] Based on one sample/0.92 acres for grid samples and one sample/approximately 30 to 50 acres for composite samples.

COST ANALYSIS OF PHOSPHORUS AND POTASSIUM 223

Table 17-4A - Soil analysis and fertilizer recommendation - - Field #4.

Management area	Acres	Soil test P	Soil test K	Rec. N	Rec. P_2O_5	Rec. K_2O	Rate error P_2O_5	Rate error K_2O	% Field
		ppm	ppm	lb/A	lb/A	lb/A			
Whole field	90	18	297	120	110	0	N/A	N/A	N/A
Zone A	38	22	317	120	50	0	+60	0	42
Zone B	42	16	305	120	145	0	-35	0	47
Zone C	10	15	195	120	160	210	-50	-210	11

17-4B - Fertilizer and management costs - Field #4.

	Fertilizer Cost/A	Fertilizer Total	Fertilizer Net Change/A	Sampling/ Analysis[1]	Map	Blend	Spread	Service Net Change/A
Whole field	$ 55	4,915	N/A	$ 62	$0	$130	$450	N/A
Variable:		5,149	+2.60	$642	$90	$140	$653	$9.81
Zone A	$ 41							
Zone B	$ 62							
Zone C	$ 96							

[1]Based on one sample/0.92 acres for grid samples and one sample/approximately 30 to 50 acres for composite samples.

Table 17-5A - Soil analysis and fertilizer recommendation - - Field #5.								
Management area	Acres	Soil test P	Soil test K	Recommendations N	Recommendations P$_2$O5	Recommendations K$_2$O	Rate error P$_2$O5	Rate error K$_2$O	Rate error % Field
		ppm		lb/A					
Whole field	104	21	273	120	65	60	N/A		
Zone A	64	24	299	120	20	0	+ 45	60	62
Zone B	27	17	247	120	130	110	- 65	- 50	26
Zone C	13	15	200	120	160	200	- 95	-140	12

17-5B - Fertilizer and management costs - Field #5.								
	Fertilizer Cost/A	Fertilizer Total	Fertilizer Net Change/A	Service Sampling/ Analysis[1]	Service Map	Service Blend	Service Spread	Service Net Change/A
Whole field	$ 53	5,544	N/A	$124	$0	$163	$520	N/A
Variable:		5,472	-0.69	$740	$104	$160	$754	$9.14
Zone A	$ 35							
Zone B	$ 75							
Zone C	$ 95							

[1] Based on one sample/0.92 acres for grid samples and one sample/approximately 30 to 50 acres for composite samples.

COST ANALYSIS OF PHOSPHORUS AND POTASSIUM

Table 17-6A - Soil analysis and fertilizer recommendation - - Field #6.

Management area	Acres	Soil test P	Soil test K	N	P_2O_5	K_2O	P_2O_5	K_2O	% Field
		ppm		lb/A			Rate error		
Whole field	145	20	277	120	80	40	N/A		
Zone A	77	25	304	120	0	0	+ 80	40	53
Zone B	52	16	282	120	145	40	- 65	- 0	36
Zone C	16	10	137	120	250	330	-170	-290	11

17-6B - Fertilizer and management costs - Field #6.

	Fertilizer Cost/A	Fertilizer Total	Fertilizer Net Change/A	Service Sampling/Analysis[1]	Service Map	Service Blend	Service Spread	Service Net Change/A
Whole field	$ 54	7,792	N/A	$ 124	$0	$221	$ 725	N/A
Variable:		8,005	+1.47	$1,034	$145	$230	$1,051	$9.59
Zone A	$ 30							
Zone B	$ 68							
Zone C	$133							

[1]Based on one sample/0.92 acres for grid samples and one sample/approximately 30 to 50 acres for composite samples.

Table 17-7A - Soil analysis and fertilizer recommendation - - Field #7.

Management area	Acres	Soil test P	Soil test K	Recommendations N	Recommendations P_2O_5	Recommendations K_2O	Rate error P_2O_5	Rate error K_2O	Rate error % Field
		ppm		lb/A					
Whole field	139	19	168	120	100	270	N/A		
Zone A	67	22	174	120	45	255	+ 55	15	48
Zone B	61	16	169	120	145	265	- 45	+ 5	44
Zone C	11	14	118	120	185	360	- 85	- 90	8

17-7B - Fertilizer and management costs - Field #7.

	Fertilizer Cost/A	Fertilizer Total	Fertilizer Net Change/A	Service Sampling/ Analysis[1]	Service Map	Service Blend	Service Spread	Service Net Change/A
Whole field	$ 91	12,700	N/A	$124	$0	$406	$ 695	N/A
Variable:		12,522	-1.28	$989	$139	$399	$1,008	$9.42
Zone A	$ 77							
Zone B	$ 98							
Zone C	$123							

[1] Based on one sample/0.92 acres for grid samples and one sample/approximately 30 to 50 acres for composite samples.

COST ANALYSIS OF PHOSPHORUS AND POTASSIUM

Table 17-8A - Soil analysis and fertilizer recommendation - - Field #8.

Management area	Acres	Soil test P	Soil test K	N	Recommendations P$_2$O5	K$_2$O	Rate error P$_2$O5	K$_2$O	% Field
		ppm			lb/A				
Whole field	155	24	199	120	20	200	N/A		
Zone A	39	34	325	120	0	0	+ 20	200	25
Zone B	62	24	167	120	20	265	0	- 65	40
Zone C	54	17	143	120	140	315	-120	-115	35

17-8B - Fertilizer and management costs - Field #8.

	Fertilizer Cost/A	Total	Net Change/A	Service Sampling/ Analysis[1]	Map	Blend	Spread	Net Change/A
Whole field	$ 64	9,853	N/A	$ 155	$0	$332	$ 775	N/A
Variable:		11,465	+10.40	$1,101	$155	$373	$1,124	$9.61
Zone A	$ 30							
Zone B	$ 73							
Zone C	$107							

[1]Based on one sample/0.92 acres for grid samples and one sample/approximately 30 to 50 acres for composite samples.

Table 17-9. Data used for calculation of fertilizer and management input costs.

Fertilizer costs Urea (46-0-0)	$232/Ton
Map (11-55-0)	$299/Ton
KCL (0-0-62)	$179/Ton
Field grid sampling	$1.25/A
Analysis of grid samples for P and K	$5.40/Sample
Composite soil sample analysis[1]	$31/Sample
Data management for fertility management map	$1/A
Variable fertilizer application	$7.25/A
Conventional fertilizer application	$5/A
Fertilizer blending charges for both variable and conventional applications	$7/Ton

[1]While no composite samples were collected for the purpose of this study, charges for analysis of such samples were included for cost comparison.

Table 17-10. Cost analysis summary for comparison of conventional to variable fertility management.

Field	Fertilizer[1] cost/A	Management[1] services/A	Total/A
1	$1.75	$9.51	$11.26
2	$4.10	$9.12	$13.22
3	-$2.96	$9.71	$6.75
4	$2.60	$9.81	$12.41
5	-$0.69	$9.14	$8.45
6	$1.47	$9.59	$11.06
7	-$1.28	$9.42	$8.14
8	$10.40	$9.61	$20.01
Average	$2.42	$9.49	$11.41

[1] (Variable fertility costs - conventional costs.)

18 Macy Farms - Site Specific Experiences

Ted S. Macy
Applications Mapping, Inc.
27 Ash Street
Frankfort, IL

Macy Farms is a grain farm operation located in Wayne County, Indiana; this is approximately 50 miles east of Indianapolis. The farm consists of 900 acres (of ground) share-rented on a fairly permanent basis. The balance of the operation consists of 100 to 1100 acres of cash rented, on short-term leases, and quite subject to change. In 1991, Macy Farms operated just over 1100 acres of corn (*Zea mays* L.) and 900 acres of soybean [*Glycine max* (L.) Merr.]. Prior to 1991, the operation had never farmed more than 1500 acres.

The farm has labor resources totalling 2.8 men during the peak times, and 1.8 men during summer operations.

The implication of this structure is that it is not possible to acquire equipment or permanent labor based upon a long-term payback period. The acreage to support such a decision cannot be guaranteed. Because the farm is usually operated with less machinery than the acreage would typically dictate, it becomes imperative that all production operations be performed as efficiently as possible to plant and harvest the crop in a timely fashion.

Macy Farms has supported spatially variable philosophies since the early 1980s. While attempts have been continually made to build an information base to implement those philosophies, production efficiencies dictated that true spatially variable material application had to wait until some level of automation was available. The farm has taken advantage of conventional controller technology since the early 1980s. It was not until 1990 that the farm first used an automatic spatially variable controller for anhydrous ammonia. In 1991, spatially variable controllers were implemented for the application of all materials used in the farm operation.

The balance of this chapter covers the areas of application rate philosophies, the physical implementation of site specific application, and a brief analysis of the financial impact on Macy Farms' operating costs.

Copyright © 1993 ASA-CSSA-SSSA, 677 South Segoe Road, Madison, WI 53711, USA. *Soil Specific Crop Management.*

APPLICATION PHILOSOPHIES

The first requirement to implement site specific techniques for the application of crop inputs is to establish philosophies to be used in the development of the target application rates. Macy Farms uses two types of information for the development of these philosophies. The first is soil survey maps prepared by the USDA. These maps provide the primary source of information for such items as target yield goals, soil texture, and organic matter characteristics. The general information contained in this database should be adjusted for personal experiences.

Macy Farms is also a proponent of intensive grid soil fertility testing. Fields are sampled on a 4-year rotational basis, with sampling done on 2.5 acre grids (330 ft X 330 ft). This sampling is done without regard to soil type.

Using these two sources of information the following philosophies have been developed:

1. Starter Fertilizer:
   Apply starter fertilizer inversely proportional to fertility test levels. Do not apply starter on high or excessively high fertility areas. Weight the P test at twice the impact of the K test.

2. Broadcast Fertilizer:
   Apply fertilizer relative to needs of each possible combination of soil type and soil testing sample location. For each of these combinations, establish a fertility recommendation taking into account the yield goal for this soil type, and the existing fertility level at that location. Generate recommendations from tables prepared by universities or consulting agronomists. Adjust this recommendation for the starter that is applied.

3. Nitrogen Fertilizer:
   Apply N relative to the yield goal associated with each soil type. Adjust for N supplied by starter fertilizer and broadcast fertilizers. Also adjust for the contribution from a previous crop. Use site-specific previous crop yields if possible.

4. Seed Population:
   Vary corn population based upon the yield goal associated with each soil type in the field.

5. Chemical Application:
   Apply preplant incorporated (PPI) chemicals and preemergence chemicals based upon soil texture traits and organic matter traits, and historical weed pressures. Record the location of spot-spraying applications for postemergence chemical application. Use this information for follow-up applications when necessary.

Macy Farms developed a geographic information system (GIS) based crop production management software package to assist in the implementation of these philosophies. USDA soil types are digitized as a layer, providing a default yield goal, texture, and organic matter characteristic. The soil test results are recorded as an additional layer, providing the information necessary to finish the fertility recommendations.

The software is capable of intersecting these two layers to provide all possible combinations of yield goal and existing fertility level. The recommendation for each of these areas is obtained from tables developed for any agronomic philosophy.

The package will also generate the application files necessary for different types of site specific machinery. These can be raster based or vector based, depending upon the device.

Example Field

The implementation of these philosophies can best be illustrated by generating recommendations for one field. Figure 18-1 is a map of the soils included in this field. The field lies in a river bottom on the west end and rises to high, light colored soils toward the east end. This is a fairly typical field for the Macy Farms operation.

Table 18-1 contains the soil type records for soils found in this field. For each soil type, a target yield goal, corn (*Zea mays* L.) seed population, and characteristic organic matter level has been created. This information was generated from USDA material and agronomic sources such as universities or agronomic consultants.

Table 18-1. Soil type characteristics.

Soil type	Corn yield	lb N / Acre	Corn population	Organic matter
Crosby silt loam	105	130	27,000	2%
Ockley silt loam	110	138	27,500	2.25%
Strawn clay loam	95	120	21,000	1.5%
Treaty silty clay loam	150	190	30,000	5%
...				

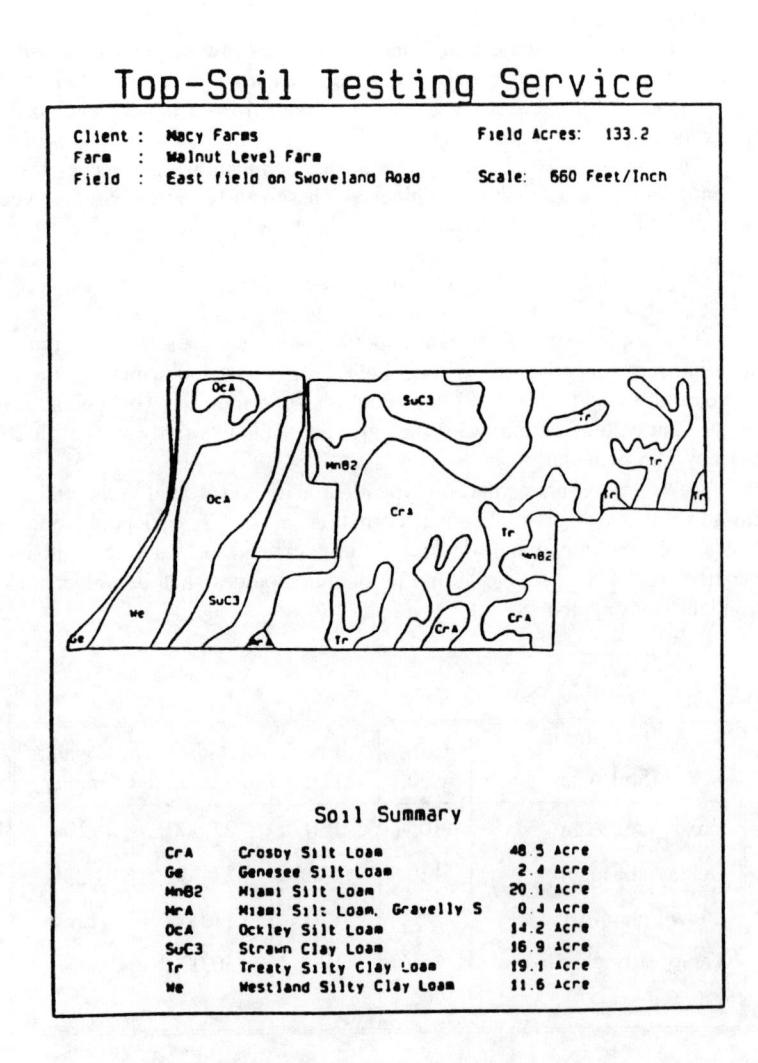

Fig. 18-1. Soil types.

SITE SPECIFIC EXPERIENCES 233

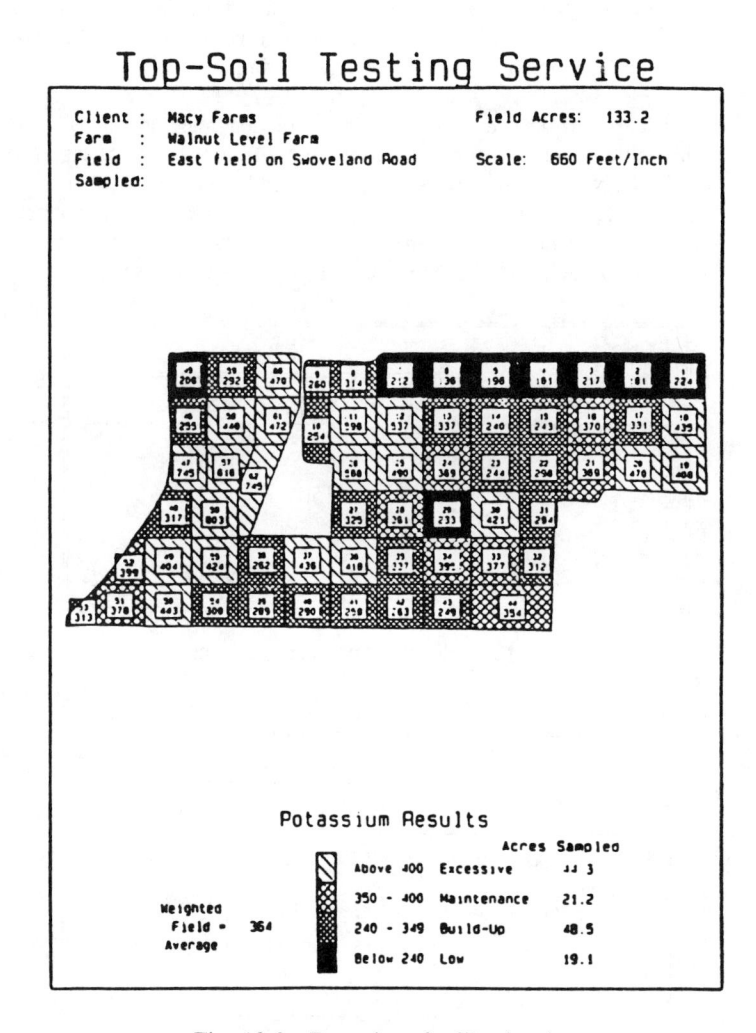

Fig. 18-2. Potassium fertility levels.

Figure 18-2 is a map of the K levels in this field. Each cell is approximately 2.5 acres. Similar information is also available for pH and P levels, as well as composited micronutrient levels.

Given the information about soil characteristics and fertility levels, execute the following steps:

1. A starter fertilizer recommendation is produced for each cell by looking at the combination of P and K levels for that cell. Table 18-2 is used to generate that recommendation for any combination of levels.

2. Broadcast fertilizer recommendations are generated from Table 18-3. This table contains recommendations for different levels of target yield goal, at given fertility levels. A recommendation is generated for each distinct region in the field. These regions are produced by intersecting the regions of the soil type map and the fertility map. A new region is created for each possible combination of soil type and sample site.

	K Test			
P Test	150	200	250	...
20	200	175	150	
30	175	150	125	
40	150	125	100	
...				

Table 18-2. Starter fertilizer guidelines.

	Corn yield goal			
K Test	90	110	130	...
100	175	181	186	
150	145	151	156	
200	115	121	126	
...				

Table 18-3. Potassium fertilizer guidelines.

SITE SPECIFIC EXPERIENCES

Each region has a target yield goal and existing fertility level. Table 18-4 is a listing of each of those regions, and their respective recommendations. These values must be adjusted for the nutrient being provided by the previously generated starter recommendation.

```
                              Fertilizer Recommendations
 Sample  Soil  Yield                  Phosphorus              Potassium
  no.    type  goal   Acres    Test   Mnt   B/U  Tot   Test   Mnt   B/U   Tot
 ---------------------------   ---------------------   -----------------------
    1    CrA   105.0  1.46      25    45    30    75    224    30    76   106
         Tr    150.0  0.52      25    64    30    94    224    42    76   118

    2    CrA   105.0  1.61      31    45    24    69    181    30   101   131
         Tr    150.0  0.59      31    64    24    88    181    42   101   143
  ...
   60    OcA   110.0  0.30     168    47     0    47    470    31     0    31
         SuC3   95.0  0.42     168    41     0    41    470    27     0    27
         We    140.0  1.39     168    60     0    60    470    39     0    39
   61    OcA   110.0  1.06     139    47     0    47    472    31     0    31
         SuC3   95.0  0.92     139    41     0    41    472    27     0    27
   62    OcA   110.0  0.92     240    47     0    47    745    31     0    31
         SuC3   95.0  1.51     240    41     0    41    745    27     0    27

 Weighted average yield goal:          113.8

 Total field recommendations:        (Lbs/Acre)         P               K
                                      Build up:        49              32
                                    Maintenance:        5              27
                                                     --------        --------
                                          Total:      54              59
```

Table 18-4. Site-specific fertilizer recommendations.

3. Nitrogen recommendations are generated from the soil type information (Table 18-1), and adjusted to reflect the N being applied by the starter recommendation and the broadcast recommendation, and the contribution from the previous crop.

4. Corn population is generated by soil type from Table 18-1.

5. Chemical rates are generated by the specific chemical label information for the characteristics associated with each soil type.

These philosophies are implemented with software developed specifically to execute these steps. Over time, an adjusted yield goal layer can be developed. This layer can reflect changes to the default soil type yield goals. These changes may be necessary due to the existence of site-specific conditions such as compaction or drainage that impact upon the general yield levels associated with that area.

Layers such as existing weed pressure, or past chemical usage can be used to fine-tune the chemical application rates. Nitrate testing can be used, and the results become a layer that is subtracted from the general N recommendation.

The development of these philosophies, consciously or subconsciously, is the first step toward site specific application techniques.

IMPLEMENTATION

Once application philosophies have been developed, and a reasonable way exists to generate recommendations reflecting the philosophies, the materials must be applied according to recommendations. The areas to receive different application rates may be marked with flags or other visual landmarks, or a means of real-time location capability may be implemented. The rate adjustment may consist of multiple passes at a fixed rate, or be manually adjusted physically, manually adjusted electronically, or automatically adjusted electronically. It is this latter option that Macy Farms has implemented.

The implementation discussion covers the design constraints for the site-specific development, the actual hardware implementation, and the operation of the application control software.

Design Constraints

Based upon the need for production efficiency, as described above, Macy Farms had to implement spatial control within the following operational parameters, generating the associated constraints:

1. Operators frequently work 16- to 20-h days. From experience, it is not reasonable to expect an operator to successfully monitor equipment operation and make the correct rate adjustments consistently. Monitoring equipment performance is more important.

    Constraint: The application rate adjustments have to be made automatically.

2. Many operations are performed after dark. This means that "landmarks" are not visible for determining location within a field.

    Constraint: Locations have to be determined automatically. An operator could not be required to indicate his location.

3. Good agronomic practice dictates the application of some materials in a pattern that is not parallel with the rows.

    Constraint: A location system that uses dead reckoning based upon a pattern of fixed width passes parallel to a straight side of a field was not acceptable. Any dead reckoning location methodology has to be able to

SITE SPECIFIC EXPERIENCES

operate on a diagonal pattern.

3A. Diagonal patterns are sometimes started at random, after dark.

Constraint: A location methodology based upon visible landmarks for establishing the angle of operation or the beginning point of the operation is not acceptable.

4. One of the operators (the author's father) is not comfortable with this level of technology. Replacing the operator is not an acceptable option.

Constraint: The operator has to be able to start an application by a limited number of commands, needs to be warned if problems are occurring, and needs to be able to ignore the existence of the equipment otherwise.

5. It is frequently desirable to apply two or more materials in the same operation, with the correlation of rates between the materials being totally random.

Constraint: Independent multi-channel control is necessary.

6. Each tractor used by Macy Farms could be interchangeably used for the application of any of the materials to be applied. It is not acceptable to install and remove controllers during times of peak use. There is not room in the tractor cabs for a different controller for every application that might be performed.

Constraint: There has to be one generic master controller capable of interfacing with application-specific control equipment on each implement.

7. Down time is expensive.

Constraint: If a problem develops with the experimental controllers, there needs to be a way to drop back to manual mode and continue conventional material application.

8. It is difficult to press keys on a keyboard or buttons on a controller when bouncing through a field.

Constraint: The controller should offer a remote switch package that can be located at a convenient place for the operator's routine access.

Application Hardware

Within the above constraints, a spatially variable control system was developed that uses a real-time location gathering device, a master controller for the determination of the target application rate from one or more material prescriptions, and conventional application controllers, modified to accept a target application rate from the master controller. The equipment used is more specifically:

1. Location:

    Both Global Positioning Satellite (GPS) receivers, and Loran receivers are used. Both devices have the ability to generate output messages at frequencies ranging from 1 to 3 s. These messages are transmitted using the RS232 protocol.

    Differential GPS is not used.

2. Master Control:

    Conventional PC's with VGA color graphics are used. These computers are interfaced with the location devices to obtain the position within the field. The software is designed to read an independent prescription for each product, and transmit a target application rate to one or more controllers. The software has a graphical interface providing the operator with constant feedback of location within the field, and the status of all control operations.

    Prescription input is done on either 3.5-in. floppy diskettes, or the SRAM cards from an Atari Portfolio.

    Micro-Type keyboards with Safe-Skins are used for the keyboard interface. These are conventional keyboards, but in a package that is only 9 in. wide and 5 in. deep.

    The master controller uses a multi-purpose interface card to interface with the radar (to determine ground speed and distance covered), analog real-time sensing devices (organic matter probe and yield monitors), and operator switches for boom control, remote run/hold, etc. The D/A function is used to interface with the DICKEY-john controllers.

    A general purpose operator switch control box was built for each application. This switch box is configured small enough to be mounted wherever the operator's hand usually rests. The switches were chosen such that the operators could easily sense switch positions and locations with his fingers. They also have enough resistance to operation to allow fingers to rest on the switches. A different box was built specific to each application configuration.

The switches on this box are connected to the interface board on the PC. They are also made available directly to a relay box that is located on the implement. This makes it possible to control valves or other devices through the master controller, or manually through the switch box leading to the ability to continue operating in the event of a master controller failure.

This hardware was installed in 1990. To test ruggedness, a policy was implemented of having the computers running whenever the tractor was being used. To date, the two computers used have logged 600 and 400 h of operational time without a hardware failure.

3. Application Tools:

Two general types of controllers are used. The first is a controller built by Macy Farms. It communicates with the master controller using the RS232 protocol. It was specifically designed to drive the stepper motor used in a Rawson hydraulic motor control package. It is installed on the piece of equipment being controlled and has no operator interface other than through the PC.

The second type of controller used is a DICKEY-john CCS-100 granular controller with a special modification to allow the input of a target application rate.

Existing implements were modified to make the following applications:

a. Starter Fertilizer:

A DICKEY-john anhydrous controller is used to regulate the application rate of starter fertilizer during the planting process. The flow sensor is an anhydrous flow meter, and the regulating valve is a conventional spray regulating valve.

b. Dry Fertilizer/Lime:

A DICKEY-john dry granular controller is used to control the application of dry fertilizer or lime from a conventional New Leader combination lime/fertilizer bed mounted on a trailer, pulled by the 7120. This is a control mechanism using hydraulic motor drives for the spreader apron and twin spreading fans. The 2670 does not have the hydraulic capacity to operate the unit. The speed of the apron is regulated by the DICKEY-john controller to maintain the desired application rate, relative to ground speed.

In 1990, an electric linear actuator was used to regulate gate height to control the application rate. That operation was satisfactory, but a DICKEY-john controller was made available for testing purposes so it was used in 1991.

c. Anhydrous Ammonia:

A DICKEY-john anhydrous ammonia controller is used for the application of N in the form of anhydrous ammonia. This unit is installed on an 11 knife DMI toolbar rented from the ammonia supplier. The bar is frequently carried by a pull-between cart, discussed below. It is used by either tractor.

d. Seed:

Seed population is controlled by a modified Systematic 2000 control system from Hydraulic Control Systems, Inc. This is a Rawson hydraulic regulator, governed by a stepper motor driven at the desired hydraulic motor speed. It comes with an operator interface box that allows the selection of a range of populations up or down from a calibrated population.

For this application, the factory interface was disconnected and replaced with a simple stepper motor controller. This controller is interfaced to the master controller through an RS232 interface. The master controller takes responsibility for determining ground speed, calculating a target motor speed dependent on target population and seed drive transmission parameters, and transmitting this target motor speed to the planter controller. The planter controller is mounted on the planter and connected to the tractor by one cable carrying both power and the interface signal.

e. Chemicals:

Chemicals are applied through a 60 ft boom for no-till, pre-and postemergence operations. They are applied in front of a field cultivator for incorporation application. They are applied through a boom mounted on the back of the anhydrous bar for the application of burn-down chemicals and anhydrous nitrogen, simultaneously. A pull-between spray cart with a three-point hitch was outfitted for all chemical regulation and application. It can pull the field cultivator or carry the 60 ft boom or the anhydrous bar.

The chemical regulation is accomplished by injecting a chemical into a water or 28% N solution carrier. The injection is done with a Raven chemical injection pump and motor, controlled by a DICKEY-john anhydrous ammonia controller. Interfacing these two units was accomplished by replacing the Raven shaft speed sensor with a DICKEY-john ammonia flow sensor, which is a proximity sensor. Several bolts were added to the perimeter of the large driven pulley on the Raven system so that the speed of the pulley could be more accurately determined. It was also necessary to implement a run/hold cutoff relay to prevent the DICKEY-john controller from trying to drive the pump speed to zero when application is not desired.

This system works, but because it is a one-material system, it is necessary to select chemicals that can be successfully mixed at full strength. That is a limitation that is acceptable, but not desirable.

Master Controller Software

The master controller software was written in-house. Due to commercial considerations, only a general outline of this application can be presented at this time.

Prescriptions for each material to be applied are prepared in the farm office. These are delivered to the master controller as files on a floppy diskette or battery-backed memory card.

The master controller polls the location device to find its most current location, and displays this position graphically on the computer screen. The operator has an opportunity to reposition the location cursor manually at any point in time. Those offsets are then applied to any subsequent location readings. The controller expects that the prescription representation is properly oriented, relative to real world coordinates. Only positional transformation is accommodated by the controller. Rotational transformation is not implemented.

Given a location, the controller obtains a target application rate from the prescription for each material being applied. It transmits that rate to the appropriate material controllers. For some material controllers, this is as simple as sending a target rate through a serial port. For other units, it requires using a digital to analog converter to send an integrated voltage to the controller. And, for anything that was built in-house, the master controller is responsible for transforming the target application rate per acre into the desired flow rate per minute, taking into account ground speed and active width. This flow rate is transmitted to the material controller.

FINANCIAL ISSUES

There are four primary financial considerations when evaluating this technology. Obviously, the cost of the equipment is a primary one. However, the costs associated with the Macy Farms application are not applicable to further installations, and this chapter will not attempt to estimate what the cost of this technology will be. The other three considerations are the savings associated with reduced crop inputs, the production benefit (or reduction), and the environmental cost.

General Input Savings

The figures in Table 18-5 are estimates of the per acre cost savings from the entire 2000 acres of crop production for 1991. They represent the change in physical inputs applied due solely to the implementation of site-specific philosophies and application technology. Where possible, these numbers represent the actual change in quantities used in 1991 and previous years. There was some new ground in the operation in 1991. Historical records were obviously not available, so historical philosophies were projected onto those acres.

These figures are savings per acre, determined by actual total savings allocated to the practices, divided by total acres in the operation. This calculates a cost savings figure that can be multiplied by the total farm acreage to help determine equipment payback.

The figures in Table 18-5 are based upon the following assumptions:

1. Corn acreage is about 60% of total acreage. Ammonia, starter, and seed savings apply only to corn acreage.

2. Reduction of actual average yield goal of nearly 18 bushels per acre on corn, based upon the acreages of each soil type, as opposed to historical goals. There was a historical tendency to shoot for yields attainable by the best soils in the field. This affects dry fertilizer and ammonia.

3. Site-specific application was only used for preplant incorporated and preemergence chemicals. This was only 25% of the 1991 acreage. The majority of the chemicals were applied as no-till burndown chemicals or postemergence chemicals. Site-specific rates were not used for these applications.

Production Considerations

Macy Farms did not have access to real-time yield monitoring in 1991. If the crop fertility tables that are in common use are correct, there would surely be increased production on those areas of a field where insufficient fertilizers were previously applied, or insufficient seed drop had occurred. These benefits can only be measured on a long-term basis, and the historical foundation may not be there to document improvement.

Similarly, there may be yield reduction in those areas where fertilizer application is reduced. Only rarely is fertilizer being applied at a rate that is beyond the economic point of marginal return. What will that reduction be?

Material	Savings per Acre	Description of Savings
Starter	$ 1.98	Historical practice: 100#/acre on all corn acreage. Savings due to application of 0 to 200 #/Acre, based upon fertility test.
Dry Fertilizer	$ 2.70	Savings due to application according to sample site, but retaining the same yield goal. This figure reflects only a small savings because the past philosophy was to apply only a little fertilizer above the average needed for all sample sites.
	$ 5.53	Savings due to modifying the yield goals throughout the field. Past procedure was to use a yield goal for the whole field that was equal to the better soils in the field.
Ammonia	$ 1.84	Based wholly upon the ability to vary nitrogen application based upon site-specific yield goal, less nitrogen from other fertilizer applications and projected carryover.
Seed	$ 1.35	Historical practice: 28,000 PPA. Savings due to populations of 18,000 to 32,000 PPA, based upon yield goal.
Chemical	$ .27	Savings due to organic matter variation on PPI chemical applications.
TOTAL:	$ 13.67	

Table 18-5. General input savings (1991).

The documentation necessary to answer the production issue is only starting to be collected. Macy Farms is admittedly operating on a "feels right" principle.

Environmental Impact

Adaptation of this technology has been shown to help reduce inputs. While it did not reduce the total N by the amount some would hope for, it did redistribute what was applied. Nitrogen use increased on some areas, but they were areas that can retain the N, and where the crop has a potential of fully using it. Applications were significantly reduced in those areas where groundwater contamination might occur more readily. If a cost is associated with the environmental impact of the overapplication of fertilizers and chemicals, then the payback to this technology is significant.

SUMMARY

Macy Farms has been able to implement site-specific materials application, in an automatic fashion. The implementation has involved the development of site-specific application philosophies, and the software to implement those philosophies. It has involved the development of a master controller that feeds target application rates to slave controllers. The master controller obtains position from a location device such as Loran or GPS receivers. Some slave controllers were built from scratch. Others were modified commercial units.

Macy Farms saw an input savings of nearly $14 per acre in 1991. This was attributable to reduction in fertilizer and seed. The reduction came from accepting a lower target yield goal in those areas of the farm that had less productive soils. Production benefits were not monitored.

There was a significant redistribution of the materials that were applied. The environmental impact was greater than the cost savings would indicate because of a redistribution of materials from areas where overapplication and waste were occurring, to areas where underapplication had been happening.

Site-specific application was implemented in a way that did not impact on the efficiency or timeliness of the operation. Most equipment used was off-the-shelf. Costs are becoming reasonable. There is little reason that this technology cannot be implemented affordably on a large share of American agriculture.

19 Working Group Report

D. Fairchild, Chair
M. Duffy, Recorder
M. Duffy, Discussion Paper

Other Participants:	R. Beck	G. Mangold
	D. Breitbach	R. Munson
	D. Buchholz	B. Nalanaphy
	T. Colvin	G. Olson
	P. Fixen	R. Olson
	M. Hammond	L. Reichenberger
	R. Joergen	J. Schulz
	C. Kohls	E. Urevig
	J. Lamb	N. Wollenhaupt

Land in its highest and best use is the way rural appraisers value farmland. Highest and best use can mean many things and include many variables. In general, highest and best is the most profitable to the owner. This usually means evaluating based on the land's most profitable agricultural use. Nonagricultural uses must also be considered for land close to developments, land that will be used as residence or land that may have recreation or other uses.

The development of modern agricultural equipment, hybrids, commercial fertilizers, and pesticides changed the way highest and best use was estimated. Soil conditions such as slope, wetness, inherent fertility and accessibility were no longer as critical as they once were. Twentieth century technology and cheap energy simplified cropping pattern decision making. This aspect of rural appraisal was simplified to determine whether there would be row crops, pasture, or a change to nonagricultural uses.

Research and technological innovations continued the move toward monocultures or very specialized crop production. Efficiency gains in terms of output per unit of input were impressive. There was a general move toward more intensive use of the land and away from the land characteristics especially in determining highest and best use. We were moving toward treating all soils as if they were identical.

Several factors have occurred that have tended to slow and perhaps even reverse this trend away from considering soil characteristics. One factor was increasing knowledge regarding the inherent productivity differences for

Copyright © 1993 ASA-CSSA-SSSA, 677 South Segoe Road, Madison, WI 53711, USA. *Soil Specific Crop Management.*

individual soils. In Iowa, for example, the projected corn (*Zea mays* L.) yield after soybean [*Glycine max* (L.) Merr.] for 140 of the major soil types averaged 131 bushels per acre. With consistent management and optimum input levels the projected yields vary from 45 to 170 bushels per acre. The standard deviation is 23 bushels per acre.

Technological innovations are another reason for the increased attention to soil specific crop management. Machinery which allows varying application rates is an example of such technology. So, too, is the use of satellites for identifying location in a field and the use of infrared photography for classifying crop development and stress areas. There are other examples of technological innovations that increase the importance of soil specific crop management.

The farm financial crisis of the 1980s served as another catalyst to help increase the interest in soil specific crop management. Anything that will add to the profitability of the farm or improve efficiency of production must be examined if the farm is going to stay in business.

Finally, increasing concern over the environmental consequences of modern agricultural practices furthered the study of soil specific crop management. Water contamination from soil-borne pesticides and fertilizers and groundwater contamination from leaching are areas of increased concern. There are other reasons but suffice it to say environmental concerns have played a major role in interest in soil specific crop management.

The recent history of U.S. agriculture can be generalized into two major areas. First, was the move toward specializing and treating all soils as identical and second is the move toward more soil specific crop management.

Before discussing the economic considerations this introduction will end by presenting two studies looking at soil specific crop management. These are not the definitive studies but rather ones that show the potential for soil specific management.

Speidel, in a creative component for a Masters of Agriculture degree at Iowa State, examined the fertilizer use issue. He randomly selected fields in Sac County, Iowa. He then rated the soils in the fields based on their yield potential. The fields fell into three general classifications; there were some with a predominance of high yield soils, some with a predominance of low-yielding soils and some fields had soils with mixed production levels. Using the assumption there was no difference in application costs (costs for different amounts of fertilizer did vary), he found that for fields with a predominance of high-yielding soils it was more profitable to fertilize for higher yields. For fields with a predominance of low-yielding soils farm profitability increased if extra fertilizer was applied to the high productivity soils. The mixed fields showed that it was better to apply fertilizer at the average rate.

Obviously this study had its limitations. It does serve to show the basic economic analysis. Some of the specific problems assumed in Spiedel's study have been discussed and must be included for a complete economic analysis.

Another analysis examined yield data to indicate the potential profitability differences by soil types. Corn yield variations are well known. Similar differences exist for other crops. For example, the expected soybean yield for

140 predominant soil types in Iowa is 44.1 bushels per acre. The range is from 15 to 57 bushels per acre with a standard deviation of 7.6 bushels.

The profitability differences vary by crop rotations. Using 1992 Iowa State cost of production estimates and $2.45 corn, $5.90 soybean, $1.80 oat (*Avena sativa* L.), and $55 alfalfa (*Medicago sativa* L.), significant differences can be observed. The range in the returns to land and management for continuous corn was $-91 to $156 with an average return of $83/per acre. The corn-soybean rotation had an average return to land and management of $125 ranging from $-51 to $204. And, a corn-oat-alfalfa-alfalfa rotation had a range from $-55 to $187 with a $105 average. When comparing different rotations on the same soil type the corn-soybean rotation dominates in terms of return to land and management. This rotation produced higher returns for all 140 soil types when compared to continuous corn. The corn-soybean also had higher returns than the C-O-M-M rotation 95% of the time. The differences between a C-Sb rotation and C-C ranged from $23 to $91 and between C-Sb and C-O-M-M the range was from $-9 to $69. The C-O-M-M rotation produced higher returns than C-C 79% of the time with a range in profitability difference from $-19 to $49 per acre.

There are other studies showing the benefits of managing by soil type. The point is that profitability differences do exist.

The remainder of this chapter will look at some of the major issues surrounding the profitability of soil specific crop management. First, will be a review of the applicable basic economic principles followed by a review of some specific areas to consider.

BASIC ECONOMIC PRINCIPLES

Marginal Analysis

In the simplest analysis an input should be used to the point where the marginal cost of the input equals the marginal revenue. For example, fertilizer should be applied until the last dollar spent just returns a dollar in additional output.

If the particular soil specific practice is of this simple form, then the analysis is straight forward. In other words if the practice does not involve additional fixed costs, then the marginal analysis would be appropriate. Varying speeds to change fertilizer rates would be an example. The revenue change would be how much yields increased or decreased by changing speeds multiplied by the product price. The marginal cost would be the value of the labor and fuel saved or added and the value of the fertilizer saved or increased.

Most soil specific management practices will not be so simple. If the practice involves new machinery or other major changes then alternative analysis techniques must be used.

Partial Budgeting

Partial budgeting is a frequently used tool. This tool should be used when the proposed practice does not involve major changes in the whole farm system. For example, partial budgeting should be used to evaluate a change in a tillage system.

Partial budgeting is relatively straightforward. Increases and decreases in costs must be identified and quantified. Next, increases and decreases in revenue must be valued. The changes in costs and revenues would involve labor, capital investments, and inputs including taxes. If the cost decreases plus revenue increases are greater than the cost increases plus revenue decreases then the practice would increase profitability.

Whole Farm Planning

Whole farm planning is another tool to use if marginal analysis is not appropriate. Whole farm budgeting is more complex than partial budgeting and should be used if a proposed practice would influence the whole farm structure. A major change in crop rotations, addition of new enterprises or major capital investments are examples of changes requiring a whole farm analysis.

Whole farm planning is a process that examines the resources available to the farm and looks at how to allocate those resources. The resources must be allocated consistent with goals of the family.

There are three basic ways to evaluate the profitability of soil specific crop management practices; marginal analysis, partial budgeting, and whole farm planning. The practice being evaluated and the degree of changes required determine which tool is most appropriate.

Regardless of the tool used the crucial key to the analysis is proper identification of all the variables impacted by the change. In some cases, there may be secondary effects that will have significant impacts on profitability and desirability. In addition to identifying the impacted variables there are several important factors that must be considered in the analysis. These factors will be covered in the next section.

FACTORS TO CONSIDER

Risk

Risk is one of the most important factors that is often overlooked in a profitability analysis. Whether the risk is real or perceived is immaterial. Farmers will react and base judgements on their perception of the risk. Risk is especially important if the perception is for high risk and the returns are relatively small.

Manure use as a fertilizer source is a primary example of risk and perception of risk. A recent Iowa State study shows only 49% of the farmers who spread manure adjusted fertilizer rates on those fields. Of those who reported adjusting fertilizer rates only 50% adjusted N rates. This means less than 25% of farmers spreading manure adjusted their N fertilizer application rate. Yet numerous studies have shown manure to be a valuable source of crop N and even a source of groundwater contamination.

One of the reasons frequently cited for not adjusting fertilizer for manure nutrients is the variability of the product. The perceived risk is having inadequate fertilizer in places due to the variability. Combining this risk with cheap fertilizer prices means farmers will simply ignore the manure value.

The subject of risk is the topic of books. Suffice it to say risks, whether real or perceived, will influence profitability and acceptance. If probabilities can be estimated then expected values can be calculated and incorporated into the analysis. If probabilities cannot be estimated the risk adds uncertainty to the analysis.

Managerial Skills

Closely related to risk is the concept of managerial skills and capabilities. Many soil specific technologies require a fairly sophisticated manager to implement. The manager must also be intelligent and willing to learn and accept new concepts.

An excellent example of this point can be found reviewing the adoption of preemergence herbicides. When farmers first used these materials they were surveyed concerning the level of weed control and whether or not they would use the material again. A significant percentage (over one-fifth) said they had good control but would not use the material again. Further questioning revealed that the farmers did not believe that a chemical could prevent weed germination. Their low weed pressure they reasoned was due to other factors. This is an example of technology that exceeded the ability for some to comprehend. Much of the soil specific technology may face similar resistance to adoption.

Nonmarket Benefits and Costs

The nature of the benefits and costs is another factor to consider when evaluating soil specific practices. Many of the benefits that accrue to soil specific technologies are nonmarket benefits. In other words, they are not traded or valued in the market. A practice that reduces water quality degradation is an example. Practices that reduce soil erosion also produce nonmarket benefits.

Valuing nonmarket benefits and costs is extremely difficult and subjective. Because they are nonmarket a farmer does not have to factor them into the decision and there is no straightforward way to value them. In many cases it is more expedient to simply note the changes rather than trying to value them. Noting that a practice lowers the level of, say, water contaminants rather than trying to value the reduction is often more useful. Simply because a good or service is not traded in the market does not mean it has no value. It only means our system does not place a value on the amenity.

Time Dimension

Another aspect to soil specific management is the time dimension. Saving soil or fossil fuels may produce benefits that are not realized until sometime in the future. These types of benefits are difficult to value.

Discounting

One approach to valuing future benefits and costs is discounting. The operating principal with discounting is that a dollar today is worth more than a dollar in the future. Estimating present value is fraught with problems. Two of the more significant ones are choosing the appropriate discount rate and the impact of technological innovations.

Similar to the other topics in this section, discounting or estimating present values is subject to debate and criticism. Regardless of the method used it is important to remember that there is a time dimension to many of our analyses. As with externalities one may choose not to estimate the value but simply note the changes. No matter which approach is used they should not be ignored.

Labor Impact

The impact on labor is the final point considered in this section. At first glance it seems relatively straightforward to estimate the increase or decrease in labor and use an appropriate wage rate to estimate the labor cost or savings. Many changes will not be so simple, however.

The appropriate wage rate may be difficult to estimate. There are published standards for hired labor but in many cases it will be the owner-operator labor. Such labor should be evaluated at its opportunity cost not a wage. The opportunity cost is the value of the time in its next best use. Saving labor may have a high or low opportunity cost depending on the individual farm. Time saved may be spent in other work activities or for leisure. The opportunity cost will vary by time of year as well as individual. Although an estimated wage rate is most frequently used it is important to remember this may or may not be appropriate.

Timeliness

Timeliness and convenience are also important labor considerations. Spending more time during a busy period may be much less profitable than during a period of less labor demand. Farmers are individuals and the convenience or ease of adopting a soil specific technology will influence its use. For example, if a farmer must mount and dismount the tractor more frequently to use a practice the profitability and acceptance will be lower.

There may be other factors in addition to the ones covered in this section. The point is that a profitability analysis must involve more than simply evaluating the market benefits and costs. Such analyses although appealing because of their simplicity can also lead to erroneous conclusions and produce undesirable social consequences.

FUTURE ISSUES

Soil specific crop management is a reemerging area using new technologies. The future developments and refinements in this area will determine the success and level of adoption.

Farmer Profitability

Farmer profitability will be the single biggest key to adoption of soil specific technology. Yield increases and the value of those increases together with the costs will determine the financial incentives to farmers.

In addition to easily measured costs farmers will also consider other aspects of the new technology. The complexity and amount of training needed are examples of additional factors farmers will consider. The more complex and "black box" the technology the more reluctance there is to acceptance especially if the marginal returns are low.

Farmers today are already faced with a massive amount of information. Sorting through this information and determining its relevance is increasingly time-consuming. Soil specific technology and its implementation will depend on farmer acceptance. Profitability is the key but not just the absolute level. It is the level of profit relative to the degree of difficulty in implementing that will determine acceptance.

University

The universities have a major role to play in studying and presenting soil specific crop management. This role should not be one of advocacy but rather generating research results and extending those results to the public.

One of the areas of uncertainty is input response rates by soil types. For some soils such information is available but for other soils it is inadequate or nonexistent.

Proper soil testing procedures is another area. Much is known but more needs to be learned. For example, impact of alternative tillage systems, depth of sample necessary, impact of sampling procedure, length of time between samples, and other factors need to be examined.

One of the roles will be to collect and analyze data that already exists. Much research has been conducted on various aspects of soil specific crop management. This research should be used and disseminated as appropriate with respect to new technology uses.

Another aspect already illuded to is information and data management systems. As the amount of data increases new and more practical systems for handling, analyzing, and disseminating need to be developed.

Some of the continuing issues with respect to universities are also present with soil specific crop management. What role should the Extension Service play, how to allocate shrinking budgets, and how to deal with the private vs. public sector issues must all be considered. These questions have been debated and the debate should continue especially to help guide future work with new and emerging soil specific technologies.

Agribusiness

The role of agribusiness will not be altered by soil specific crop management. They will still try to package a saleable product that will be acceptable to the farmers.

Similar to the considerations for farmers two major factors should influence the agribusiness approach. They must be able to devise a system that is profitable to both themselves and their customers. Additionally, agribusiness must develop systems that are simple and easy to understand.

Similar to universities, agribusiness should look towards information systems to promote and sell. Information and information management have often been viewed as free goods or enticements. Soil specific management techniques require information and such information should be in the fee structure.

This change in philosophical outlook necessitates a need for a sound marketing plan. Agribusiness should also maintain and strengthen ties with the universities. Information and knowledge change rapidly. If agribusiness is going to market information, it must be current.

Public

As discussed earlier in this chapter, society has a stake in the development and use of soil specific crop management. Society can benefit in many ways. For example, a cleaner environment, safer food supply and more orderly use of nonrenewable resources.

Agricultural policy has already begun to move towards soil specific legislation. The conservation reserve program, the wetlands program and conservation compliance are all geared towards specific soil types or locations.

The future direction of agricultural policy is unknown. It is likely, however, that future policies will continue the trend toward site specificity. Environmental benefits are a public good. Society through legislation could force adoption of soil specific technology or they could offer inducements that would improve the profitability.

Soil specific crop management and technologies are just emerging. The profitability and useability of this technology will determine its acceptability. Careful economic analysis will be required to ensure that future directions in soil specific technologies are acceptable to farmers and society.

CONCLUSIONS

When considering a profitability analysis be sure to use the correct tool. Three tools were discussed in this chapter and anyone may be used depending on the nature of the technology considered.

In any economic analysis identifying the impacted variables is the key. Beyond correct valuation it is important to remember the impacts that are more difficult to quantify. Risk and uncertainty, nonmarket goods and services, and the timing of benefits and costs are all important considerations.

Soil specific technologies represent an increase in our knowledge and technological capabilities. An economic analysis of such technologies can be relatively simple or extremely complex depending on the nature of the technology under study. The analysis can focus on profitability, the environment, resource conservation or all three. Regardless of the technology or analysis farmers must see and comprehend the benefits if it is to be adopted.

SECTION V

ENVIRONMENT

20 Best Management Practices for Efficient Nitrogen Use in Minnesota

G. W. Randall
Southern Experiment Station
University of Minnesota
Waseca, MN

Passage of the Comprehensive Groundwater Protection Act of 1989 by the State Legislature significantly altered the direction of water resource protection with regard to N fertilizer management in Minnesota. This was a result of three separate but related components of the Act: (i) the development of a groundwater protection goal, (ii) the enhanced regulatory authority for fertilizer practices within the Minnesota Department of Agriculture (MDA), and (iii) the responsibility for development of a Nitrogen Fertilizer Management Plan (NFMP) by the Minnesota Department of Agriculture.

Because of the complexity of N fertilizer effects on water resources, and the controversial nature of associated management decisions, the Legislature authorized the MDA to establish a Nitrogen Fertilizer Task Force to make recommendations to the Commissioner of Agriculture on the structure of the NFMP. Legislative direction stated "the task force must include farmers, representatives from farm organizations, the fertilizer industry, University of Minnesota, environmental groups, representatives of local government involved with comprehensive local water planning, and other state agencies, including the Pollution Control Agency, Department of Health, Department of Natural Resources, State Planning Agency, and Board of Water and Soil Resources."

The primary goal of the NFMP developed by the N Task Force was to manage N inputs to crop production so as to prevent degradation of Minnesota water resources while maintaining farm profitability. The central tool for achievement of this goal is the adoption of **best management practices** (BMPs) which were based upon the concept of **total nitrogen management**. Because of the ability to manage and control plant nutrients, the primary focus of the BMPs is N fertilizer. However, consideration of other N sources and agronomic practices was necessary for an effective and practical total N management system.

Best Management Practices are defined in Minnesota Statute as:
"... voluntary practices that are capable of preventing and minimizing degradation

---

Copyright © 1993 ASA-CSSA-SSSA, 677 South Segoe Road, Madison, WI 53711, USA. *Soil Specific Crop Management.*

of groundwater, considering economic factors, availability, technical feasibility, implementability, effectiveness, and environmental effects."

The N BMPs recommended by the task force are based upon research, particularly that which has been conducted at the University of Minnesota and other land grant universities, and upon practical considerations. This ensures that the BMPs are technically sound and, at the same time, likely to be adopted by growers (Minnesota Department of Agriculture, 1990).

THE THREE-TIER BEST MANAGEMENT PRACTICE STRATEGY

The task force developed a three-tier structure of BMPs for Minnesota. The first tier is a set of BMPs, which are not crop or region specific, to be adopted throughout the state. The second tier consists of five sets of regional BMPs, each designed to be adopted in one of the five general regions of the state. The third tier consists of BMPs for special situations that exist and present unique environmental or management concerns.

STATEWIDE BEST MANAGEMENT PRACTICE (Tier 1)

Statewide BMPs can be considered to be "generic" in that they apply to all areas in the state. The succeeding tiers refine the statewide recommendations. In general, statewide BMPs are applicable to all cropping systems and agronomic practices. The statewide BMPs were based upon the concept that accurate determination of crop N needs is essential for profitable and environmentally sound N management decisions. The statewide BMPs are:

1. Develop realistic yield goals.
    Selection of yield goal and the subsequent N application rate has a profound effect on groundwater quality. Limited research has indicated that growers tend to set unrealistic goals, commonly missing them by 10 to 30% and, as a result, application rates are higher than necessary to maximize yields and maximize economic returns. The Nitrogen Fertilizer Task Force strongly recommends the "running average" concept for yield goal selection. Yield goals are based on the past 5-year average, excluding the worst year. This approach will provide a sound basis for a field specific or soil specific N rate that is environmentally and agronomically sound. Tools such as soil testing, and to a lesser degree plant tissue sampling, play a valuable role in determining application rates once a yield goal is established. Soil testing, in the appropriate portions of the state, is highly recommended.

2. Develop and use a comprehensive record keeping system to record field specific information.
    Accurate field records should be kept by farmers for use in their crop management decisions since they are essential to the development of realistic yield goals and attainment of maximum profitability. This farm

specific or soil specific information should be used to evaluate past experience and plan future N management programs. At a minimum, farmers are encouraged to accurately and systematically keep information on crop yields, N fertilizer, and manure applications, and soil test results. The information can be used to monitor and adjust N management in a precise fashion for profit and environmental benefits.

3. Adjust N rate according to soil organic matter content, previous crop, and manure application.

    Mineralization of soil organic matter releases N that is useable by crops. Nitrogen recommendations in Minnesota should be adjusted for soil organic matter content as determined by a soil test.

    Legumes in a crop system can supply substantial amounts of N to subsequent crops. For example, first year credits for N can range from 40 lb/A for soybean to 150 lb/A for alfalfa. Similarly, N application rates can be significantly adjusted to account for manure application.

    By failing to account for these sources of N in determining the correct application rate, a surplus of N may be created; this surplus can potentially leach to groundwater.

4. Use a soil nitrate test when appropriate.

    The use of a deep soil test to measure residual soil nitrate in the root zone can substantially improve N recommendations in regions of Minnesota where average annual precipitation is approximately 25 in. or less. University of Minnesota research indicates that soil nitrate test results are inconsistent in predicting N needs in more humid regions of Minnesota. With appropriate interpretation, soil nitrate sampling may provide useful information where average annual rainfall is greater than 25 in. This interpretation must be tailored to each field based on exact past and current conditions.

5. Use prudent manure management to optimize N credit.
    a. Test manure for nutrient content.
    b. Calibrate manure application equipment.
    c. Apply manure uniformly throughout a field and do not apply over the recommended rates.
    d. Injection of manure is preferable, especially on strongly sloping soils.
    e. Incorporate broadcast applications whenever possible.
    f. Avoid manure application to sloping, frozen soils.

6. Credit second year N contributions from alfalfa and manure.

7. Do not apply N fertilizer above recommended rates.

    Nitrogen application rates higher than current University of Minnesota and other land grant universities' recommendations have been shown to significantly increase nitrate-leaching losses and subsequent contamination of groundwater. A high degree of confidence can be placed in University of Minnesota and neighboring land grant universities' recommendations because they are based on long-term field research. Environmental impacts will be reduced if recommendations are followed.

Soil specific nutrient application rates based on productivity of the soil in question is one of the best ways of matching application rate to crop need and thus reducing the likelihood of overapplication. Combining an accurate soil map that contains up-to-date soil productivity information with variable-rate technology can significantly improve application rates so that they are more environmentally and economically sound. Soil nitrate tests in some areas and perhaps other available soil N tests in more humid zones should enhance more precise application rates in the future.

8. Nitrogen applications should be timed to achieve high N-use efficiency.

Nitrogen application timing can significantly affect the efficiency of N use and the potential for nitrate contamination of groundwater. Generally, the greater the time between application and crop uptake, the greater the chance for N loss. However, if sidedressed N is applied too late for crop use and/or to dry soils (wherein root activity is limited), N not used by the crop can leach to groundwater. Regional or soil specific recommendations should be used to achieve high N-use efficiency.

REGIONAL BEST MANAGEMENT PRACTICE (Tier 2)

In order to achieve a goal of minimizing environmental impacts while optimizing agricultural profits, BMPs must account, to some extent, for local variation in soils, hydrogeologic conditions, and climatic conditions. In this interest, the state has been divided into five regions based upon general climatic conditions, soil characteristics and the resulting sensitivity to groundwater contamination. Figure 20-1 depicts the locations of the five regions in the state. The regional BMPs refine the prescriptions of the statewide BMPs.

Southeastern Minnesota

Southeastern Minnesota is characterized by permeable, silt loam soils with underlying fractured limestone bedrock. This karst region is very susceptible to groundwater contamination. Average annual precipitation in the region is greater than 30 in. Cropping systems include corn, forages, oat, and soybean. Livestock production consists primarily of dairy, beef, and hogs.

1. Do not apply fertilizer N in the fall.

The risk of leaching loss of nitrate from fall N application is heightened in southeastern Minnesota due to the high average annual precipitation, the well-drained and permeable nature of the soils, and the presence of karstic terrain. Spring preplant or sidedress N applications provide for more efficient use.

2. Anhydrous ammonia or urea sources of N should be used in spring preplant applications. Broadcast urea should be incorporated within 3 days of application. Consult the Soil Conservation Service for further information if soils are high erosive.

3. Sidedress applications to corn should be applied prior to the V4 stage of development.

4. Sidedress applications of urea and UAN-28 should be injected or incorporated to a minimal depth of 4 in.

5. A nitrification inhibitor should be used with preplant N applications if soils are poorly drained and soil moisture levels are high in the upper portion of the profile. Check label for registered crops.

6. Minimize direct movement of surface water runoff to sinkholes.

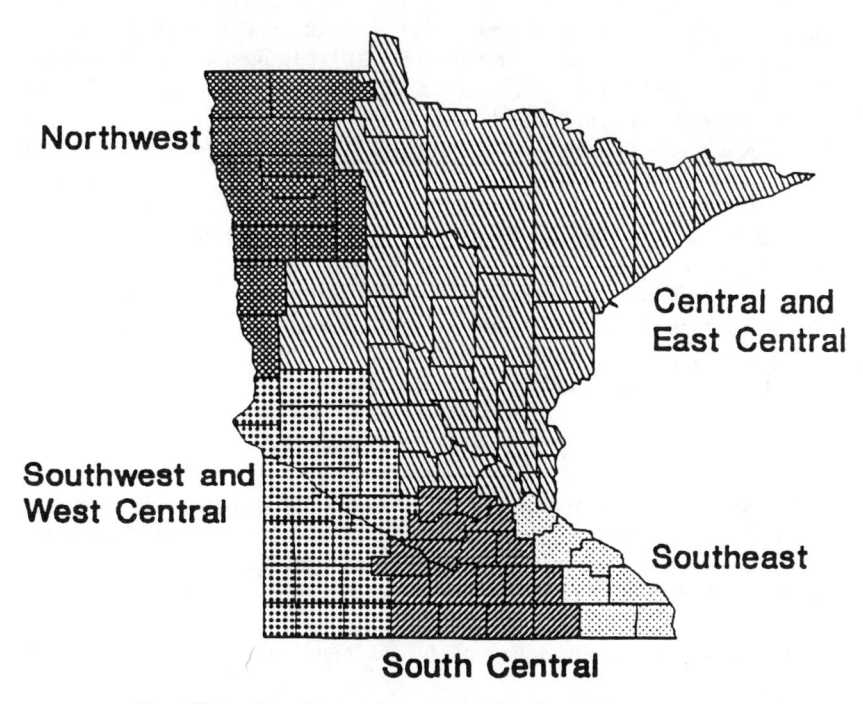

Fig. 20-1. The five regions for which Best Management Practices are formulated.

Table 20-1. Effect of N treatments on the 1990 and 4-yr average corn yields and NO_3-N concentrations in the soil water at 5 ft in Olmsted County.

Tillage	N rate lb/A	Time/Method	Grain yield 1990 bu/A	Grain yield '87-90 bu/A	Nitrate-N[1] Conc. in water mg/L
Chisel	0	--	76	84	1
"	75	Spr. preplant	145	156	11
"	150	"	155	172	29
"	225	"	156	167	43
"	150	Fall	145	169	43
"	150	Fall + N-Serve	148	169	50
"	75 + 75	Spr. + SD	154	168	47
No till	150	Spr. preplant	140	168	20

[1] In ceramic cup samplers on 5 Sept. 1990.

Results from University of Minnesota trials in southeastern Minnesota during the last few years reinforce the above BMPs. A continuous corn study started in 1987 clearly demonstrates that N should not be applied at above recommended rates and not in the fall (Table 20-1). Highest 4-yr average yields occurred with the 150-lb N rate; however, NO_3-N concentrations in the soil water at 5 ft also began to climb rapidly at this rate. Perhaps a N rate of about 120 lb/A would have optimized yield and profitability to the farmer while minimizing nitrates in the groundwater.

Fall application (13 Nov.) of anhydrous ammonia with and without N-Serve gave yields in 1990 that were 7 to 10 bu/A less than with the same N rate applied in the spring before planting (Table 20-1). Moreover, NO_3-N concentrations in the soil water were 50 to 70% higher with the fall applications. Split application of anhydrous ammonia (50% preplant + 50% sidedress at 8- to 10-leaf stage) did not improve yields over the preplant treatments but did result in higher NO_3-N concentrations in the soil water.

South-Central Minnesota

South-central Minnesota is characterized by fine-textured soils formed in glacial till and sediments. Most soils have naturally poor-to-moderate internal drainage and are tiled to improve drainage. Average annual precipitation in the region is 25 to 30 in. Cropping systems are predominantly corn and soybean.

1. Spring preplant applications of N are highly recommended over fall applications.

2. If the N is fall applied, delay application until the soil temperature is below

50°F at a 6-in. depth. Anhydrous ammonia should be used for fall applications.

3. Anhydrous ammonia or urea sources of N should be used in spring preplant applications. Broadcast urea should be incorporated within 3 days of application. Consult the Soil Conservation Service for further information if soils are highly erosive.

4. Sidedress N to corn should be applied prior to the V4 stage of development.

5. Sidedress applications of urea and UAN-28 should be injected or incorporated to a minimal depth of 4 in.

6. A nitrification inhibitor should be used with fall or preplant N applications in poorly drained soils that have high moisture levels in the upper portion of the profile.

7. Carefully manage N applications on soils characterized by a high-leaching potential.

    a. Do not apply fertilizer N in the fall to coarse-textured soils.

    b. When soils have a high-leaching potential, application of N in a sidedress or split application program is preferred. Use a nitrification inhibitor with early sidedressed N on labeled crops.

Results obtained at Waseca corroborate these BMPs. Fall application of ammonium sulfate in early November resulted in significantly lower corn yields than spring applications (Table 20-2). Moreover, NO_3-N lost out of the tile lines was markedly higher with the fall applications.

Table 20-2. Corn yields and NO_3-N lost thru the tile lines as influenced by N rate and time of application at Waseca.

N Treatment Rate	Time	1978-82 Yield Avg.	Average annual NO_3-N lost thru tile lines
lb N/A		bu/A	lb/A
0	---	66	7
120	Fall	131	27
120	Spring	150	19
180	Fall	160	34
180	Spring	168	26

Table 20-3. Nitrate-N losses through tile lines at Waseca during 1991 and corn yield as influenced by time of N application to corn following soybean.

Parameter	Time of application			
	Fall No N-S	Fall N-S	Spr ------ No N-S ------	Spr + SD
Drainage (acre-in.)	17.6	16.7	16.2	18.0
NO_3-N Loss (lb/A)	75	72	56	57
Flow-weighted NO_3-N Conc. (mg/L)				
June	34	27	17	16
Corn yield (bu/A)	122	143	151	153

Since the data in Table 20-2 were obtained using ammonium sulfate, many farmers questioned whether the same results would occur if anhydrous ammonia (AA) was used. Results obtained on a Webster clay loam at Waseca in 1991 show the highest NO_3-N concentrations in the tile water and greatest NO_3-N losses with the fall-applied AA without N-Serve (Table 20-3). The addition of N-Serve to the fall AA reduced NO_3-N concentrations slightly but not to the level of the spring and split applications without N-Serve. Corn yields did not vary greatly among the four N treatments.

Southwest and West-Central Minnesota

The Southwest/West-Central region of Minnesota is characterized by soils of medium-to-fine texture which were formed in glacial till. Many soils in the region have naturally poor to moderate internal drainage and are consequently tiled to improve drainage. Average annual precipitation is less than 26 in. Cropping systems are dominated by corn, soybean, and small grains.

1. Use a soil nitrate test with a 2 to 4 ft depth to determine N needs. Soil samples should be taken in the fall, after the soil temperature is below 50°F at the 6-in. depth, or in early spring.

2. Anhydrous ammonia or urea are recommended in spring preplant applications. Broadcast urea and preplant applications of UAN-28 should be incorporated within 3 days of application. Consult the Soil Conservation Service for further information if soils are highly erosive.

3. In situations where fall N applications are used, delay application until the soil temperature is below 50°F at a 6-in. depth. Use AA or urea sources of N. UAN-28 should not be fall-applied.

4. Sidedress N to corn should be applied prior to the V4 stage of development.

5. Sidedress applications of urea and UAN-28 should be injected or incorporated to a minimal depth of 4 in.

Benefits from using the soil NO_3 test to provide fertilizer N recommendations in western Minnesota can be readily seen from data collected at Lamberton in 1990 (Table 20-4). Soil samples taken in the spring to a 5-ft depth and analyzed for NO_3-N show substantially more residual NO_3 in the profile following corn compared to following soybean. Based on NO_3-N in the top 4 ft and a yield goal of 140 bu/A, fertilizer N recommendations of 31 and 116 lb/A for continuous corn and corn following soybean, respectively, were applied. This amount for continuous corn was 120 lb N/A less than the recommendation without the soil NO_3-N test. Corn yields for the two cropping systems were not different. Nitrate-N concentrations in the tile water reflected the accumulation of NO_3 in the soil profile for continuous corn especially in May.

Table 20-4. Residual soil NO_3-N, corn yield and NO_3-N losses from tile lines as influenced by continuous corn compared to a corn-soybean rotation at Lamberton in 1990.

Parameter	1989 Crop: 1990 Crop:	Corn Corn	Corn Soybean	Soybean Corn
Nitrate-N in top 5 ft of soil (lb/A)		168	136	75
N recommended and applied based on soil NO_3-N (lb/A)		31	0	116
N rate recommended without soil test (lb/A)		150		120
1990 Corn yield (bu/A)		120		123
May-June				
Tile flow (in.)		0.8	1.1	0.7
NO_3-N loss (lb/A)		5.3	6.2	3.6
NO_3-N Conc. (mg/L)				
May		27	20	16
June		34	33	27

East-Central and Central Minnesota

The Central/East-Central region of Minnesota is characterized by soils of coarse-to-medium texture. Most soils in the region were formed in glacial till. Outwash plains are common in this region. Many central/east-central soils are moderately to excessively drained. Average annual precipitation in the region is greater than 25 in. Cropping systems are dominated by corn and forages.

1. Carefully manage N applications on soils that have a high-leaching potential.

2. Anhydrous ammonia or urea sources of N should be used in spring preplant applications on fine and medium-textured soils. Broadcast urea should be incorporated within 3 days of application. Consult the Soil Conservation Service for further information if soils are highly erosive.

3. Sidedress applications of urea and UAN-28 should be injected or incorporated to a minimal depth of 4 in.

Northwest Minnesota

The Northwest region is generally characterized by fine-textured soils formed in lacustrine deposits. The annual average precipitation in the region is less than 24 in. The major cropping systems in the region are small grain (*Triticum aestivum* L.), soybean [*Glycine max* (L.) Merr.] and sugar beet (*Beta vulgaris* L.).

1. Use a soil nitrate test to a 2- or 4-ft depth to determine N needs. Soil samples should be taken in the fall after the soil temperature is below 50°F at the 6-in. depth or early spring.

2. Delay fall N application until the soil temperature is below 50°F at a 6-in. depth. Anhydrous ammonia or urea sources of N should be used for fall applications.

   UAN-28 should not be fall applied. Broadcast urea and spring preplant applications of UAN-28 should be incorporated within 3 days of application. Consult the Soil Conservation Service for further information if soils are highly erosive.

3. Nitrification inhibitors are not recommended on fine-textured soils but are recommended on coarse-textured soils with high-leaching potential.

SPECIAL SITUATION BEST MANAGEMENT PRACTICES

The third tier of BMPs are referred to as Special Situations BMPs. The special situations are a result of certain combinations of management and environmental conditions that may render an area or site more susceptible to groundwater contamination than would be predicted by the general characteristics of the surrounding region. The third tier accounts for those management situations or sites that are interspersed throughout the state.

Irrigated Soils

Irrigation, especially on coarse-textured (sandy) soils and shallow-rooted crops, may increase the leaching potential of applied N. Irrigation increases the soil water content of the root zone, thus enhancing mass and diffusive transport of nitrate in the subsurface past the zone of effective crop utilization.

Irrigated soils in Minnesota were typically formed in outwash plains or alluvium and are consequently of coarse texture. Localized areas of irrigation occur throughout the state.

Water use in these areas is variable depending upon soil and geologic conditions and average yearly precipitation. Commonly irrigated cropping systems include corn and potato.

1. Do not fall-apply fertilizer N to soils in the following textural classes: sandy loam, loamy sand, and sand.

2. Follow proven water management strategies to provide effective irrigation and minimize leaching.

3. Test irrigation water for N content and adjust N fertilizer rates accordingly.

4. Use sidedress or split applications of N on irrigated soils. Do not rely on fertigation for delivering more than one-third of the required N. (Fertilizer chemigation rules are being developed by the Minnesota Department of Agriculture at this time. Backflow prevention, well head safety and other techniques are recommended until rules are adopted.)

5. Use a nitrification inhibitor when the bulk of the N is applied in a single preplant or early sidedress application. For corn, N treated with a nitrification inhibitor should be applied prior to the V4 growth stage. Check label for registered crops.

6. Include a small amount of N in starter fertilizer in most situations (10 to 20 lb/A).

7. Do not delay N applications past optimum uptake period.

8. Establish a cover crop following early harvest of crops. Consult the Soil Conservation Service for further information if soils are highly erosive.

Coarse-Textured (Nonirrigated) Soils

Coarse-textured soils need special management to prevent leaching losses. Coarse-textured soils are present in many different regions and can be found throughout the state in outwash plains, alluvial river valleys, and ancient beach ridges. These soils have considerable leaching loss potential due to rapid infiltration characteristics and low water-holding capacities which can easily be exceeded. Furthermore, these soils are often associated with unconsolidated sand and gravel aquifers that may have water tables that are near the soil surface.

1. Do not apply N fertilizer in the fall to coarse-textured soils.

2. Apply N in a sidedress or split application program

3. Use a nitrification inhibitor with early sidedressed N.

SUMMARY

Statewide and region-specific BMPs developed for Minnesota should greatly improve fertilizer N efficiency, economical return, and environmental quality if they are implemented by crop producers. Some of these BMPs, e.g., realistic yield goals; record keeping systems; adjustment of nutrient application rate based on soil OM, soil test, previous crop and manure; and proper time of application, can be implemented into a soil specific farming system that will enhance profitability while minimizing negative impacts on water quality. Technology is now available to package these BMPs into profitable crop production systems for today's farmer.

REFERENCES

Minnesota Dep. of Agriculture. 1990. Recommendations of the Nitrogen Fertilizer Task Force on The Nitrogen Fertilizer Management Plan. MDA, 90 W. Plato Blvd., St. Paul, MN.

21 Social Issues Related to Soil Specific Crop Management

Peter J. Nowak
Department of Rural Sociology and
Environmental Resources Center
University of Wisconsin-Madison
Madison, Wisconsin

Science at the most elementary level is the study and explanation of variation. Modern agriculture is the process of managing both predictable and unforeseen variation in a profitable manner. Science contributes to agriculture by initially translating unexpected variation into expected variation, and then by creating procedures for managing this diversity. Soil specific crop management is a case in point. Input efficiencies, yields, and environmental impacts can vary significantly within a field management unit due to variation in landscape and soil. While most farmers have intuitively understood this relationship, and some have even attempted to adjust management accordingly, the systematic application of science to this situation has been lacking. The scientific community is now on the threshold of developing systems of soil specific crop management.

Most recognize that much of the science underlying soil specific crop management is not new. Rather it is the integration and synthesis of this knowledge base into farm-scale application that is innovative. These applications are being developed in both the private and public sectors. At issue is the opportunity to direct or guide these incipient development efforts in such a way to maximize social, economic, and environmental benefits while minimizing any negative consequences.

Directing the development and application of a new technology is based on understanding how various consequences emerge. In other words, research priorities, design specifications, and implementation of technology transfer efforts need to be planned relative to some probability of achieving preferred outcomes. These desired outcomes can be defined in terms of a combination of agronomic, environmental, economic, and social criteria.

Copyright © 1993 ASA-CSSA-SSSA, 677 South Segoe Road, Madison, WI 53711, USA. *Soil Specific Crop Management.*

Relative to the focus of this chapter, there are four antecedent conditions that will directly influence adoption rates, diffusion patterns, and consequences of soil specific crop management techniques. These are: (i) the nature or form of the technological package representing soil specific crop management; (ii) the attributes of the physical resource base in which the technology is applied; (iii) the distribution of farm firm and human capital characteristics in areas targeted for the technology; and (iv) the nature of the institutional support network.

ANTECEDENT CONDITIONS AND SOIL SPECIFIC CROP MANAGEMENT

Research priorities, design specifications, and implementation of technology transfer efforts relative to soil specific crop management need to be guided by a rational process if beneficial outcomes are to be optimized. It has been noted that adoption rates, diffusion patterns, and consequences of soil specific crop management techniques will be directly influenced by certain antecedent conditions. Each of these antecedent conditions need to be addressed prior to discussing probable consequences.

Nature of the Technology

Soil specific crop management is a very heterogeneous bundle or package of techniques, tools, and management requisites. This technological package making up soil specific crop management can be classified along three general component areas or dimensions. These are the information base, positioning techniques, and application processes.

The **information base** refers to the comprehensiveness, scale, accuracy, affordability and accessibility of data on soils, topography, field position, climate, and agronomic information. Current variation is due to the lack of uniformity in the presence, accuracy, accessing requirements, and scale of relevant information. The more complex forms of soil specific crop management will be very dependent on a comprehensive and accurate information base. Each informational item can vary in terms of needed precision and the span of the temporal period in which the information will remain valid. The latter factor, the temporal dimension, is exemplified by the soil characteristics that are relatively permanent while pest cycles, climate fluctuations, or crop rotations can change information demands on an annual basis.

A soil specific crop management package must also address positioning the production tools within a field. **Positioning techniques** refers to the means of establishing a spatial location in order to guide the application processes. This can vary between on-board computers communicating with satellites to "rules of thumb" and other experiential-based decision rules of farmers. The precision in positioning is directly related to the efficiencies gained. That is, the greater the precision, the greater the potential to increase production efficiency. There is, however, an upper limit to this relation. This is where the costs of additional precision in positioning exceed the benefits gained through using soil specific crop management.

Application processes refer to the data-dependent tools and techniques used in the input and management operations. Knowledge of the variability in soils and landscape is combined with the position on that landscape to make the requisite managerial decisions. This application process can be guided by sensors, output from on-board computers, satellite imaging, predetermined decision rules or the experience of the farmer. Application does not just refer to agricultural inputs. Another application of soil specific crop management is knowing the efficiency of this technological package. That is, does the difference in output justify the additional expenses and effort required? Consequently, an important application process is the measuring of yields or some other gauge of output. This output measure, of course, can also refer to changes in the loadings or impact on the environment.

Relation of Technology to Adoption Processes

The exact nature of soil specific crop management can vary significantly with various combinations of the three component areas just discussed. There is no one soil specific crop management technology. Instead, the components of the three dimensions of soil specific crop management just discussed can be integrated into a variety of packages applicable to different areas of the county and agricultural production systems. A critical issue is how this substantial number of alternative technological packages can be evaluated along a common set of dimensions.

There is an extensive research literature that has examined the adoption and diffusion of new agricultural technologies (Rogers, 1983). A generalization from this literature is that the nature of the innovation has a direct impact on adoption processes (Nowak, 1987). For example, one agricultural innovation may diffuse rapidly among a population gaining widespread use while a different innovation can be characterized by slow adoption rates or even rejection by the same population of farmers. The nature of the innovation is one of the central factors explaining this difference in adoption rates. The ensuing implication of this research generalization is that it is critical to be able to classify or engage in a technology taxonomy of agricultural innovations.

Several different criteria have been proposed to classify new technologies. Suggested characteristics of technology have included cost, returns to investment, efficiency, risk or uncertainty, communicability, compatibility, complexity, scientific status, point of origin, and susceptibility to successive modification among others (Zaltman et al., 1973). Research, however, has reduced this list to five central characteristics (Rogers, 1983). These include relative advantage (both economic benefits and social status), compatibility to the current farming operation, complexity of the technique, trialibility or ability to be tried on a small scale and the observability of the results or use of the technology (Rogers, 1983). The relation of each of these characteristics of new technologies to adoption rates should be evident.

Yet one additional qualification to technology classification needs to be added. The apparent nature of a technology can change across the diffusion process. That is, the above characteristics of a innovation are inherent in the target population, not the innovation itself. For example, the relative advantage of a new production technique may initially be high. As the number of farmers who adopt increases, supply also increases which decreases price received. Thus early adopters receive a higher "adoption rent" than later adopters. The relative advantage of the innovation decreases across time. Another example would be where trialibility decreases in importance as more farmers adopt. Because many of one's neighbors are already using the practice, establishing the "workability" of a practice is no longer as salient an issue as it was for the early adopters. The permanence of the perceived characteristics an innovation has across the diffusion process is a research topic that has received little attention.

The relative advantage of soil specific crop management will depend on the current level of production efficiency and the amount of investment required to purchase and maintain the system (or rent the services, i.e., custom). Lower levels of current efficiency combined with systems of low investment, maintenance or rental costs will have the highest relative advantage.

Determining the compatibility of soil specific crop management will depend on certain farm firm characteristics and the physical setting. It appears likely that soil specific crop management, at least in many of the forms currently being discussed, will require higher levels of managerial expertise. Especially relative to the more complex, hardware-dependent components. For capital and managerial intense techniques, a scale bias will also operate. That is, it will only be cost effective on larger operations where there is a sufficient degree of expertise or specialization in management. Relative heterogeneity in landscape and soils will also increase the compatibility of soil specific crop management up to a point. This will be discussed in more detail under the antecedent condition of the physical resource setting. In sum, it appears that the initial forms of soil specific crop management will only be compatible to a limited number of farm operations.

Complexity has already been discussed to a certain degree. The complexity of a soil specific crop management system can vary tremendously. The higher the complexity of the system, then the smaller the segment of farm operations on which this technology will be appropriate. Reducing the complexity of a soil specific crop management system will depend on modifying the technology or increasing the opportunity to enhance requisite management skills.

The feasibility of soil specific crop management should be fairly high. Farmers should be able to try a system on a limited acreage where the plausibility of the results depend on the accuracy of input and output records. Even in those situations where there is a high initial investment cost, it should be possible to emulate programs developed with residue management systems allowing farmers to try the system at little expense or risk. It also appears that some forms of soil specific crop management will not be adopted by individual farmers, but offered as a custom service. The availability of custom services within a trade area should also enhance the trialability of this technology.

The observability or visibility of the results of a soil specific crop management system will be very low. This will retard adoption and diffusion rates. The increased efficiency resulting from a soil specific crop management system becomes evident only through being able to accurately quantify inputs relative to outputs in accord with a refined record keeping system. Thus far firm output measures and records, however, are rarely available to the public or other farmers considering the technology. Consequently, efforts to accelerate the adoption of this practice will depend heavily on making results visible through "signing," testimonials and other visibility-enhancing efforts.

A way of summarizing this discussion is to examine two different soil specific crop management systems; one based on "high technology" (e.g, satellites, on-board computers, etc.) with significant capital requirements, and the other on "low technology" (e.g., a grid system imposed on field soil maps, variable rate applicators, etc.) with minimal capital requirements. For most farms, the high technology system will have a low relative advantage, compatibility and observability with high complexity and trialibility. This will make this type of soil specific crop management system initially applicable only to small, niche markets. The low technology package will have low observability while being high on the other four dimensions. The potential market for this second package will be much larger, but the profit margins will lower resulting in less emphasis on marketing by the private sector, especially equipment manufacturers. Complicating this process is the role played by the private sector. It is expected that agrichemical input dealers and suppliers will make the investment in many of the high technology packages to offer as custom services. Few farmers will need to invest in the high technology option, but will rely on custom forms of soil specific crop management provided by the agriservice sector. Most farmers will invest in the low technology end of this continuum, probably after experiencing the advantages of soil specific crop management under custom arrangements. The adoption rates of these two contrasting technological packages will be very different in any given area.

Attributes of the Physical Resource Setting

The second antecedent condition that will influence consequences of soil specific crop management are the attributes of the physical resource setting in which the technology is applied. Agriculture in the United States is built upon a very diverse physical and climatic base. There are 189 Major Land Resource Areas, six principal climate types ranging from superhumid to arid, and eight categories of land capabilities where the amount of land suitable for continuous cultivation is approximately equal to the amount of land not suited for cultivation (USDA, 1981). At the national level there is more than 10,500 soil series with 4500 soil families. This diversity at the national level may also be found to varying degrees in local settings including within farm operations. The explanation for this diversity can be attributed to geologic (wind, water, ice, and gravity) climatic, and other pedogenic factors.

The very nature or function of a soil specific crop management system is to manipulate inputs relative to diversity in the physical setting to optimize outputs. A critical question is whether this technological package is applicable across varying levels of diversity. That is, does the applicability of a soil specific crop management system remain constant across the variability of the

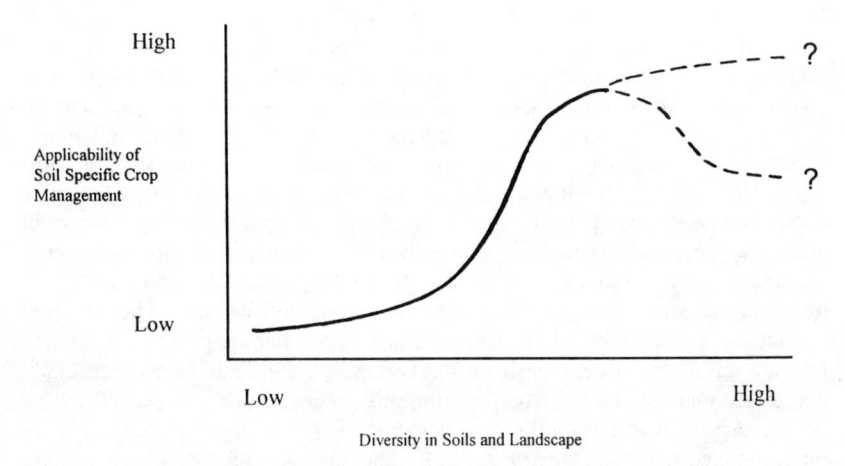

Fig. 21-1. Applicability of soil specific crop management relative to diversity in soils and landscape.

physical resource setting? The probable answer to this question is graphically represented in Fig. 21-1.

Here it is hypothesized that the applicability of a soil specific crop management system is restricted in very diverse and very homogeneous settings. In very homogeneous and very heterogeneous settings (major variation in soils and landscape) the applicability or utility of soil specific crop management will be restricted. While the reason for this restriction in a homogeneous setting is self-evident, more explanation is required for a very heterogeneous setting.

A very diverse setting in terms of soils and landscape imposes real constraints on the land manager. These constraints have been recognized with various land classification and capability rating schemes (Tivy, 1990). In these "constrained" settings we often find small, fragmented farms and fields due to concerns over erosion, water movement, safety of machinery operations, machinery limitations all of which define what constitutes a "manageable" acreage. A good example of this type of setting is found in the upper Midwest in the unglaciated region surrounding the Mississippi River. Here we find significant diversity in both soils and landscape. Based on present definitions of what comprises a soil specific crop management system, there would be little applicability for such a system in this region. This is not meant to deny the need for the principles underlying soil specific crop management, but only to state that current designs of these systems would face major constraints in this type of physical setting.

Relation of the Resource Base to Adoption Processes

The physical resource base will influence adoption processes by delineating the appropriateness of the new technology. For example, research has found that the adoption of various residue management systems is highest in parts of the country with the most erodible soils under cultivation (van Es and Notier, 1988). The same relationship was found within a region (Nowak and Korsching, 1985). It is expected that the appropriateness of soil specific crop management systems will be influenced by both the diversity in soils and characteristics of those soils. Soil specific crop management will have little applicability across uniform soil mapping units that have a consistent management history. Diversity of soils within field boundaries or within mapping units will be important in defining appropriateness of the technology. Also important will be the characteristics of the soil (i.e., organic carbon) that influence nutrient cycles and ability to attenuate contaminants. Finally, the interaction between past management and the characteristics of the physical base will also influence appropriateness. Fragile soils, soils more capable of "holding" nutrients and areas that have experienced serious deposition or sedimentation will respond differently to this technology.

Distribution of Farm Firm and Human Capital Characteristics

It was noted earlier that there is significant diversity in the national, agricultural physical resource base. The same diversity also applies to farm structure, social organization of the farm firm, managerial capabilities, market and assistance networks available to work with the land user. It is into this diverse socioeconomic setting that new technologies such as soil specific crop management are introduced. Ignoring socioeconomic diversity when introducing new technologies will have the same potential negative consequences as ignoring the diversity in the physical resource base.

The term *farm structure* is often used to characterize the features of this diverse socioeconomic setting. Rasmussen (1989, p. 1-2) noted that it includes "the number and sizes of farms by commodities and regions, the degree of specialization in production and the technology employed, the ownership and control of the productive resources, barriers to entry and exit in farming, and the social, economic, and political situations of farmers." Farm structure, therefore, will also influence applicability and appropriateness of soil specific crop management.

Table 21-1 (Ahern et al., 1990) present data illustrating the importance of farm structure to corn, a crop often mentioned relative to application of soil specific crop management.

The illustrated variation between farms producing corn is significant for several factors that could influence the applicability of soil specific crop management such as size, specialization, and ability to invest in these new techniques. Moreover, if the size bias discussed earlier is salient, then only certain parts of the nation will see wide-scale development of soil specific crop

Table 21-1. Characteristics of corn farm operations for the USA, 1987.

	Less than 25	25 - 99	100 - 499	500 or more	All farms
Number of farms:	124,683	207,256	135,244	15,317	482,500
Share of all farms:	25.8%	43.0%	28.0%	3.2%	100%
Corn characteristics:					
Avg planted acres	12	53	190	675	101
% of acres rented	33.1%	44.9%	59.9%	67.2%	53.2%
Yield (bu/acre)	90.6	106.1	121.3	127.6	118.2
Variable expense					
as a % of total	45.2%	45.4%	45.1%	49.6%	46.2%
Production specialty:					
Cash grain	35.3%	47.9%	61.0%	76.9%	49.2%
Other crop	12.1%	10.0%	3.8%	1.8%	8.5%
Beef, hog, sheep	32.6%	19.7%	20.4%	13.9%	23.0%
Dairy	18.8%	21.8%	14.1%	7.4%	18.4%
Other livestock	1.2%	0.7%	0.6%	0.0	0.8%
Operator age distribution:					
<25	1.9%	2.7%	0.1%	0.0	1.7%
26-49	34.7%	39.7%	53.4%	61.2%	42.9%
50-65	38.4%	37.4%	37.8%	33.6%	37.6%
>65	25.0%	20.2%	8.8%	5.1%	17.7%
Major occupation:					
Farming	61.1%	83.1%	92.3%	99.0%	80.5%
Other	38.9%	16.9%	7.7%	1.0%	19.5%
Farm organization:					
Partnership	5.0%	5.9%	15.2%	39.6%	9.4%
Individual	94.5%	92.0%	78.0%	40.3%	87.1%
Corp and Coop	0.5%	2.0%	6.8%	20.0%	3.6%
Economic class:					
$250,000 or more	1.2%	3.8%	23.2%	81.3%	11.0%
$100,000-$249,999	2.7%	21.2%	43.1%	17.3%	22.4%
$40,000-$99,999	12.5%	33.6%	26.5%	1.4%	25.1%
$0-$39,999	83.7%	41.4%	7.2%	0.0	41.4%
Percent receiving any					
gov. payments	45.3%	68.4%	90.6%	97.8%	69.6%

Large Farm Counties*

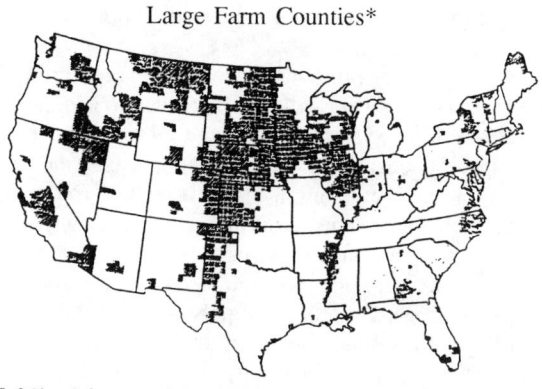

*Less than 59.2% of farms with 1982 gross farm sales of less than $40,000.

Fig. 21-2. Location of large farm counties, 1982.

management, at least in its current forms. Figure 21-2 (Carlin and Saupe, 1990) presents the location of the U.S. counties where large farms are concentrated. It would be expected that soil specific crop management packages would be promoted and concentrated within these areas.

Another dimension of this spatial distribution is the variation between USDA production regions. Soil specific crop management, at least as is currently being discussed, is limited in application to a small set of cash grains and specialty crops. As Table 21-2 (U.S. Congress, OTA, 1986) illustrates, various agricultural commodities tend to be concentrated in certain production regions. This concentration, as well as the factors discussed, all indicate that the adoption and diffusion of soil specific crop management systems will be clustered in certain areas of the nation.

Table 21-2. Percent of U.S. sales of each commodity by region, 1982.[1]

Commodity groups	North-east	Southern	North Central	Western	Total USA
Cash grains	1.3%	21.4%	66.2%	11.0%	100.0%
Cattle and calves	1.3	28.2	45.6	24.9	100.0
Dairy	19.9	18.4	41.2	20.5	100.0
Poultry and eggs	8.6	62.2	17.0	12.2	100.0
Hogs and pigs	2.0	15.3	80.1	2.5	100.0
Fruit and tree nuts	6.9	23.4	5.2	64.5	100.0
Vegetables	7.6	25.8	9.0	57.6	100.0
Cotton	0.0	54.9	1.4	43.7	100.0

[1] Note: Totals may not add due to rounding.

Relation of the Farm and Farmer to Adoption Processes

The size, enterprise mix, managerial capacity (human capital), and planning horizon of the farm firm will also be related to adoption rates, diffusion patterns, and social consequences of soil specific crop management techniques. Depending on specific characteristics of the package, the size or scale of operations will influence the adoption of soil specific crop management. Soil specific crop management systems requiring significant capital investments will be more applicable to larger farm firms or agribusiness. The more specialized the enterprise mix of the farm on crops will also influence adoption rates. More diversified farms with regard to crops, animals, horticultural, and other specialty crops may find little utility in a technology that only applies to a small part of the management unit. Managerial capacity of the farm firm will also influence adoption if the soil specific crop management system has high managerial requirements. Finally, the planning horizon of the farm firm will influence adoption rates if the soil specific crop management system has high initial investment costs or has a longer period over which learning and transition costs have to be assimilated.

Nature of the Institutional Support Network

To paraphrase the English poet John Donne, "no farmer is a production unit onto them self." The modern producer is very dependent on external sources not only for production inputs and market outputs, but also for the production process itself. This dependence takes the form of information, expertise, support services, as well as the generation of new production products and processes. The adoption and diffusion of soil specific crop management within any geographical area will partially depend on the nature and strength of these external sources.

One trend is clear when looking into the history of agriculture and projecting this process forward; the increasing importance of the private sector in defining technology transfer processes. Although not well documented, it appears that agricultural private sector firms are surpassing public sector organizations in determining who adopts what practices, where and when this occurs. Much of this has to do with the generation of basic research within the public sector which is then translated into applied products and processes by the private sector. Regardless of the causes of this trend, the importance of the private sector in the adoption of agricultural innovations is increasing.

Relation of the Institutional Support Network to Adoption Processes

Lawrence Brown (1981) in his book on innovation diffusion discusses two general models. What he calls the traditional model focuses on the demand side of the equation. That is, those personal, situational and farm firm factors that impels an individual to adopt an innovation. Much of the traditional adoption research has emphasized these factors. This is contrasted with what he calls the market-infrastructure model that focuses on the supply of innovations. The loci of this model are institutional characteristics, both private and public, that affect access to the innovation as well as establish various constraints under which individual decisions are made.

These institutional characteristics will influence the adoption of soil specific crop management in two general ways. First, "the location of the agencies and the temporal sequencing of their establishment determine where and when the innovation will be available and provide the general pattern of the spatial pattern of diffusion" (Brown, 1981, p. 51). Second, the specific promotional or implementation strategy selected by each agency will "contribute further detail to the spatial pattern of diffusion by creating different levels of access to the innovation depending on a potential adopter's economic, locational, demographic and social characteristics" (Brown, 1981, p. 51).

Both of these processes are already at work as evidenced by popular reports in the farm press. For example, Reichenberger and Russnogle (1989) reported that Soil Teq Inc. had sold 26 computer-controlled, variable-rate applicators to custom applicators. It is common to find situations reported where a limited number of variable rate technological packages (equipment, software, mapping, and positioning capabilities) are being made available in certain portions of certain states (Pocock, 1991). The trade area for each of these custom applicators or dealers will largely determine if local farmers will have access to this technology.

A common theme for the current promotional or implementation strategies is the opportunity to save on input costs. This has been concentrated on crop nutrients, largely P_2O_5 and K_2O. While none of these popular reports give data on the customers adopting these practices, the testimonials often give the impression of larger farms on some of the more productive soils.

Two outcomes are probable when considering the role of the private sector relative to soil specific crop management. First, it is expected that the private sector will engage in market analyses to examine many of the factors discussed earlier to determine the profitability of investing in soil specific crop management. These investments will be made only in areas or products where there is a good potential for profit. Second, farmers will not have equal access to this technology and the various supporting systems and services. Several farmers will never have the opportunity to assess this technology as it simply will not be available or supported in their local trade areas.

SOCIAL IMPACTS

It was stated that four antecedent conditions will directly influence adoption rates, diffusion patterns and consequences of soil specific crop management. Each of these four antecedent conditions were briefly described: (i) the nature or form of the technological package representing soil specific crop management; (ii) the attributes of the physical resource base in which the technology is applied; (iii) the distribution of farm firm and human capital characteristics in areas targeted for the technology; and (iv) the nature of the institutional support network. A brief summary of current research as well as future research needs were summarized for each antecedent condition. Based on this preliminary and speculative discussion, social impacts can be estimated for three areas; individual farms, the private and public sector, and the agricultural sector as a whole.

Individual Farm Level

Early adopters will enjoy increased profitability due to enhanced efficiency and adoption rent. Consistent with the history of the adoption of other agricultural innovations, early adopters will accrue certain economic benefits. This results from their competitive advantage in the larger market relative to non-adopters.

Current research, however, is not clear as to whether this enhanced efficiency can be maintained. The issue is whether soil specific management will inhibit some of the initial, large within-field differences in nutrients or pesticides across time. The answer will depend on the degree these differences are due to past management practices as opposed to underlying pedogenic, drainage, or other physiographic factors. Differences caused by past management will reduce the effectiveness of soil specific crop management as use of this technology will nullify or dampen this variation across several crop cycles.

There will be increased labor demands at critical planting and harvesting windows. This anticipated social impact is more difficult to document due to the uncertainty of the exact nature of soil specific crop management techniques. However, at present there are two certainties. First, soil specific crop management needs to be driven by an intense soil sampling process. While fall sampling may be possible for P_2O_5 and K_2O, this may not be feasible as soil specific crop management is applied to N, herbicides, and insecticides. Consequently, some form of preplant sampling and scouting will be necessary in the spring. Second, the operating philosophy behind soil specific crop management is efficiency. That is, changes in input and output ratios so as to increase farm profitability. The implication of this is the need for documented output (yield variation within a field). While much of the technology is being designed for "on-the-go" monitoring within the combine or harvesting unit, there is still a need to analyze this data relative to inputs. Both of these situations, preplant sampling and scouting along with postharvest data analysis will increase labor demands. While much of this is currently being contracted with off-farm services, it can be anticipated that as soil specific crop management develops and becomes more scale neutral, more of these labor demands will be assumed by the farm unit.

Related to the last point will be **increased demands for modern record-keeping and analytical systems.** Unless soil specific crop management is contracted for as a custom service, there will be increased demands placed on the farm operator. These demands relate to the process of guiding or directing this technology to enhance efficiencies. Efficiency is not inherent in soil specific crop management itself. Rather, efficiency results from adapting and adjusting the system to fit equipment inventories, managerial abilities, and the quality and quantity of available information. As noted, calculating efficiency is dependent on the ability to analyze detailed input and output records. Input and output values have to be analyzed relative to some combination of variation in field soil mapping units and the sampling protocol employed. This will require some combination of judgement based on experience, and access to formal decision rules. This latter item, formal decision rules, may appear in the form of software packages.

The two previous social impacts emphasize the **need for enhanced management skills**. While calling for enhanced management skills is fairly common, the sequence in which these skills must be learned will require significant adjustments among farm managers. In essence, soil specific crop management involves the application of management and technology to variation within a field boundary. Yet many farmers currently fail to differentiate management and technology between fields. For example, a field producing corn is treated like all other fields growing corn within a farm operation.[1] Consequently, it is going to be difficult to get farmers to adopt soil specific crop management techniques until management recognizes and responds to between-field differences.

High-capital, complex soil specific crop management packages will be inappropriate to many small to medium sized farms. Soil specific crop management will not be an economic and conservation panacea for all of American agriculture. Current forms of soil specific crop management are capital intensive and complex. Unless offered as a local service in a competitive fashion, many farms will find this technology beyond their managerial, labor, and investment potential.

As soil specific crop management becomes more formalized, routinized and available, there is going to be significant promotion from both the private and public sectors for adoption. As a result, **mismanagement and inefficiency due to "pressured" adoption will occur**. That is, some farmers will be persuaded into adopting this technology when financial or managerial resources are inadequate. This situation will defeat the purpose under which soil specific crop management is being promoted. These financial or managerial constraints will result in significantly less efficiency from the system relative to its potential.

If soil specific crop management begins to more fully address N and pesticide issues, and if this system is managed to address environmental concerns (e.g., leaching, run-off, etc.), then **nonadopters will face increased vulnerability to environmental regulation**.

---

[1] 1991 data from Wisconsin found that 71% of 836 farmers surveyed relative to nutrient management practices did not differentiate rates of commercial nutrient application between corn fields.

Equity concerns will arise in areas where soil specific crop management technologies are inappropriate. Parts of the country where there are small, fragmented farms will find the current versions of soil specific crop management inappropriate. The current hardware components of soil specific crop management have a clear economy of scale relative to field size. Current versions would be inappropriate for farms composed of small (e.g., 10 acres) fields. This is going to limit the applicability of this technology as a best management practice for environmental objectives. What remains to be seen is if the concepts and processes underlying soil specific crop management can be made more scale neutral.

Private and Public Sector

Those who develop recommendations and products for the agricultural sector will be under **increasing pressure to develop input "packages" rather than individual products**. This is consistent with the trends associated with farming systems, whole farm analysis or integrated farm management.

There will be **an increased need to develop, promote, and support accurate record-keeping systems**. Several record-keeping systems already exist in agriculture. These vary from the very simple "boot box" systems to complex computer software packages. Most, however, are not designed for recording within-field variation as found under a soil specific crop management system.

There will be **an increased demand for more accurate within-field yield-estimating techniques**. As noted, the philosophy behind soil specific crop management is to enhance efficiency by varying inputs and management strategies within a field management unit. Whether this efficiency is achieved is dependent on a valid and reliable measure of variation in output (yields). Continued innovation is needed in engineering to develop these cost-effective techniques that are compatible to record keeping and analysis systems.

Increased demand for public support of mapping, positioning, and modeling data or technologies. Soil specific crop management is very information dependent. A completed as well as validated modern soil survey will be a pre-requisite for use of soil specific crop management in many areas. In addition, accurate positioning data (e.g., global position satellites, GPS) or access to satellite imagery will also dictate the applicability of this management system. Soil specific crop management may also require the development of models that predict crop responses, pest cycles, or environmental impacts for efficient use. All of these requirements will necessitate public expenditures for development and support. In an era of budget constraints this may require trade-offs relative to other agricultural funds or development of a "user pays" strategy.

There will be **demand on the public sector to provide support of soil specific crop management in areas where the private sector has bypassed due to profitability concerns**. This situation will also induce more farmer-innovation and adaptation as farmers attempt to create "localized" versions of soil specific crop management. This will make the public sector task of developing uniform recommendations and procedures more difficult.

If soil specific crop management is developed as a best management practice, and if research can specify the site-specific conditions under which it protects or enhances the environment, then **there will be a decreased need for environmental regulation in applicable areas with high adoption rates.** Yet to be resolved is whether soil specific crop management can be developed to fit the definition of a best management practice.

Agricultural Sector

Soil specific crop management, at least in its present capital-intensive forms, will **sustain the trend toward fewer, larger farms controlling a greater share of commodity production.**

There will be an economy of scale operating within farms that adopt soil specific crop management. That is, the investment in equipment, management skills, learning and transition costs, and access to supporting information will pressure farmers to apply this technology to as many acres as possible within a farm. Consequently, this will **increase the pressure toward farm specialization rather than diversification.**

There is increasing discussion and probability that these large "factory" farms will be treated as point sources of potential pollution and regulated accordingly. Soil specific crop management could be developed as an "environmental practice" to meet this regulatory pressure.

Due to the many complexities and requirements associated with soil specific crop management, it can be expected that **professional land management firms will increase acres controlled due to specialization, scale of economies, and human capital factors.**

Soil specific crop management should allow adopters of this technology to operate more efficiently on rented land. Therefore, wide-scale adoption of this technique should **enhance the existing trend in the proportion of rented to owned acres on farms due to specialization, scale of economies, and human capital factors.**

CONCLUSION

It should be clear that there are a variety of complex factors that will influence the consequences associated with the adoption and diffusion of soil specific crop management techniques. Four sets of antecedent conditions will largely determine these consequences. One of the major limitations in delineating these consequences is related to the first factor; the lack of specificity in current applications of the technology. At the present stage of development, soil specific crop management remains ambiguous at best. Application of P_2O_5 and K_2O is fairly well defined and being practiced in certain areas. However, broader applications for N, pest control, or environmental considerations remain largely undeveloped. Moreover, regional variation in physical conditions and farm structure will influence the types of technology appropriate for a soil specific crop management system. Due to these many legitimate sources of variation, ambiguity surrounding the consequences of soil specific crop management is likely to remain.

There has to be an overall optimistic view of soil specific crop management. Just the basic acknowledgement that any new agricultural technology will have both positive and negative consequences is an important step. An acknowledgement that was missing on many past technologies until well after the fact. This fact is complimented by an opportunity for managing consequences to maximize environmental, economic, and social objectives. As the specific components and procedures associated with soil specific crop management continue to evolve, there is a need for a constant questioning of how this technological package will impact prevailing environmental, economic, and social goals. The evolution of soil specific crop management needs to be guided by a broader vision than simply more efficient crop production.

REFERENCES

Ahern, M., G. Whittaker, and D. Glaze. 1990. Cost distribution and efficiency of corn production. p. 33-50. In Determinants of farm size and structure. Dep. of Economics, Iowa State University, Ames.

Carlin T., and W. Saupe. 1990. Structural change in agriculture and its relationship to rural communities and rural life. p. 103-117. In Determinants of farm size and structure. Dep. of Economics, Iowa State University, Ames.

Brown, L. 1981. Innovation diffusion: A new perspective. Methuen, New York.

Nowak, P. 1987. The adoption of agricultural conservation technologies: Economic and diffusion explanations. Rural Sociol. 52:2, 208-220.

Nowak, P., and P. Korsching. 1985. Conservation tillage: revolution or evolution? J. Soil Water Conserv. 40:2, 199-201.

Pocock, J. 1991. Put your money where it counts. Prairie Farmer, January: 12-14.

Rasmussen, W. 1989. Agricultural structure and the well being of society revisited. p. 1-10. In Determinants of farm size and structure. Dep. of Economics, Iowa State University, Ames.

Reichenberger, L., and J. Russnogle. 1989. Farm by the foot. Farm J. Mid-March: 11-16.

Rogers, E. 1983. Diffusion of innovations. Free Press, New York.

Tivy, J. 1990. Agricultural ecology. Longman Scientific and Technical, New York.

U.S. Congress, Office of Technology Assessment. 1986. Technology, public policy and the changing structure of American agriculture. OTA-F-285. U.S. Gov. Print. Office, Washington, DC.

U. S. Department of Agriculture. 1981. Soil, water and related resources in the United States: Status, condition and trends. USDA, Washington, DC.

van Es J., and P. Notier. 1988. No-till farming in the United States: Research and policy environment in the development and utilization of innovation. Soc. Natural Resour., 1:93-107.

Zaltman G., R. Duncan, and J. Holbek. 1973. Innovations and organizations. John Wiley and Sons, New York.

22 Use of Soil Property Data and Computer Models to Minimize Agricultural Impacts on Water Quality

David I. Gustafson
Rhone-Poulenc Ag Company
2 Alexander Drive
Research Triangle Park, NC

As described by its most vocal proponents, Soil Specific Crop Management (SSCM), is a technology for optimizing net returns to the grower. By applying farm chemicals in a spatially varying manner according to localized needs and yield potentials, the grower can place these materials in the portions of the field that need them most. It is now widely assumed that such an approach will carry environmental benefits as well, although this has yet to be demonstrated in any scientifically valid way. Conceptually, it would seem that making chemical applications more efficiently in tune with localized crop and soil needs would reduce waste and therefore the net outflows of chemicals from the field. However, there is no guarantee that this will always be the case, and if it is desired to derive environmental benefits from this technology then there will have to be an environmental component built in to the SSCM application strategies. A prescription for accomplishing this is proposed in this chapter.

Following a brief historical perspective on the issues surrounding pesticide impacts on water quality, the important environmental impacts that are to be avoided are enumerated. This is proceeded by a listing of the soil properties that would need to be monitored for SSCM reduction of such impacts. A discussion is then presented of the modeling techniques available for describing pesticide movement through soil, including a list of the key pesticide properties that would need to be part of any computerized system for prescribing application strategies. The final section of the chapter is devoted to an implementation strategy based on the introduced soil, pesticide, and modeling concepts. The recommended tactic is to use the USDA-SCS screening/modeling method that is now being developed for field-averaged soil properties. With only a few modifications, this SCS procedure could be adapted to the SSCM paradigm with presumed beneficial effects on the environment.

Copyright © 1993 ASA-CSSA-SSSA, 677 South Segoe Road, Madison, WI 53711, USA. *Soil Specific Crop Management.*

HISTORICAL PERSPECTIVE

Ever since the publication of Rachel Carson's (1962) prototypic environmental anthem, *Silent Spring*, the American public has become increasingly sensitive to the occurrence of pesticides in the environment. Though undeniably useful and an integral part of the technology responsible for the minimal costs we pay for food, pesticides are undoubtedly one of the easiest targets of the burgeoning Green Movement now sweeping the world.

As analytical methodologies have become more precise and monitoring networks more extensive, the occurrence of pesticides in drinking water supplies has been more readily and widely documented. Scientists in academia, industry, and government have sought to develop accurate models for predicting how, when, and where pesticides move into drinking water supplies.

The models which have been developed may be broadly categorized as either screening or simulation models. As a practical matter, it is not always possible or cost-effective to simulate, in intimate detail, every aspect surrounding the use of a particular pesticide on a particular field. Screening models make it possible to weed-out the use scenarios that are very unlikely to result in off-site movement of the applied pesticide, and thereby focus the finite modeling resources onto the smaller set of circumstances which have a real potential for contamination. These screening models could be easily adapted to the SSCM paradigm, through the introduction of enabling technologies.

However, it should be remembered that there is still considerable debate in the scientific community over the question of exactly which screening and simulation models are the most appropriate to use, as a fairly large number of different methods have been proposed. This cacophony of conflicting models is present today despite the fact that several regulatory agencies throughout the world, including the U. S. Environmental Protection Agency and Germany's BBA have adopted particular screening methods and simulation models based on the physical properties of pesticides to serve as triggers for determining whether movement into groundwater is possible.

ENVIRONMENTAL IMPACTS

The environmental impacts of pesticides used in agriculture may be broadly classified into two categories: human health effects and ecotoxic effects. In the USA, human health effects are avoided through the promulgation and use of Health Advisory Levels (HAL's) and Maximum Contaminant Levels (MCL's). These are specific concentration levels for pesticides in drinking water which the U. S. Environmental Protection Agency has deemed to be safe. Concentrations above this level are not necessarily dangerous, but concentrations at this level or below have been determined to not present an unreasonable risk to human health, even when conservative allowances are made for the possibility of water consumption by the most sensitive members of the population.

Ecotoxicity of pesticides refers to the effects that some chemicals can have on so-called nontarget biota, such as birds, fish, and microscopic aquatics.

The toxicity study end-points most commonly used to describe the effects of pesticides on such organisms are LD50's (the dose killing 50% of the exposed members of the species), acute NOEC's (the 48- or 96-h exposure level at which no adverse effects are observed, and chronic NOEC's (generally the 21-d exposure level at which no adverse effects are observed).

SOIL PROPERTIES

Of all the soil properties affecting pesticide movement, none is more important than the capacity of soil to sorb molecules of the applied chemical and thereby remove them from the microscopic rivulets of leaching water in which transport takes place. This effect is rivaled only by the importance of permeability, which can become pre-eminent in sandy or other similar soils where only limited sorption can occur due to the lack of sufficient organic matter and clay-bearing materials.

Both permeability and slope affect the rate at which water leaches downward or is directed into surface runoff. These two pathways of water flow are responsible for the movement of pesticides into ground and surface water, respectively. These physical characteristics of the field should be part of the two-dimensional description of the field in the SSCM strategy.

Most simulation models of pesticide transport through soil are based on the use of a retardation factor approach for determining the velocity of the pesticide wetting front relative to the velocity of the aqueous wetting front as water moves through the soil. The retardation factor method assumes that pesticide molecules reach instantaneous, linear equilibrium between their sorbed and dissolved forms as the water moves through the soil matrix. As detailed later, the partition coefficient for pesticide exchange between these two forms is generally assumed to be proportional to the organic matter content of the soil. The retardation factor is given by:

$$RF = 1 + (BD \, K_d / FC) \qquad (1)$$

in which RF (unitless) is the retardation factor, BD (kg/L) is the dry soil bulk density, K_d (L/kg) is the linear partition coefficient for the pesticide in the soil, and FC (unitless) is the soil's field capacity, often defined to be the water held by soil when drained to a matric potential of 0 K_p.

The partition coefficient is given either as a multiple of the soil organic carbon content:

$$K_d = OC \, K_{oc} \qquad (2)$$

or the soil organic matter content:

$$K_d = OM \, K_{om} \qquad (3)$$

in which OC (unitless) is the soil organic carbon content, OM (unitless) is the soil organic matter content, and K_{om} (L/kg) and K_{oc} (L/kg) are the pesticide soil sorption coefficients normalized by organic matter and organic carbon, respectively. In cases where one is forced to convert between these two constructs and an assumption must be made concerning the relationship between soil organic matter content and soil organic carbon content, a factor of 1.724 has been recommended (Lyman et al., 1982).

Besides the soil properties already mentioned, the pH of soil is also important in the special case of ionizable pesticides, such as the sulfonyl-ureas, whose behavior is known to be a strong function of soil pH (Brown, 1990). In virtually all such cases, the mobility of such organic acids increases with higher pH. Often, there are also strong effects of pH on the degradation rate of such materials.

One important use of correlations between pesticide behavior and soil properties has been the generation of maps and lists of soils on which undesirable environmental impacts are more likely to occur. For instance, such correlations and more sophisticated modeling techniques were used to enumerate, on the product-use label for Temik (aldicarb), those soils on which applications were to be restricted due to leaching concerns.

MODELING TECHNIQUES

The dizzying array of modeling techniques available to describe agricultural applications of pesticides serves to blur one critical issue reducing their utility -- in a large number of cases, pesticides enter drinking water in ways not accounted for by the models. These include: applications near wells, surface water bodies, or sinkholes, back-siphoning, abandoned wells, glasshouses, applications to lawns, golf courses, glasshouses or nurseries, poorly constructed wells, leaks at storage facilities, agricultural drainage wells, spray drift during application, and spills -- both the large, accidental variety and the small, intentional kind.

Despite this shortcoming, models and screens have been demonstrated to provide useful information in the management of pesticides to minimize environmental impacts.

The most successful philosophy behind the development of effective screening methods is to first select a group of pesticides that have been well-characterized with respect to their potential for occurring at detectable levels in drinking water. The next step is to select the key physical properties of those pesticides affecting their potential for contamination. The two properties generally accepted to be most important are persistence, as measured by the soil half-life (days), and mobility, as measured by K_{oc} or K_{om}. An X-Y plot is then formed in which the key physical properties are used as the axes and different symbols are used to represent the contaminants and the noncontaminants. The plot is then examined to see if any patterns exist such that reasonable prediction of contaminant classification could be accomplished through the introduction of simple mathematical criteria.

This is the methodology that was employed in the development of the groundwater ubiquity score (GUS), a numerical index that has proved useful in predicting whether certain pesticides are likely to be contaminants of groundwater (Gustafson, 1989). Upon examination of classes of contaminants and noncontaminant assembled by the California Department of Food and Agriculture in 1986, it was found that the GUS index, defined as

$$GUS = \log(\text{soil half-life}) * [4.0 - \log(K_{oc})] \qquad (4)$$

gave complete separation of the two contaminant classes at a GUS value of 2.3. On the basis of this observation, a categorization scheme was proposed in which those pesticides having GUS values in excess of 2.8 were said to have a high leaching potential, and those with GUS values less than 1.8 were said to have a low leaching potential.

This screening approach was used by the U. S. Department of Agriculture Soil Conservation Service (SCS) in the development of their screening procedure for recommending to growers whether the application of a particular pesticide on a particular field would likely result in off-site movement into groundwater (Goss and Wauchope, 1990). In addition to classifying pesticides, the SCS method classifies the leaching and runoff potential of the soil according to its permeability, organic matter content, and field capacity. A matrix is then formed with soil classification on one side and pesticide classification on the other. In the SCS method, field-averaged values for soil properties are generally used or reliance is placed on soil databases such as the published county survey. Once the growers choice of pesticide is placed in the proper cell of the matrix, a decision is made as to whether further investigation is to be performed.

The follow-up analysis (also known as Tier 2 or Tier 3 investigations) might include a consideration of the toxicity of the chemical, mitigating factors such as application method or a low application rate, and the proximity and importance of the nearby drinking water sources. Inevitably, higher-tier testing will involve the use of more sophisticated simulation models such as GLEAMS or PRZM. These simulation models incorporate many more factors than are considered in the screening models, such as detailed daily weather data, crop management parameters, and more extensive soil and pesticide property data. Such models, however, are beyond the scope of what could be incorporated into a practicable SSCM minimization of environmental impacts.

IMPLEMENTATION STRATEGY

An ideal approach to implement SSCM minimization of environmental impacts would be to modify the SCS screening methodology in such a way so as to use -- instead of field-averaged values for the soil properties -- the specific values for the various spatially distinct areas in the field. The SCS matrix approach could then be separately applied to each mapped portion of the field.

As a practical matter, this might result in the use of different pesticide products on different portions of the field, in those rare instances where a real potential for contamination exists at particularly vulnerable spots within the application area. The costs associated with such prescription applications must be carefully weighed against the potential benefits, but it would seem likely that such sophisticated management techniques will become more practicable as the enabling technologies associated with SSCM become more widely available.

REFERENCES

Brown, H. M. 1990. Mode of action, crop selectivity, and soil relations of the sulfonylurea herbicides. Pestic. Sci. 29:263-281.

Carson, R. L. 1962. Silent spring. Riverside Press, Cambridge, MA.

Goss, D., and R. D. Wauchope. 1990. The SCS/ARS/CES pesticide properties database-II using it with soils data in a screening procedure. p. 471-493. In Pesticides in the next decade: The challenges ahead. Proceedings of the Third National Research Conference on Pesticides. 8-9 November 1990.

Gustafson, D. I. 1989. Groundwater ubiquity score: a simple method for assessing pesticide leachability. Environ. Toxic. Chem. 8:339-357.

Lyman, W. J., W. F. Reehl, and D. H. Rosenblatt. 1982. Handbook of chemical property estimation methods. McGraw-Hill Book Co., New York.

23 Nutrient and Pesticide Threats to Water Quality

Robbin S. Marks
Natural Resources Defense Council
1350 New York Avenue
Washington, DC

Justin R. Ward
Natural Resources Defense Council
1350 New York Avenue
Washington, DC

Agricultural activities are a major contributor to water pollution. As discussed in this section, groundwater and surface water contamination by pesticides and fertilizer nutrients is a matter of growing public concern.

NUTRIENT POLLUTION OF WATER SUPPLIES

American farmers use massive quantities of fertilizer to enhance crop production. Of particular environmental importance, N fertilizer use in U.S. agriculture increased nearly fourfold from 1960 to 1990. In 1990, farmers applied 11 million tons of N fertilizer to boost crop production (Taylor, 1992).

Numerous studies have documented widespread nitrate contamination of water supplies, traceable largely to agriculture. Citing findings from the U.S. Department of Agriculture (USDA) and Resources for the Future, the National Research Council's 1989 Alternative Agriculture report noted that approximately one-half to three-fourths of the nutrients (predominantly N and P) reaching the nation's surface waters derive from agricultural fertilizers and livestock waste (NRC, 1989, p. 99). Nutrient pollution accelerates the process of eutrophication, whereby water is robbed of the oxygen necessary to sustain aquatic life.

With respect to groundwater, a 1985 study from the U.S. Geological Survey (USGS) found nitrates above natural background levels in well samples taken in every state (U.S. Geol. Surv., 1985). A 1987 report by the U.S. Department of Agriculture estimated that more than one-third of all counties nationwide -- and a much higher fraction of counties in agricultural regions such as the Midwestern grain belt -- are highly vulnerable to groundwater nitrate pollution (Nielsen and Lee, 1987, p. 14). National surveys by the U.S. Environmental Protection Agency (EPA) in 1990 and 1992 detected nitrates in over half of the rural domestic and community drinking water wells tested (U.S. EPA, 1990, 1992).

Copyright © 1993 ASA-CSSA-SSSA, 677 South Segoe Road, Madison, WI 53711, USA. *Soil Specific Crop Management*.

Various farm states that have conducted groundwater surveys have found nitrate contamination in levels that threaten human health.[1] For example, more than 20% of private wells tested recently in Kansas and South Dakota exceeded the health standard for nitrates (Fedkiw, 1991, p. 24-25 and p. 27). A 1988-1989 state-wide survey of rural water quality throughout Iowa found that more than 18% of private wells exceeded the standard (Iowa DNR, 1990). Water samples from Iowa's agricultural Big Spring Basin from the 1960s to the 1980s indicated a strong correlation between increases in cropland applications of N fertilizer and nitrate concentrations in groundwater (Hallberg, 1989). Nebraska's 1990 analysis of well-water samples conducted in areas of suspected pollution showed nitrate levels that exceeded the federal health standard in 21% of all wells, and 31% of the irrigation wells tested (Fedkiw, 1991, p. 11-12).

Nitrate contamination presents significant public health implications, in that groundwater provides drinking water for approximately 40% of this country's overall population and for nearly 100% of rural residents. In high concentrations, nitrates can cause infant methemoglobinemia, commonly known as "blue-baby syndrome." This condition impedes oxygen transport in infants' bloodstreams and has led to at least one reported death (Johnson and Bonrud, 1988). Other public health concerns surrounding nitrate contamination of groundwater include a link between nitrates and certain cancers, birth defects, high blood pressure, and developmental problems in children (Nielsen and Lee, 1987, p. 22). Within the human digestive process, nitrates can form potent carcinogens known as *nitrosamines* (Nielsen and Lee, 1987, p. 22).

PESTICIDE CONTAMINATION OF STREAMS, LAKES AND WELLS

The nation's agricultural system employs enormous volumes of chemical pesticides to control weeds, insects, and plant disease. In 1989, U.S. farmers applied more than 800 million pounds of pesticides within their operations (Aspelin et al., 1991).

Water pollution by agricultural pesticides has been well documented. Recent USGS findings revealed widespread herbicide contamination of rivers and streams throughout the Mississippi River Basin (Goolsby et al., 1991, p. 24-30). The agency conducted sampling during the spring of 1991, a period of intensive herbicide use in production of corn and other crops within the region. All of the water samples in this USGS survey revealed detections of atrazine, one of the most widely used herbicides in American agriculture. One-fourth of the samples

---

[1] EPA has set a drinking water health standard for nitrates of 10 milligrams per liter. This level is primarily designed to prevent "blue-baby syndrome," and does not reflect cancer risks or other chronic health effects that may result from nitrate exposure.

exceeded federal health levels for atrazine.[2] The results of the 1991 study were consistent with previous USGS research that measured the concentration of herbicides in streams throughout the Midwest (Goolsby et al., 1991, p. 2-6). Most municipal drinking water systems lack pesticide-removal technology to treat water that reaches household taps.

Pesticides present additional threats to groundwater quality. In the late 1980s, EPA compiled a database of individual state monitoring results for pesticides in groundwater. This compilation documented a total of 46 pesticides in the groundwater of 26 states from normal agricultural use (U.S. EPA, 1980, p. 1-2). For example, the herbicide alachlor, a potential human carcinogen that has been banned in Canada but remains in wide use in the USA, has been found in the groundwater of 12 states (U.S. EPA, 1988, p. 1-2).

ENVIRONMENTAL POTENTIAL OF FARMING BY SOIL

Soil specific crop management technology, also known as "farming by soil (FBS)," is an emerging, sophisticated technique for matching the field application of fertilizers and pesticides to soil types. The technology is distinguished from typical farming methods in which fertilizers and pesticides are applied according to the average soil characteristics of farm fields.

Farming by soil relies on data collected through aerial color-infrared photographs or extensive soil testing on the ground. Soil type may be measured as frequently as five times within a three-acre grid. This testing information provides the basis for computerized, color-coded maps that indicate fertilizer and pesticide application rates commensurate with farmers' crop and yield goals (Elliot, 1987; Pocock, 1981, p. 12).

New, high-technology chemical spreaders are equipped to "read" these computerized maps and vary the fertilizer and pesticide doses and blends as the machines move across farm fields (Pocock, 1991, p. 12). Measured chemical parameters, such as levels of N, P, pH and K are major determinants of the environmental fate of applied pesticides and fertilizers (Munson and Runge, 1990, p. 17). The soil tests can also reveal information about physical characteristics of soils, including the potential for leaching of chemicals to groundwater (Larson and Robert, 1991, p. 108). Related site-specific tools under development include machines that, variously, test soil for N content and simultaneously apply inputs (Munson and Runge, 1990, p. 70), change from one tillage method to another according to soil characteristics (Larson and Robert, 1991, p. 111), and seed in patterns and planting depths based on soil type and moisture retention (Gaultney, 1991, p. 7).

Soil specific crop management holds potential to help reduce pollution at the source. As it develops, the technology may complement existing

---

[2] EPA has classified atrazine a "Group C" possible human carcinogen and has set a "maximum contaminant level" (mcl) of three parts per billion based on the risk of liver and kidney damage.

techniques by which farmers can save money through more efficient management of nutrients and pesticides used in crop production. Substantial reductions in the waste stream of nitrates, for instance, can be accomplished through testing of soils and preparation of nutrient "budgets" to prevent fertilizer overapplication and set appropriate yield goals, as well as through improved timing and placement of fertilizer on croplands.

The financial and environmental benefits of farm management adjustments such as these are well documented. For example, Iowa farmers who credited N from alfalfa (*Medicago sativa* L.) planted in a crop rotation achieved an average 27% reduction in N fertilizer inputs over an 8-year period (Iowa DNR, 1990). These producers incurred no appreciable reductions in yields, and saved an average of approximately $1,000 per farm annually. A study by Iowa's Leopold Center for Sustainable Agriculture projected that farmers in that state could reduce their use of N fertilizer use by up to 29% and save as much as $86 million annually through variable, better-timed fertilizer applications in lower amounts (Kanwar and Baker, 1992, p. 7-9).

Soil specific crop management may eventually supplement farming practices that enable reduced pesticide inputs. For example, the new technology could complement agricultural techniques with demonstrated success as alternatives to high pesticide use in Midwestern corn (*Zea mays* L.) and soybean [*Glycine max* (L.) Merr.] production, including ridge tillage, strip intercropping, planting crops in narrow rows, and applying chemicals in bands on farm fields (Curtis et al., 1991).

POLICY CONSIDERATIONS

As the technology advances, farming by soil may provide producers with a useful tool to meet their obligations under state and federal laws for water pollution control. For example, Nebraska's 1986 Groundwater Management and Protection Act requires N use reductions in areas with serious, persistent nutrient pollution of underground water supplies (The Groundwater Management and Protection Act. Nebraska Rev. Stat. Sec. 46-656-674 [Cum Supp. 1986]). Minnesota's 1989 Groundwater Protection Act sets a nondegradation goal for water quality, and contemplates mandatory pollution control measures in cases where health levels for nitrates or pesticides are exceeded (Minn. S.P.A., 1989).

At the federal level, 1990 amendments to the Coastal Zone Management Act (CZMA) require implementation of runoff control measures to meet water quality standards in areas vulnerable to pollution. Draft EPA guidelines to the states for CZMA implementation, although highly discretionary, emphasize nutrient and pesticide source reduction strategies that could encourage or force adjustments in farming practices (EPA, 1991). Pending federal Clean Water Act legislation includes strong source reduction provisions that seek to reduce runoff pollution from agriculture and other "nonpoint" sources.

The potential environmental benefits of FBS will not be realized without policies that promote, rather than discourage, judicious use of chemical inputs in agriculture. Sound policy in the following areas will be key to the success of soil specific crop management technology and other source reduction techniques:

Financial and Technical Assistance

Notwithstanding potentially significant cost savings from input reductions, the high cost of the new technology will be daunting for many small- and medium-scale farming operations. Government cost sharing and technical assistance could help enable wider adoption of soil specific crop management. One prospective vehicle for this type of assistance is the Water Quality Incentives Program authorized by the 1990 farm bill (Food, Agriculture, Conservation and Trade Act [FACTA] of 1990, Pub. L. No. 101-624, Sec. 1238, 104 Stet. 3590-3594, 16 USC 3838). Under this program, producers in water quality problem areas may receive technical assistance, as well as up to $3500 in annual cost sharing to implement source reduction measures included in multi-year plans. Unfortunately, this program has been hampered by inadequate funding levels (Center for Res. Econ., 1992, p. 25). Under the 1990 farm bill, USDA is obligated to provide technical assistance to farmers who wish to develop water quality protection plans to assist in their compliance with state and federal laws (FACTA of 1990, Sec. 1238D, 16 USC 3838d).

Commodity Program Rules

Commodity support programs administered by USDA reward intensive production of a few commodity crops (NRC, 1989, p. 10-11). The program rules frustrate some farmers' efforts to diversify their production systems and reduce chemical inputs.

Environmental reform could help ensure that the commodity programs work for, rather than against, the prospective benefits of soil specific crop management. In this regard, the 1990 farm bill's Integrated Farm Management Program Option (IFMPO) represents an important step toward allocating federal farm payments for resource-conserving crop rotations and other environmentally beneficial farming alternatives (FACTA of 1990, Sec. 1451, 7 USC 5822). Additional reforms are needed to correct the commodity programs' bias toward monoculture and surplus production.

Pesticide Use Information

Environmental success of farming by soil could be greatly enhanced by accurate, site-specific data concerning chemical applications. Pesticide use information is essential to farmers, decision makers, and the public to identify problem areas and develop source reduction strategies. Unfortunately, most states have not instituted requirements for pesticide use record-keeping, and USDA has been slow to implement a record-keeping provision for restricted use pesticides mandated by the 1990 farm bill (Center for Res. Econ., 1992, p. 30).

CONCLUSION

Controlling agricultural water pollution from nutrients and pesticides is an urgent environmental priority. Soil specific crop management, while no panacea, represents an encouraging element of a comprehensive strategy to stop pollution at the source. Public policy favoring source reduction and sustainable agriculture is necessary to achieve environmental benefits from farming by soil and other environmentally promising technologies.

REFERENCES

Aspelin, A. L., A. H. Grube, and V. Kibler. 1991. Pesticide industry sales and usage: 1989 Market estimates. U.S. Environmental Protection Agency, Office of Pesticide Programs, Washington, DC.

Curtis, J., L. Mott, and T. Kuhnle. 1991. Harvest of hope. Natural Resources Defense Council, New York.

Elliot, C.. 1987. Fertilizing-blending and spreading on-the-go computerized maps and radar guidance. 1. International Off-Highway and Powerplant Congress and Exposition, Milwaukee, Wisconsin. 14-17 September 1987. Pocock, J. 1991. "Put Your Money Where It Counts", P. 12, Prairie Farmer. January 1.

Fedkiw, J. 1991. Nitrate occurrence in U.S. waters (and related questions), A reference summary of published sources from an agricultural perspective. p. 24-25 and p. 27. U.S. Department of Agriculture, Washington, DC.

Gaultney, L. 1991. Prescription farming based on soil property sensors. p. 7. In New technology in agriculture - Proceedings of the 1991 Crop Production and Protection Conference. Iowa State University, Ames, Iowa, 3-4 December 1991.

Goolsby, D. A., R. C. Coupe, and D. J. Markovchick. 1991. Distribution of selected herbicides and nitrate in the Mississippi River and its major tributaries, April through June 1991. p. 24-30. U.S. Geological Survey, Denver, Colorado, Water-Resources Investigations Rep. 91-4163.

Hallberg, G. R. 1989. Nitrate in ground water in the United States. In R. F. Follett (ed.) Nitrogen management and groundwater protection. p. 54. Elsevier Science Publishers, Amsterdam, the Netherlands.

Iowa Department of Natural Resources. 1990. A progress review of Iowa's agricultural-energy-environmental initiatives: Nitrogen management in Iowa. p. 5-6. Iowa City, IA.

Iowa Department of Natural Resources, Geological Survey Bureau. 1990. Iowa State-Wide Rural Well-Water Survey, Summary of Results: Nitrate and Bacteria. p. 35.

Johnson, C. J., and P. Bonrud. 1988. South Dakota Department of Health. "Methemoglobinemia: Is it coming back to haunt us?" p. 3-4. In Health and Environment Digest. Vol. 1, No. 12.

Kanwar, R. S., and J. L. Baker. 1992. Effect of split N-fertilizer applications on drainage water quality and NO_3-N leaching. In Progress Report. Leopold Center for Sustainable Agriculture, Ames, Iowa, Vol. 1.

Larson, W. E., and P. C. Robert. 1991. Farming by soil. p. 108. In Soil management for sustainability. Soil and Water Conservation Society, Ankeny, IA.

Minnesota State Planning Agency. 1989. The Minnesota Groundwater Protection Act of 1989: A Summary.

Munson, B., and C. F. Runge. 1990. Improving fertilizer and chemical efficiency through "High Precision Farming". p. 17. Center for International Food and Agricultural Policy, St. Paul, MN.

Nielsen, E. G., and L. K. Lee. 1987. The magnitude and costs of groundwater contamination from agricultural chemicals. p. 14. USDA Economic Research Service, Agricultural Economic Rep. 576.

National Research Council. 1989. Alternative agriculture. p. 99. National Academy Press, Washington, DC.

Pocock, J. 1991. Put your money where it counts. p. 12. Prairie Farmer.

Taylor, H. 1992. Agricultural economist, Economic Research Service, U. S. Department of Agriculture, personal communication.

U.S. Environmental Protection Agency. 1988. Pesticides in groundwater data base - 1988 Interim Rep. p. 1-2.

U.S. Environmental Protection Agency. 1990 and 1992. National Pesticide Survey: Phase I Report. PB91-125765, Fall 1990; National Pesticide Survey: Phase II Report.

U.S. Environmental Protection Agency. 1991. Proposed guidance specifying management measures for sources of nonpoint pollution in coastal waters. p. 2-41, 2-57.

U. S. Geological Survey. 1985. National water summary 1984. Water Supply Paper 2275.

24 Working Group Report

W. E. Larson, Chair
G. Malzer, Recorder
R. S. Marks and J. R. Ward, Discussion Paper

Other participants:	R. Barnes	B. Montgomery
	T. Dao	L. Mulkey
	D. Gustafson	C. Onstad
	J. Huffman	G. Randall
	P. Hunt	W. Riedell
	B. Khakural	J. Spetzman
	R. Knutson	J. Ward
	M. Jetland	R. Wilbur

Soil specific crop management (SSCM) refers to management on a site-specific basis based on the spatial variability of soil, terrain, plant, or other properties within a field. It implies applying inputs on a microscale rather than a field scale. While some information is available, the relationship between SSCM and environmental parameters needs further study.

Research in SSCM thus far has concentrated on management of chemical inputs. However, site-specific management can also be applied to selection of crops, planting rates, tillage, and erosion control practices.

Intuitively, a mass-balance approach appears appropriate for analyzing chemical input-output relationships in the soil-water-crop-atmosphere continuum. If chemicals are supplied at an amount to meet only the needs of the plant (crop or weeds) and the interactions with the soil (sorption, degradation, etc.), then excess chemicals are not available for transport. Additionally, if chemicals are supplied to each soil unit at an optimum amount, without excess, economic crop production will be enhanced.

Under some conditions, trade-offs between economics and environmental concerns may be necessary. If so, SSCM may be used to optimize the trade-off. While soil-specific management has been applied most often to crop production management, the principles can also be applied to other uses of the land.

Soils have three primary functions, namely to (i) provide a medium for plant growth, (ii) partition and regulate water in the environment, and (iii) act as a buffer in the environment. Whatever function soils perform, management should be carefully matched with the characteristics of the soil on a site-specific basis.

LAND MANAGEMENT ISSUES

Soil Resources

The soil is a complex physical, chemical, and biological body that must be described in terms of its many attributes such as water-holding capacity, texture, structure, pH, nutrient content, and organic matter. The various attributes are usually interrelated so that a change in one attribute will change other attributes. For example, a change in the organic matter content may alter the availability of nutrients, the water-holding capacity, and structure. Because of the interrelations among attributes, the behavior of some attributes can be predicted from others.

Larson and Pierce (1992) have suggested that for assessment and for quantifying the sustainability of soils, a minimum data set (MDS) be selected and monitored. They further pointed out that the MDS could be used for predicting other soil attributes by the use of pedotransfer functions (PTF). Pedotransfer functions are means to predict an attribute from values of other attributes. The relationships are often statistical.

Soils vary greatly in their vulnerability to damage, even in the same landscape. Some soils because of the soil and landscape characteristics are prone to severe erosion. If the granular material is of limited depth over consolidated material, the soil will be damaged quickly from erosion. Similarly, mismanagement can easily damage soils' ability to perform the three primary functions.

Water Resources

The soil acts as a primary body in partitioning and regulating water flow in the environment, as a sink for chemicals, and as an incubation chamber for microbial transformations of organic materials. The soil attributes determine the effectiveness of how well the soil performs these functions.

Nearly every drop of rain that falls on the soil and the conditions at the soil surface determine whether or not the water infiltrates the soil or runs off. If it infiltrates the soil it may be stored for use by organisms, move downward to groundwater, or move laterally to appear later in springs or seeps. While in the soil it can pick up chemicals and transport them to other parts of the environment.

Water as it moves through the environment transports chemicals. It can move agriculture chemicals such as (a) plant nutrients from applied sources (fertilizers or human and animal wastes); or (b) naturally occurring salts. On the other hand, the soil is a very effective filter that, in many cases, improves the quality of water as it moves. Larson and Robert (1991) estimated that the hydraulic conductivity varied by five-fold, depending on soil mapping unit in a 50 acre field in Jackson County, Minnesota. Hence the potential for differential leaching of chemicals is large.

Both water quantity and quality are being increasingly looked at as important resources that are being endangered. The management and protection of water and soil resources as they relate to the environment is an important national and international issue.

Landscapes

The soil and its characteristics is an integral part of the landscape. Both the soil and the landscape must be considered together in management of the environment. The physical characteristics of the landscape (degree and length of slope, drainage characteristics, etc.) along with weather, soil, plants, and management affect the amounts of erosion, for example.

The environment surrounding a parcel of land may influence the desired use and management of land. For example, drainage of water-containing chemicals from an agricultural area into wetlands or sensitive wildlife habitats may have important consequences. Drainage of water from land into a lake or impoundment for municipal water supplies may require a particularly careful land management treatment. In all cases the desired management of the land must consider the soil and landscape on a site-specific basis.

SOIL PROCESSES

The interactions of agricultural management practices with the soil resource are indeed complex. Agricultural management may directly or indirectly influence many surface and subsurface processes that regulate the quality and quantity of ground and surface water resources. The dynamic soil-water ecosystem in an agricultural setting may in turn have a significant impact on atmospheric exchange of N, C, and other compounds.

Surface Processes

Agricultural management practices imposed onto a landscape will interact with the various soil attributes across that landscape to provide a range of potential environmental consequences. For example, the manner of tillage and residue management may influence the rate of water infiltration, thereby impacting the runoff from the area. The amount and timing of runoff will directly influence soil loss, and the pesticides and nutrients that may be contained in the runoff water or attached to the soil particle via erosion. Soil attributes, tillage/residue management and landscape position, therefore, become important factors that should be considered in assessing the impacts of soil specific management.

Fertilizer and pesticide management practices also need to be considered when evaluating the environmental impacts associated with soil specific management. Agricultural practices which result in increased agricultural chemical concentrations at the soil surface (surface-applied herbicides, or immobile nutrients) result in a higher potential for loss due to surface runoff processes. Conservation tillage practices reduce the quantity of runoff, but frequently demand surface applications of nutrients and pesticides. Reduced runoff may also lead to differential percolation of water into the soil increasing the potential for leaching of nutrients and pesticides.

Subsurface Processes

Differential infiltration and percolation of water into and through the soil will be created by landscape position and soil characteristics. Agricultural management practices will also interact with the above factors to influence leaching of chemicals. For example, the form, amount, and time of N fertilizer application will affect the amount of nitrate N that is present in a soil profile at a given point in time. The loss of nitrate via leaching through the soil, however, will depend not only on fertilizer management practices, but also on the soil and climatic attributes which also influence water movement.

FIELD MANAGEMENT

Crop production practices, including rotations, use of waste products such as manure, and anticipated production levels must also be considered in assessing the impacts of soil specific management. Production systems that include animals frequently have legumes in rotation which fix atmospheric N contributing to the soil N reserve. The availability of this N reserve for plant uptake or loss to the environment will be dependent on the interaction of the soil attributes with landscape and climatic characteristics. Other inputs to the soil system such as manure, residues, and waste products, along with synthetic inputs will be counterbalanced by the crop demand and soil processes that modify, degrade, or transform the original material placed into the soil system. Larson and Robert (1991) estimate that the corn yield potential in a 50-acre field in Jackson County, Minnesota varies by about 100%, depending upon soil mapping unit. If N is applied at a rate to maximize corn yields on the most productive soils, the potential for leaching in the less productive soil is obvious. The dynamics of each of these mechanisms will vary with the soil conditions present.

Randall (chapter 20 in this book) at this workshop outlined statewide best management practices (BMP) for N management for preventing degradation of Minnesota water resources while maintaining farm profitability. The statewide BMP are (1) selection of realistic yield goals, (2) use a comprehensive record keeping system to record field (or site) specific information, (3) adjust N rate according to soil organic matter content, previous crop and manure application, (4) use a soil nitrate test when appropriate, (5) use prudent manure management, (6) credit second year N contributions from alfalfa and manure, (7) do not apply N fertilizer above recommended rates, (8) nitrogen applications should be timed to achieve high N-use efficiency. The Minnesota statewide BMP's are appropriate over a broad spectrum of soils and farming conditions. However, many of the eight BMP's need application on a soil specific basis to be most effective. For example, yield goals must be estimated by soils within a field to be most effective. Likewise, organic matter estimations and nitrate tests are best applied on a soil specific basis. Similarly, BMP's for other management decisions (e.g., tillage, erosion control, herbicide application, etc.) can be most effective if applied on a site specific basis.

RESEARCH

Using a mass balance approach, it appears reasonable to believe that by optimizing management on a site-specific basis within fields, the needs of a crop can be met without damage to the environment. In the case of N input, for example, N can be applied to provide adequate amounts for the crop without providing excess for leaching. For erosion control, tillage and crop residue management can be varied on a soil and topographic basis to bring erosion amounts within acceptable levels while optimizing crop yields. While soil-specific crop management offers great promise in environmental protection, little information is available.

Impacts

Research needs to be directed toward the economic, environmental, and efficiency impacts of soil-specific crop management.

Environmental impacts should include such items as surface and groundwater quality, erosion, carbon sequestering, effects on nearby fragile ecosystems, wildlife, and aesthetics.

Mass Balance

A mass balance approach should be considered. Input-output relationships should be considered, where feasible, to account for the eventual disposition of the inputs. This approach appears particularly attractive where chemicals are applied. It may also be used in the case of erosion where the source of the sediment and the eventual deposition of the sediment can be identified.

The mass balance approach is needed on an ecosystem basis where input-output relationships can be studied on all parts and all facets of the ecosystem. Risk-benefit studies on an ecosystem basis should be carried out.

Three-dimensional Assessments

Environmental assessments are needed on areas large enough to include all segments of the ecosystem. The area needs to consider all facets of the environment, including agricultural, wildlife, hydraulic, and surrounding critical areas. Areas suitable for study may be defined by a watershed, aquifer, soil catena, etc.

Aerial databases should include topographic, soil, drainage, aquifer, plant, and land use information. Geographic information systems and similar systems offer powerful tools for ecosystem assessments.

Long-term Experiments

Long-term (10 or more years) experiments are needed to quantify many changes in the soil-water-atmosphere continuum. Long-term quantitative data are extremely scarce. The residence time of chemicals in the soil and vadose zone is often considerable. Thus, it may require many years to detect the chemicals in groundwaters. Likewise, the soil is a large sink for carbon, but because of natural variability the change in C content of the soil is difficult to measure over short periods. The C content of the soil is a critical parameter in soil management and has important implications for global climate change.

Spatial Quantification

Soil attributes vary spatially and while many attributes are related, the quantification of specific attributes is needed for the various applications of management. Grid maps particularly suited for environmental assessments are needed. For example, a grid map of a field showing soil test values for P is most useful for applying P fertilizers and for assessing potential P in runoff water. In the case of N grid maps of nitrate amounts, along with mineralization potentials and realistic crop yield estimates are needed. Additionally, an estimate of leaching potentials may be useful. For application of lime, pH and soil texture maps are desirable. For application of erosion control methodologies, the physical attributes of the soil and the landscape characteristics are needed. Obviously, preparation of the large number of grid maps is expensive.

In many cases, a minimum data set may be desirable from which simple models can be used to estimate critical parameters for the maps. Research is needed on what grid maps are most meaningful for the various applications, what field sensors and laboratory analyses are needed to prepare the maps, and what applications can be made from the various maps.

Models for Using Spatial Data

Computer-based models are needed that can take the spatial data as input and make predictions on an areal basis such as a watershed or hydrologic unit. These models can be useful for prediction of natural resource behavior within the area of study, but also to predict the impact on adjacent sensitive areas such as wetlands and wildlife areas.

Some models are available but need validation under various conditions. Other models need development.

Environmental Information Transfer

The environmental benefits of SSCM appear to be considerable. As these benefits are substantiated a strong effort will be needed to transfer this knowledge to both crop producers and the concerned public. The environmental benefits and/or risks should be an integral part of SSCM best management practice recommendations.

REFERENCES

Larson, W. E., and F. J. Pierce. 1992. Conservation and the enhancement of soil quality. In Proceedings of International Workshop on Evaluation for Sustainable Land Management in the Developing World. Chiang Rai, Thailand.

Larson, W. E., and P. C. Robert. 1991. Farming by soil. In R. Lal and F. J. Pierce (ed.) Soil management for sustainability. Soil and Water Conserv. Soc., Ankeny, IA.

SECTION VI

TECHNOLOGY TRANSFER

25 Prescription Farming

William Holmes
RR #2
Oran, MO

In March of 1989, Holmes Bros. Farms, Inc., Southeast CO-OP of Advance, and Delta Growers Association of Charleston joined in an effort to learn and understand new technology that would aid in the more efficient use of agricultural inputs.

Our first objective was variable rate application of plant nutrients. Within 2 weeks we had signed up 40 producers with approximately 10,500 acres for the project. At first we intended for each producer to collect his own soil samples for lab analysis and we would digitize the information into a map suitable for use in an on-board computer in a fertilizer truck. The computer would recognize the different areas within the field and make the necessary changes in rates as it crossed each area of the field.

After a good start, we became concerned about the consistency of the data if many different people were taking the soil samples. We also recognized the complexity of the task undertaken and began to look for other resources to aid in our project.

By mid-June, we had contacted the University of Missouri-Columbia (MU), the Soil Conservation Service (SCS), and the Missouri Department of Natural Resources (MDNR) and asked for assistance. After a review of the project proposal we received confirmation from all agencies for assistance. The MU provided technical assistance and a reduced lab fee for soil test analysis, SCS provided a four-person team of soil scientists for 6 weeks for the purpose of conducting a more detailed soil survey on the project acres, and MDNR provided a grant of $93,060 to aid in the project. Additionally, the local farmers have provided a substantial amount of funding support through acreage sign-up.

PROGRESS TO DATE

The project was initially a 3-year project. One of the main goals of this project was to determine whether there is enough variability within fields to warrant variable rate technology (VRT). We are maintaining a large database with information from lab analysis of soil samples from each field. Soil samples are taken on a 330 x 330 ft. grid and the soil analyses are loaded into the computer. Interpretation and fertilizer recommendation maps are developed based on MU soil test interpretations. The field is divided into zones for fertilizer application and the data stored in a digital map. Five fertilizer blends most optimum for the field will be spread by the computer-driven spreader truck using this digital map. To date, we have grid sampled 9000 acres and 6000 acres have been fully processed for VRT application. More than 4000 acres have had fertilizer applied using a Soil Teq[1] spreader truck.

It is clear that many of these fields will benefit from variable rate fertilizer application. It is not uncommon to see P levels range from <20 to >200 and K levels range from <100 to >400 in the same field. We believe the same benefits will be evident for VRT pesticide application. In fact, much of the data generated in this project affecting fertility recommendations will be important factors when making pesticide recommendations. Information about soil type, texture, organic matter, soil pH, and cation exchange capacity (CEC) will be key factors when dealing with many of the soil sensitive chemicals. Other factors will include matching the proper chemical with the pest that exists in various parts of the field. We hope to build on the knowledge and data gained in this project as we move into variable chemical application. In our project we are using geographic information system (GIS) methods capable of using data from infrared photography, aerial photo base maps, and SCS soil survey maps to define differences in soil type, texture, and water-holding capacity. This system uses several stand alone programs operating under a series of shell scripts. Many operations are integrated but some are not. Much efficiency is lost due to data translations required for data to pass from one program to another. This can be remedied by using a true GIS relational data base environment. We are consulting with the Space Remote Sensing Center (SRSC) located at the Stennis Space Center in Mississippi to develop this type system. We are in the process of converting all data from the project to this format.

EXISTING CHALLENGES

One of the big problems we face is managing the massive amounts of data that are generated by prescription farming. Being able to interpret the amount of data and make good agronomic and environmental recommendations is difficult. The collection and organization of data is time consuming and

---

[1]Soil Teq., Inc. Waconia, MN 55387.

expensive. If data that is routinely collected by the normal activities of the Soil Conservation Service (SCS), Agriculture Stabilization and Conservation Service (ASCS), and other government agencies can be used by a compatible GIS, the cost will be reduced and the technology will be adopted by a larger number of producers for prescription farming purposes.

In the future, SCS will be developing layers of data with the Geographic Resources Analysis Support System (GRASS), which is the GIS that will meet their needs. Some of this data will be invaluable for the purpose of prescription farming (if in a compatible format) and will reduce the cost of implementing prescription farming for the producer.

Another valuable source of information could be used from the ASCS activities. These include accurate field boundaries, crop history, and a universal method of naming (i.e., farm no., tract no., and field no.) fields. Additional layers of data will need to be developed that may not be available from other sources, but will be critical for prescription farming recommendations.

FUTURE PLANS

We plan to extend the acreage to include another 25,000 acres in the Missouri Bootheel. Data collection will be automated in every way possible and practical. We hope to incorporate global positioning systems (GPS) equipment with real time differential correction to acquire soil sample and field boundary locations. The system could include a bar-code reader at the field level and lab locations to facilitate the process.

We plan to pursue an automated way of collecting yield data within the field. This can be accomplished with yield monitors mounted on the combine receiving position data from the GPS. Research involving on-board yield monitors and GPS systems is being conducted by Dr. Steve Borgelt (MU Agricultural Engineering Department) and Dr. Ken Sudduth (USDA-ARS). Also, it may be possible to estimate yields, within a reasonable degree of accuracy, over a large geographic area using satellite imagery. Satellite data and other forms of remote sensing may play an important role in agriculture in the future. Remotely sensed information such as yield estimates within the field, plant vigor[2], and maturity[3], could aid the producer in the overall management of the crop if supplied on a timely basis. SRSC has committed many of their resources to aid our efforts over the next few years to develop remote sensing techniques, data collection systems, and GIS to bring these management tools to the agricultural producer.

---

[2]Relating to areas of stress indicating plant diseases, insect infestation, or drought which could aid in irrigation scheduling.

[3]Used for timing of growth regulators in some crops like cotton and scheduling harvest in many specialty crops such as fresh vegetables.

We plan to develop a powerful GIS and integrated relational database system that is capable of managing all pertinent data layers and assist in prescribing the proper fertilizer or pesticide at the correct rate for each area of the field. The system will use universal fertility recommendation equations[4] that incorporate the use of a coefficient table for Missouri. This will allow the system to be easily modified by substituting a different coefficient table, to match recommendations being made in other states. Chemical rates could be recommended for each area of the field based on manufacturers' label recommendations and also use a new Soils Pesticides Interaction Screening Procedure (SPISP) being developed by SCS. This procedure will rate pesticides applied to a particular soil with a 1-3 rating. A rating of 1 would indicate little chance of contamination to ground or surface water. A rating of 2 would indicate a possible problem and a 3 rating would indicate a likely chance that the pesticide could cause problems when used. Recommendations for the pesticide could also be biased by a mask of all environmentally sensitive areas provided by MDNR or EPA (Environmental Protection Agency). The system would accommodate the compilation of historical records of pesticide application for future compliance requirements by EPA, MDNR, or the Department of Agriculture. Historical data (i.e., yield and fertilizer applied) could be used to model fertility data allowing lab analysis to be amortized over a longer period of time. This system will need to assess, at the onset of field analysis, the economic and/or environmental advantage of VRT. The first assessment should be made prior to grid fertility sampling. This can be accomplished by using some form of remote sensing data to determine the amount of variability that exists within the field. It would seem logical to assume the greater variability in biomass and vigor that exists within the field, the better candidate a particular field would be for VRT. If the decision is made to proceed after the first assessment is made then a thorough grid fertility sample should be made. The grid sample results would be loaded into the GIS for further analysis before the decision to use VRT is made. This could be accomplished using a theoretical technique to establish yield response to variable spreading which employs the use of Missouri's fertility index equation. If after this analysis, the system indicated enough economic potential for the producer to proceed, the proper VRT could be recommended. The system should be capable of monitoring the effectiveness of the prescription (i.e., yield, pest control) and optimize inputs for future applications.

ADOPTION CONSTRAINTS

How fast and to what degree this technology and these systems are used will depend upon several factors:

---

[4]These equations are being developed by Dr. Dale F. Leikam (Farmland Industries, Inc.) in consultation with Dr. D. D. Buchholz (Univ. of Missouri).

1. The ability to automate data collection.
2. The development of VRT equipment to apply prescriptions.
3. The user friendliness of these systems.
4. Cost of systems per unit served.
5. The extent of shared data by government agencies.
6. Economic factors affecting agriculture and the economy.
7. Technology transfer and education to agribusiness.

26 Illini FS Variable Rate Technology:

Technology Transfer Needs from a Dealer's Viewpoint

John Mann
Illini FS
1509 East University
Urbana, Illinois

Illini FS is a member cooperative of the GROWMARK system which does business in Illinois, Iowa, and Wisconsin. Illini FS operates in Champaign, Douglas, and Edgar counties in east central Illinois. We currently have about $36,000,000 in annual product sales and those are broken down - 70% plant food and 30% energy (petroleum) products.

Our involvement with variable rate technology (VRT) started when I was a retail fertilizer plant manager with Corn Belt FS. One spring day, an older gentleman who was a good customer of ours came into my office and put the "Farming by the Foot" article on my desk and said, "This makes a lot of sense. Why can't we do this on my farms?" At that point I didn't give it a lot of thought, although we did start an aggressive campaign of soil testing, flagging, and spot spreading fertilizers.

In July of 1989, I was promoted to area manager at Illini FS. That spring day when the farmer came into my office with the magazine article never left my mind. Our marketing staff meets twice each month to discuss marketing ideas and I thought it was a good time to throw out the idea of site-specific fertilizer applications. The idea met with overwhelming acceptance and the decision was made to continue to gather information about this futuristic possibility.

At the next marketing meeting we invited Roger Knutzen of Soil Teq and his staff down to further explain the VRT process and what would be involved. One of the other key ingredients to our success is that from this initial meeting and on into our growth with the program, we involved the University of Illinois.

The success that Illini FS has had with VRT was not an accident. The achievements were part of a systematic action plan to achieve a set of predefined marketing objectives.

Copyright © 1993 ASA-CSSA-SSSA, 677 South Segoe Road, Madison, WI 53711, USA. *Soil Specific Crop Management.*

Illini FS has been able to offer the farmer a cost-effective solution to the site-specific fertilizer application question. We have our program broken into two phases, Phase 1 and Phase 2. The first phase is everything it takes to get us to the point where his field can be spread by the VRT system. We purchased two all terrain vehicles (ATVs) and equipped them with a marine compass to keep us on a north-south or east-west course, and an acre counter to keep track of distances between grid samples. The first map is an example of the field that has been sampled on a systematic 3.3 acre grid (330 by 440 ft). The first step is to measure the perimeter of the field to determine how many 3.3 acre grid points we need per field. Once that is done, we mark all of the grid points on a hand-drawn map and start sampling. We are careful to take our five core samples that make up each data point from precisely the correct spot. This care ensures us that we have the most representative soil test possible and that we can replicate this same testing procedure when we retest in 4 years so that the results from year one and year four correlate.

Once the samples are taken and delivered to a University of Illinois cooperating lab, they are analyzed for pH, P, K, and organic matter. The only difference is that, in addition to the traditional information, the X and Y coordinates for each sample are attached at the soil lab. Once the samples are done they are sent into our computer center at Urbana via phone modem. Once the results reach Urbana they are put through nine different software programs. The results of this Phase 1 program are: 1. the soil test; 2. the hand drawn map showing where the samples were taken; 3. the three-dimensional Kriged map showing a side view of the variations in P test levels throughout the field; 4. the three-dimensional Kriged map showing an overhead view of the variations in P test levels throughout the field; 5. the three-dimensional Kriged map showing a side view of the variations in K test levels throughout the field; 6. the three-dimensional Kriged map showing an overhead view of the variations in K test levels throughout the field; 7. the three-dimensional Kriged map showing a side view of the variations in pH test levels throughout the field; 8. the three-dimensional Kriged map showing an overhead view of the variations in pH test levels throughout the field; 9. the digitized map showing the management zones to be spread; 10. the fertilizer recommendation sheet showing the different fertilizer rates to be applied on each of the management zones.

****

To get the VRT program off the ground and successful, we had to have the cooperation of several different groups. First was the marketing staff. Each of the members had to accept that this was the correct direction our company should be taking and to demonstrate that belief by committing a large percentage of our capital asset budget to make it all possible.

Another very key group that had to be sold on the concept was the Board of Directors. This is a board made up of farmers that are geographically representative of our company. The board at Illini FS is a forward thinking group that is not afraid to spend money in pursuit of a way to better the company. Our board is well educated and able to see the marketing direction that we have laid down with the VRT program. After the board was presented with the concept, its advantages, and the capital that would be necessary to make a go of it, they gave us a unanimous level of support for the program.

The next group to be sold on the concept was our own crops salesmen. The concept was not hard to sell to our people. Most of the concerns with the salesmen were operational in nature. Some of the questions that we faced were: What will VRT do to my tonnage? What will VRT do to my income? How will you service my plant? How much VRT can you do? What is the turn-around time for getting my results back? How can I charge that much for this program? How can I be sure that VRT is done correctly? These are just a few of the questions. One of the biggest advantages to the VRT program is that it forces the crops salesman to go out to the farmer and work up a fertilizer plan before season is upon us.

The final and perhaps the most important group to be sold on the concept of VRT are the farmers. Although most of the selling of the program is done by our crops salesmen with one-on-one farm interviews, we felt it was necessary to pull together a group of farmers for a focus group meeting to tell us what they liked and disliked about our program. The information that we gained from this frank discussion continues to help us shape VRT into something the farmers are asking for.

****

Obviously, the VRT program cannot survive without the economics in place. Going into the program we analyzed it in three different scenarios, a worst case, middle case, and best case scenario. After being into the program now for almost 3 years, I can say that we are about on track with the middle case scenario. We have not made a large amount of money with this program, but it has allowed us to be a market leader. Going into the program, we underestimated the amount of expense it would take to operate. We also underestimated the demand the VRT program would generate. We also overestimated the amount of business we would be doing as the mapping entity for the central part of the Midwest.

*****

With the demand for VRT better than we had hoped for, our need for additional qualified people and enhanced computerization became apparent. Up to this point we had been operating our VRT department with an assistant controller who liked to dabble in computers. His expertise was essential in the beginning but we had outgrown his level of expertise. It was at that time that we began to look for a programmer who had a farming background, and could be an effective communicator to our crops salesmen. Once that individual was located and hired we decided that we needed to install a LAN system to speed up the processing and offer us some enhanced operating efficiencies. This system became operational in January of 1992. Since the employment of our Manager of Information Systems, he has tied independent programs together, automated several mundane processes, and increased our mapping output threefold.

With the LAN in place we saw the need for an additional person to market the VRT concept to farm managers and perhaps market future mapping services to farmers. This person was also identified and employed further adding to our staff and our commitment to the VRT process.

*****

The next step for our company was to take the VRT program out of its pilot stage and open the program up to the entire company. With this goal in mind we devised some marketing programs and strategies to achieve 40,000 acres signed up across Illini FS in fiscal 1991-92. Other goals included educating the salesmen on some of the more technical points of the program. Getting the salesmen to sell the program a season ahead of when it was to be applied so as to smooth out the seasonality of our computer operations. Get the focus of our VRT program onto getting new business as opposed to using it on established business. Getting the VRT message communicated to all of the farmers in our trade territory, and just simply making the VRT selling process fun and exciting. To achieve these ends we came up with different marketing programs, targeted to address each of the areas discussed above. The VRT program now stands in the spring of 1992 with 70,000 acres under contract, of which 30% is new business.

****

Farm managers control about 30% of the acres in our trade territory. We have made a conscious effort to court their business and bring them along with the innovations in fertilizer technology that we have been marketing. Our past experiences tells us that farm managers very rarely get called on by the retail fertilizer plant salesman. When we introduced another product new to the marketplace (Soft Lime) in the early 1980s, we did not concentrate on the farm managers and it has subsequently been a challenge to get the farm managers to use the product. Not wanting the same fate to befall the VRT program, we hired an individual that spends a large percentage of her time explaining the VRT system. She will also put on grower/farm manager meetings and go over VRT results with the farm managers. It is our intent to expand our mapping entity to include record keeping and this individual will plan an important role as we move forward with this technology.

*****

Another key aspect of our VRT program is our advertising and promotion. From the onset we decided that we were going to take a proactive stance on getting the word out on what we were doing to be good stewards of the soil and groundwater. When we started with the VRT system there was very little promotional literature available on the subject. We went to work on an 11-min video that tells about the program, why we got into it, and where we are going with it. We were also contacted by several magazines and asked to do articles on the VRT process. These articles have looked at what we are doing from several different angles. Some of those are to spend money on fertility only where needed; precision application; grid soil sampling and its effectiveness; and the mapping of soil fertility levels with different types of soil testing methods. It is our feeling that important support came from the University of Illinois' handbook for 1991-92. In the handbook they stated that the farmer will receive an increase in yield and profit by using the site-specific fertilizer method.

*****

With acres being enrolled, we had to constantly take a look at our equipment needs and allocation to ensure that they were fairly distributed across our trade territory. After an excellent summer and fall selling season we realized that we had more application acres than two fertilizer trucks could handle. The decision was made to approach the Board for a third truck. It was voted on and accepted in January of 1992.

In monitoring acres we have found that we are changing the ratio of P to K in the recommendations that we are making. The amount of P and K recommended depends on the particular field and the fertilizer program that has been used in the past. Overall, we are putting on less P and more K. This is true for our fields in east central Illinois and may not apply elsewhere. Our average fertilizer recommended was 446 lb per acre. This broke out to be 176 lb per acre of phosphate and 270 lb per acre of potash. When this average analysis of 446 lb is compared to our typical recommendation of 36-92-120 (200 lb of phosphate and 200 lb of potash), we are actually putting on 11.6% more pounds per acre. However, the mix is shifting from phosphate to potassium which is making overall price per acre for the farmer lower. One must keep in mind when looking at averages that any particular field will be different than the average based on what has happened in the past with regard to nutrient application and withdrawal.

*****

One of the future directions that Illini FS will be taking is with the implementation of the global positioning system (GPS). This satellite tracking system will give us the ability to variably apply products that we are not able to run through our present machines. The first of these applications will be with our Soft Lime machines. We will be taking Kriged soil test maps of pH and putting them into the GAPS interface to be able to display the map of the field and its corresponding management zones for limestone applications over the top. With the GAPS system we will be able to tell when we change management zones for lime rates and manually change our rate on the go as we are applying lime. This will offer our farmers another service and round out the program to include another of the essential ingredients.

The GAPS system, when fully integrated with Soil Teq's system, will enable us to replace the dead reckoning system of navigation with more accuracy. This will allow us to more accurately spread contours, terraces, and point rows.

The GAPS system will also allow us to variably apply rates of sprayed herbicides in much the same manner that we will be applying the Soft Lime.

*****

Variable Rate Technology mapping is a seasonal business. In order to balance out the seasonality of this business and to spread the fixed costs of our enhanced computerization, we have gotten into what we call Ag Information mapping. This is where the farmer or farm manager provides us with the raw information he has about his farm and we then take this information and produce a map. This map can be used for communication with landowners, farm managers, fertilizer suppliers, floating drivers, seedsmen, hired help, etc. The information can also be down-loaded into a database and in the case of large acreage (farm managers) different reports on yield, inputs, etc. can be generated. This technology also puts us in the lead for keeping track of pesticide applications that will be needed to comply with the forthcoming regulations.

*****

The following are some limitations in the area of technology transfer from the research and development of the original equipment manufacturer to the distributor to the fertilizer dealer to the farmer:

1. There is very little actual yield research to prove the system works. All of the work that we can find has been done with theoretical yield response.
2. This program costs money. I believe that you can make money and pick up new business with the concept but those rewards come after a substantial outlay of capital.
3. Assets make things possible, but people make things happen. People will make or break your program, and that includes people at all levels.
4. OEM's need to be brought up to speed with the technology as much if not more than the dealers that are marketing it. Our experience has been that the two of us have learned at about the same pace.

5. You cannot wait in the office for the farmer to order his VRT. This is a program that must be sold and sold preferably a season ahead of application.
6. It takes a considerable amount of perseverance to get this program in place, marketed, and delivered. Do not get discouraged in the early stages of the program.

27 Computerized Recordkeeping for Variable Rate Technology

John S. Ahlrichs
CENEX/Land O'Lakes
Box 64089
St. Paul, MN

Variable rate technology (VRT) requires sophisticated, computerized recordkeeping and data management. Of the groups that could implement VRT, few have the computer systems and support staff in place to do so. Most farmers, dealers, and consultants are still keeping paper records. They have to enter and struggle with the computer age before they can implement VRT. This process will not happen overnight.

Today detailed crop production recordkeeping is a necessary part of running a profitable farming business. However, not everyone has adopted this attitude as reflected by occasional application errors and less than optimum profit. Rotation, application rate, and plant-back restrictions of the newer crop protection products are one example where poor recordkeeping can cause serious problems. Records also provide a good view of the past, a critical step when trying to improve production practices for the future.

With implementation of the 1990 farm bill, farmers not already keeping production records will be forced by law into keeping track of what, where, when, and how much for applications of restricted use pesticides. Over time that will evolve into cradle-to-grave tracking of pesticide applications. Current California recordkeeping laws provide a case study. Specifically, growers must get authorization from the County Ag Commissioner to apply most pesticides. This approval must be given to their dealer before the farmer can purchase the products. The dealer must report all pesticide sales to government agencies. Farmers must also report each pesticide application. The numbers must match. The state of Washington currently is more aggressive than the 1990 Farm Bill by requiring that farmers keep records of restricted use pesticide applications for 7 years. The volume and complexity of this type of recordkeeping requires a computer-based solution. However, few of the affected parties know how to use computers and fewer have access to or have purchased software to help manage these records.

Copyright © 1993 ASA-CSSA-SSSA, 677 South Segoe Road, Madison, WI 53711, USA. *Soil Specific Crop Management.*

Keeping records of plant food applications is no different. Nebraska's Central Platte (river) Natural Resource District contains 200,000 ha dominated by irrigated corn (*Zea mays* L.), where the timing and rate of plant food applications are tightly controlled to minimize groundwater nitrate levels (Schepers et al., 1991). These laws require detailed records to monitor plant food application timing and rates to avoid civil fines. Similarly, the Clean Water Act currently being negotiated in Congress will probably require recordkeeping and input planning by farmers to prove that they are totally tracking and managing their production inputs to reduce groundwater contamination. Again, a computer-based solution is a virtual necessity to manage the data.

Mandated recordkeeping is changing the way agriculture does business. This is not all bad. These laws and regulations on managing production inputs, along with financial and political pressures, help pave the way for the next generation of crop production management-variable rate applications. Variable rate technology requires a sophisticated integration of computerized equipment, software, databases, and decision aids. Today, few groups have the computer experience or personnel to really implement this complex technology. I believe most of the development in this area will occur at the dealer or crop consultant level. Individual farmers do not have the time or capital required to learn, implement, and support all of the technology required for a fully usable information management system. There are a few exceptions, such as large or corporate farmers, farm managers, etc.

General surveys of Cenex/Land O'Lakes Cooperatives indicate that less than 1% of their farmers faithfully use computer software for agronomic recordkeeping or as an agronomic decision aid. A larger number use computers as a financial analysis tool. Farmers are not allocating the time for managing computer-based production records and this trend does not seem to be changing. In addition, the complexity of modern laws and products is forcing farmers to turn to their dealers, consultants, and Cooperative Extension Service for agronomic knowledge and recommendations. This creates a great opportunity to offer production recordkeeping/management services. Dealers and consultants will probably be the most aggressive developers of these services and users of agronomic information systems because of their vested interest in the farmer. They want to offer unique services that will keep the farmer from purchasing products/services elsewhere.

Currently, few dealers have fully implemented computer-based management of production records. They are, however, starting the process. Consultants have a slightly greater degree of computerization. Before dealers and consultants successfully implement computer-based management systems, the following obstacles need to be overcome:

1. They need to make a purchase decision. Good software is not cheap.

2. They need to learn (by trial and error) how computer-based information systems fit into their business.

3. They need to allocate human resources to ensure success.

4. They need to figure out how to charge for information services.

5. They need to develop a certain degree of comfort and proficiency with computer systems.

6. They need to make many mistakes in order to learn what not to do.

This is a long learning process. As an example, Cenex/Land O'Lakes began implementing AgriSource, the Information System for Crop Technology™ in June 1988. This was one of the first comprehensive, integrated information systems designed specifically for the agronomic decision maker (primarily agronomy department managers and their sales and support staff). Of these direct users of AgriSource, 70% had no previous computer experience. Many were not even keeping grower records. Each account using AgriSource worked through all of the above process. The average time period to integrate the AgriSource Agronomic Management System into a dealer has been about 2-1/2 years with free regional and on-site training, in addition to telephone support and marketing assistance.

The first success for a dealer implementing an information system often includes tissue and soil test results being transmitted from a lab directly into electronic grower records (Fig. 27-1). Computerized farm planning that automatically tracks planned and actual applications for the whole field is the next step. This often involves integration with point of sale software that ties to their accounting system. The new business and inventory management tools provided by a computerized point of sale system often temporarily dilutes the focus toward agronomic information management. After dealers automate their recordkeeping system and have farmer participation (both physically and financially) in the process, they are positioned to fully use the decision aids, expert systems, and technology that would all be included under a heading of an Information Management System (Fig. 27-2). By this point, dealers are starting to use the computer systems to optimize the management of a whole field. Many AgriSource accounts are finally at this level today.

Dealers have to be near this comfort level before they are ready to move to the next step: managing variable rate information. This is supported by the observation that we are just beginning to get questions about using computers for mapping and VRT. These questions are starting to come from accounts that have realized that VRT is not a black box with many unsolvable problems, rather it is a logical addition to an existing information system. They also recognized that VRT will require even more changes in decision-making processes, procedures, software, and recordkeeping to manage and track production inputs.

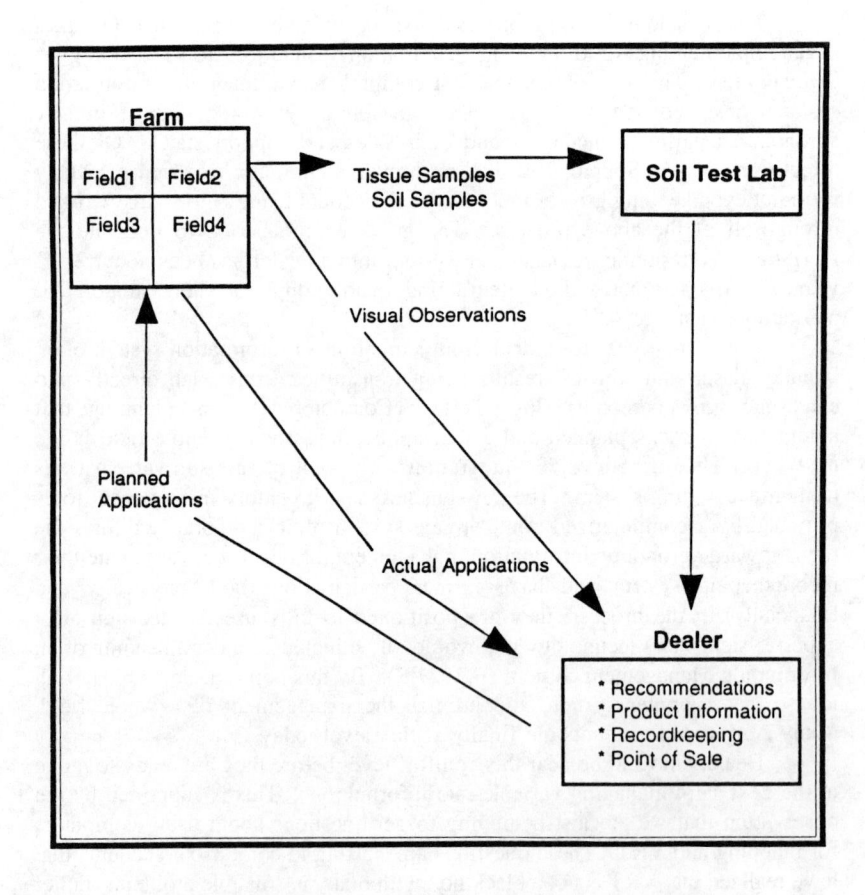

Fig. 27-1. Information flow for dealers just starting to use agronomic information management systems.

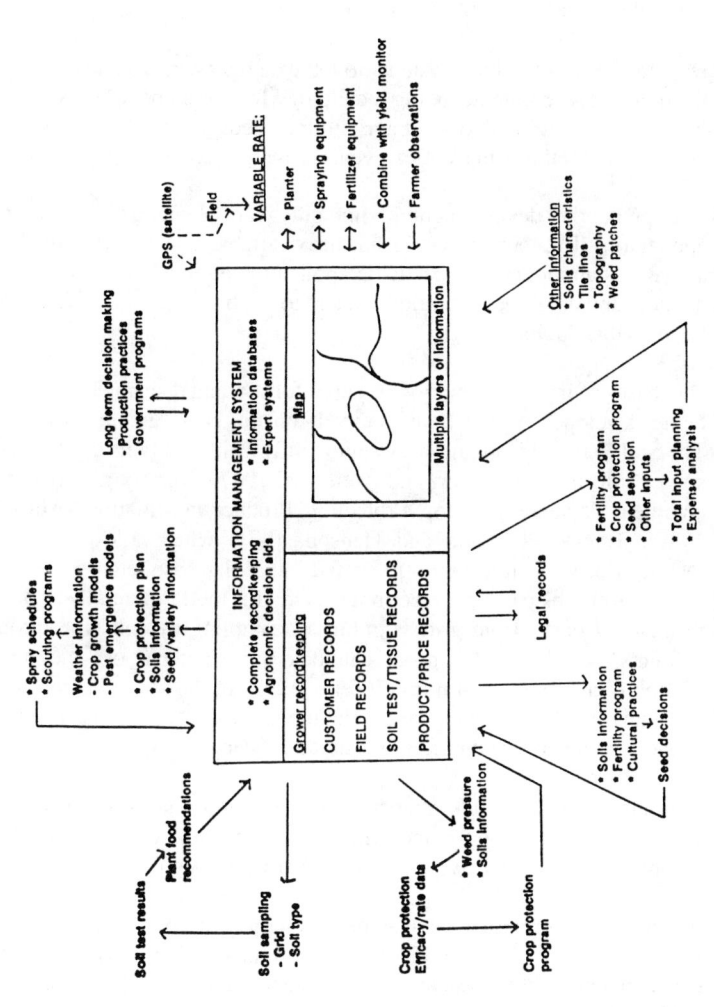

Fig. 27-2. Information flow for advanced agronomic information systems.

Six interdependent factors are converging to help that occur:

1. Laws requiring "Best Management Practices" (BMP) for all inputs are being developed.

2. Development of tracking systems to tell exactly where you are in a field. Inexpensive, accurate decoding technology for global positioning system (GPS) is being refined to meet agricultural needs (Mangold, 1992). Other positioning methods may offer even cheaper solutions (Palmer, 1991).

3. Entrepreneurial development of not only variable rate dry plant food applicators (Brunoehler, 1991) but also variable rate liquid plant food, anhydrous ammonia, and pesticide application equipment. Variable rate planters and combines with flow meters to continuously track yields may be available soon.

4. Databases of input information required for running decision software are being developed or can be developed "on the go". Examples include: The Soil Survey Information System (SSIS) which contains digitized soil survey information (Robert and Anderson, 1987), AgriSource contains a complete database of Crop Protection Product information (Ahlrichs, 1989), Track Net (Tofte and Hanson, 1991) which allows the field locations of weed infestation to be fed into a database, etc.

    Soil labs can now electronically return the large numbers of test results that occur from grid sampling or sampling by soil type through modems. In-the-field sensors can determine percent organic matter (Shonk et al., 1991) and nitrate (Zenk, 1992) in the soil and store them in databases, thus allowing pesticide application rates to vary according to the parameters which affect pesticide efficacy.

5. Computer decision aids that manage this information and tell the equipment what to apply and where, are being refined. This includes mapping software like AgMapp from AgriCad (Degnan, 1991).

6. Agronomic information management systems are now being developed to integrate all of these pieces of crop production information and decision aids. These systems will be able to track the planned and actual inputs, field observations, and decisions of farmers and consultants.

Each of the six factors have their own obstacles to "success." When they are integrated, the result will be a new ability to manage a field as a set of continuously changing variables instead of applying a single solution to the whole field.

With others focused on the positioning, application, and information collection technology, the biggest obstacle to full implementation of managing the decisions and recordkeeping for VRT becomes how to completely integrate all of these processes into a paperless flow of data that can be used by the agronomic decision makers (farmers, the applicator, and agronomists).

Cenex/Land O'Lakes is addressing this problem through AgriSource, the Information System for Crop Technology ™. This communication/information system is a library of integrated databases and decision aids located at the farm cooperative and designed for a dealer and farmer to use together.

The AgriSource system routinely collects information from outside sources (soil test results from soil labs, pesticide product use information from manufacturers, marketing/pricing information from suppliers, and product efficacy data from researchers) and in the near future will pass information to and collect actual application rate and location information from variable rate application equipment. All of this information is linked together and stored in databases for use by computer decision aids and people engaged in farm planning, making input recommendations, and keeping application records.

The movement to VRT has been and will continue to be gradual. Having a solid base of computer users that understand the merits of value added services is required to accelerate the process. Each of the AgriSource accounts that implements VRT will already be managing whole field information that is independent of VRT (Fig. 27-2). Examples include inventory items (crop protection and plant food product/price records) fertility recommendation equations, farm planning programs, base field information (location, crop, etc.) and databases detailing crop protection products use restrictions, crop protection product application rate equations, etc. Database search and reporting programs could also be included in this category. Variable rate technology will be a logical enhancement to dealers already managing whole fields and will be achieved by adding mapping software that allows planned and actual variable rate information to be visually displayed and managed.

Complete integration of these traditional decision aids with mapping software should be a straight-forward task that will permit full management of variable rate information. The following examples illustrate the fact that there is no one software solution and that all pieces need to integrate.

CROP PROTECTION PRODUCT EXAMPLE

Areas of weed infestation in a field are delineated in a mapping program for small area (spot) spraying. The computer adds up the area and passes this information to pesticide decision making software. The computer recommends crop protection products and application rates (based on pH, organic matter, cation exchange capacity, growth stage, infestation, etc.) to be approved by the agronomist and farmer. The computer generates the mix sheets for the application and the state or federally required application records. The product is invoiced and the DOT shipping document is generated to make product delivery legal. The location, product, and rate information is electronically transferred to the variable rate application equipment. The application equipment tracks (with the aid of positioning software) the actual locations and rates of pesticide application. This actual application information is saved back in the computer in the mapping software for later decision making. The application records are all linked to inventory and sales records for complete reporting.

This process allows the state regulators and the dealer to track what was sold and to whom and the dealer, farmer, and chemical rep (in case there are complaints) to track product application information: what, when, how much, and where. Without automatic, precise recording of pesticide application information, variable rate pesticide recordkeeping will produce a legal, regulatory, and management nightmare.

PLANT FOOD EXAMPLE

The dealer grid samples a field or samples by soil type a field noting (or encoding from a positioning system) the exact coordinates of each sample. The samples are sent to a lab and as the analysis of the samples is completed, they are electronically transmitted to the dealer. These results can be displayed in the mapping program and are stored in the field records. They can also be exported to Kriging/Surfing software to determine the different fertility management zones. Fertility recommendations are automatically determined for each zone and can be displayed by zone in the mapping program. The total amount of plant food for the field is calculated, priced, and invoiced. The application information is loaded into the variable rate spreader and the plant food is accordingly applied. Variable rate chemical impregnation can be added to this scenario. The actual application rates are saved back to the mapping program, which then links all products with the inventory and sales records for complete tracking and reporting of product use.

Similar concepts are being applied to determine variable seeding rates for planters and continuously recording harvest yield. The result will be a completely managed crop input/output system.

Recordkeeping as discussed above is of value to meet regulatory requirements and track what was applied. It will be forced as VRT is implemented. However, the real power of information management for farmers will come from the ability to modify future production practices and optimize future inputs based upon past experiences. In the first scenario, there might be a need to expand the area of weed control or move to a different pesticide program to control the specific weed mix in a field or portion of a field. In the second scenario, tissue testing and crop monitoring may show that plant nutrients were over- or under-supplied in one area. Some areas of the field may have nutrient/disease/soil problems that need to be managed. Computer-based records allow projection of the effect of changes in inputs or practices on profitability and production.

Unfortunately, this type of computer technology is far beyond the comprehension of most of our customers. We have not brought them with us as we entered the computer age. The best of them are just discovering the advantage in automated basic recordkeeping. Until computers and computerized recordkeeping systems are an integral part of a dealers agronomic business, we will not be able to implement our "obvious" technological solutions.

The challenge for production agriculture is profitably managing each part of each field in an environmentally sound manner. Variable rate technology will be part of that. The lag time for implementing VRT will continue to be beyond expectations because our customers are still struggling with the technology to enter the information age. However, the long term savings in time, expenses, and input optimization that will come with VRT and the decision tools we develop for it will be well worth the effort of educating our customers.

REFERENCES

Ahlrichs, J. S. 1989. AgriSource: The information system for crop technology. p. 353-355. In Freshwater foundation (ed.) Agrichemicals and ground water protection: Resources and strategies for state and local management. Conference, St. Paul, MN. 24-25 Oct. 1988. Freshwater Foundation, Navarre, MN.

Brunoehler, R. 1991. Grid fertilizing rig finally hits pay dirt. Farm Ind. News 24 (12): 22-23.

Degnan, J. 1991. Field mapping by satellite. Farm Ind. News 24 (11): 42-44.

Mangold, G. 1992. Satellite power raises new questions. Soybean Digest (Jan.) 1992: 34-36.

Palmer, R. J. 1991. Progress report of a local positioning system. p. 403-408. In Automated Agriculture for the 21st Century. Proc. of the 1991 Symp. Chicago, IL. 16-17 Dec. 1991. Am. Soc. Agric. Eng., St. Joseph, MI.

Robert, P. C., and J. H. Anderson. 1987. A convenient soil survey information system (SSIS). Appl. Agric. Res. 2 (4):248-251.

Schepers, J. S., M. G. Moravek, and R. Bishop. 1991. p. 641-647. In W. F. Ritter (ed.) Irrigation and Drainage Proc. 1991 National Conference. Honolulu, HI. 22-26 July 1991. Am. Soc. Civil. Eng., New York.

Shonk, J. L., L. D. Gaultney, D. G. Schulze, and G. E. Van Scoyoc. 1991. Spectroscopic sensing of soil organic matter content. Trans. Am. Soc. Agric. Eng. 34:1978-1984.

Tofte, D., and L. Hanson. 1991. Networking monitors, servos and memory for manual and automatic machinery control. p. 409-417. In Automated Agriculture for the 21st Century. Proc. of the 1991 Symposium. Chicago, IL, 16-17 Dec. 1991. Am. Soc. Agric. Eng., St. Joseph, MI.

Zenk, P. 1992. Soil doctor has nitrogen fix. Farm Ind. News. 25(2): 32-33.

28 Working Group Report

D. Buchholz, Chair
D. Leikam, Recorder
G. Miller, Discussion Paper

Other participants:	T. Ables	J. Mann
	J. Ahlrichs	A. McQuinn
	J. Anderson	P. Novak
	R. Brady	L. Reichenberger
	T. Hall	B. Schmidt
	W. Holmes	I. Sether
	P. Hurtis	D. Wesley
	T. Macy	

Successful transfer of technology is dependent upon the acquisition of knowledge by the intended recipient. Knowledge can be described as one's understanding by actual experience. One's knowledge is acquired through access to information and participation in educational activities. The acquisition of knowledge is demonstrated by a change in the recipients' actions and activities.

PERSPECTIVE

During this century, some technology available to agricultural producers has gained near universal on-farm implementation within a very short period of time following initial availability. Examples of rapid adoption of new technology include: (i) hybrid seed corn in the 1930s and 1940s; (ii) commercial fertilizers in the late 1940s and 1950s; (iii) pesticides in the 1950s and 1960s; and (iv) large equipment in the late 1960s and early 1970s. However, other technologies have not gained universal adoption. For example, estimates are that less than 15% of the agricultural producers located in the upper Midwest actually own or use home computers for their business. The change to no-tillage systems has also been resisted by producers. In fact, statewide surveys conducted in Iowa suggest that for every three farmers that experiment with no tillage, two reject the practice (Padgitt, 1989). Integrated Pest Management (IPM) is another strategy that has failed to achieve widespread adoption. A statewide survey of Iowa farmers revealed that 16% subscribe to a commercial crop scouting service and another 42% always systematically scout their own fields. One in six (17%) walk their fields with sufficient regularity to meet the standards of adequate scouting (Padgitt et al., 1990).

Copyright © 1993 ASA-CSSA-SSSA, 677 South Segoe Road, Madison, WI 53711, USA. *Soil Specific Crop Management.*

Why do some practices become rapidly adopted and other practices require longer incubation and evolution? At the first glance one may conclude that increased productivity, reduced labor, and increased profitability or perceived profitability may be the overriding factors. Cases in point include hybrid seed corn and commercial fertilizers. On the other hand, the complexity of a practice and the need for increased levels of management skills can be an impediment for rapid adoption of a new technology. Case in point for these conditions may include home computers, no-tillage systems, and IPM. These technologies are potentially profitable for agriculture producers, but require the acquisition of new technical and management skills. A similar case can by argued for soil specific crop management (SSCM).

A TECHNOLOGY TRANSFER MODEL FOR TARGETED PROJECTS

An analysis of Iowa informational activities for agricultural producers resulted in the development of a four-phase implementation model. This model focused on a direct marketing approach for target clientele. Its intent is to accelerate the adoption of new and refined technologies in all areas of farm production.

Phase I consists of inventory, assessment, and evaluation. This phase provides for the establishment of baseline data concerning clients' farming practices, physical inventory, attitudes, perceptions, and suggested need for change. Data collection for this phase is accomplished by developing project specific survey instruments. Survey instruments may be delivered to the target audience by mail, face-to-face interviews, small group interviews or a combination of these methods. In addition to the target farmer cooperator audience, a similar survey instrument was delivered to neighbors and other farmers on a regional or statewide basis. Evaluation of the baseline data information collected in these initial surveys allows for the development of specific marketing activities and informational programs.

Similar follow-up survey instruments were delivered to the same targeted project cooperators, neighbors, and other farmers within the region or state midway through the project (for 5-year projects) and at the conclusion of the project (for 3- and 5-year projects).

Phase I results provide for the documentation of change by project cooperators, neighbors, and other farmers. Because of variability in weather and buffering in natural systems, environmental effects of changed farming practices may be difficult to document by biological, chemical, and physical measurements of the targeted soil and water resource over the typical 3- to 5-year project. Phase I methodology provides for an indirect measurement of change by documenting the change in clientele's practices, attitudes, and perceptions.

Phase II involves localized demonstration of existing technology. This phase provides "hands-on" application of practices with cooperating farmers. Some demonstrations can be designed with multiple replicated treatments to include a treatment that mirrors the cooperator's current management practice. Replicated treatments allow for data collection and statistical treatment of the data. In this phase, technology is transferred by the process of observing, participating, evaluating results, and processing information. Additionally, neighbors are exposed to the technology.

Phase III consists of information marketing. This phase involves multiple communication strategies that ensure that project cooperators, neighbors, and other farmers on a regional or statewide scale are consistently aware and informed of the technologies being transferred, project activities, and results. Success stories featuring cooperators published in project newsletters and news releases to local media generate interest in the demonstrated technologies among neighbors and other farmers.

Phase IV involves networking among project cooperators and neighbors and sharing of hands-on experience about the demonstrated technologies. When cooperators learn from each other what works and what does not work, the learning process is supported and accelerated. Networking activities can be accomplished through cooperator tours, field days, informational meetings, conferences, workshops, and annual meetings to review project results and social events, as examples. Networking among cooperators who are targeted for technology transfer and who in turn will potentially implement change provides a vehicle to overcome the barriers to effective use of information (McLaughlin and McLaughlin, 1989).

DEFINING SOIL SPECIFIC CROP MANAGEMENT TECHNOLOGY

Soil specific crop management technology consists of managing appropriate levels of vertical technology to identifiable sampling points on the landscape. Vertical technology is defined as the introduced inputs layered on the natural system by a manager.

Higher	Global positioning systems "On-the-go" mapping Field sensors	Higher
Specialization	Variable rate technology Automated mapping Automated in-field controllers	Technology
	More intensive data collection Computer Automation Geographic information (GIS)	
Lower	Soil maps Map unit sampling Sketches Non-automated implementation	Lower

Soil specific crop management technology may consist of varying levels of specialization and sophistication applied across varying levels of sampling points. Conceptually, some existing SSCM technology can be categorized as shown previously.

In this example, the user defines the level of specialization or sophistication as well as the level of sampling intensity or data management best suited to all factors affecting the individuals capabilities and wishes. As a result, there is no single SSCM technology that is either adopted or not adopted. Instead, adoption consists of identifying SSCM technology specific to the needs and capabilities of the individual producer.

The vertical technology used to assist the manager in decisions of SSCM will be driven by the availability of site specific information for the creation of necessary input layers. For example, to vary fertilizer inputs, soil test levels and crop yield estimates are generally accepted database requirements. For herbicides, weed populations, soil texture, soil pH, soil organic matter, crop rotation, and cropping intentions may all be necessary to employ the SSCM system. Other uses of SSCM strategies will require other site-specific information or databases for effective implementation.

The vertical technology will also be driven by the availability of appropriate engineered sensors, controllers, and positioning devices used on fertilizer and agricultural chemical applicators, as well as tractors, combines, and other various farm implements. The interface of all these devices is likened to the need for a universal power take-off or hitch for rapid, wide-spread, potential adoption.

THE TARGET AUDIENCE AND ULTIMATE USER

An important issue is identifying the targeted user. The technological complexity of some SSCM systems will limit the ultimate users to those capable of putting together the necessary team of human resources and capital required to adopt the specialization. As we attempt to define the ultimate user of SSCM technologies, in many respects we are attempting to define the commercial use of these technologies. The ultimate user of SSCM is the crop producer. However, current SSCM technologies will likely come through retail agricultural supply dealers or crop consultants. Current and future SSCM technologies would define target users to be retail agricultural input supply dealers, crop consultants, commercial farmers, and University/public agency personnel.

The retail agricultural input supply dealer and crop consultant are clearly critical to the adoption of SSCM technologies as known today. Most recognized SSCM systems involve adjustment of fertilizer and herbicide inputs. A majority of these inputs are applied by the dealer. Development of the database dictating the variable application of these inputs may often come through the retail dealer and crop consultant. Therefore, strategies for education of these two targeted groups will be discussed in a later section of this chapter.

A select group of commercial farmers will adapt certain SSCM technologies on their own. These farmers will likely come forward through some self-selection process. However, as a specific targeted group, selection will be focused to the innovators in this early stage of development. Obviously, the types of engineered controllers and the complexity in the development of the information or database to drive the controller will dictate the breadth of adoption of SSCM technologies at the farm level.

As a primary support base to the agricultural industry, the public sector specialists and agency personnel must be aware and informed to the availability and acquisition of appropriate databases and technology required in successful adoption of SSCM technologies.

ADOPTION CONSTRAINTS/OBSTACLES FOR SSCM TECHNOLOGY

While SSCM makes intuitive sense to many farmers, agribusiness people and university personnel, the technology will be questioned by others. As is true for the transfer of any new technology to the end user, there are factors that will limit the rate of adoption of SSCM by agriculture.

The speed of adoption of new technology is related to the ease/cost of adaptation, the required level of specific knowledge and management skills, the perceived risk/benefit of the technology and the ability to quantitatively/visually identify the effects of adoption. While the perceived benefit of SSCM is high for some, it is perceived as being high risk by others. And depending on the level of SSCM being considered, the cost of adoption is often perceived as being relatively high and the adoption process relatively complex. Finally, because of the nature of SSCM, the benefits of this technology will likely not be highly visually apparent and will be difficult to quantify in terms of profitability, yields or environmental impact.

Because of these characteristics, there will be many obstacles to overcome before SSCM technology is widely adopted. Presented below are a few of the most often mentioned obstacles to SSCM adoption. They are categorized for the different groups that will be involved in the transfer of this technology to agriculture. While several of these objections are common to farmers, retail dealers, consultants, and university personnel, the relative importance may be different. Other obstacles are confined to a single group.

<u>Farmer</u>
Tradition
Perceived cost
Skepticism
Complexity
Real cost
Increased management skills

<u>Dealer</u>
Tradition
Perceived cost
Complexity
Higher level of management
Cost (capital and resources)

University	Consultant
Tradition	Tradition
Perceived cost	Perceived cost/viability
Data/research needs	Profitability
Benefit/skepticism	Proprietary information
Complexity	Loss of market

Obstacles Common To All Groups

Possibly the most universal obstacle to overcome is the change in mindset that will be required. Many will insist that the potential benefits of SSCM need to be documented with research relative to the profitability, yield, and environmental benefits of the technology. It is important that SSCM be recognized as simply an extension of what we have previously identified as being important for best management strategies. Evaluating SSCM from a farmer viewpoint rather than scientist viewpoint only will be needed. The mindset needed is that the only difference between conventional crop management and SSCM is that we are now delineating smaller "fields" within a field defined by roads and fence rows.

Some have the mindset that SSCM is an all-encompassing strategy consisting of grid sampling, VRT equipment, GPS, etc. However, it must be kept in mind that there are many SSCM strategies of varying degrees of sophistication and complexity.

Another mindset requiring change is the perceived roles of all of the players involved in production agriculture. Farmers, dealers, consultants, and university personnel will all need to accept the role of researcher, educator, and student. As the different components of SSCM are conceptually advanced, developed, evaluated and refined, all will have experiences and knowledge to pass to one another.

Movement away from traditional, entrenched ideas/practices is difficult and will initially be a major obstacle to SSCM adoption. Farmers, however, may well be the group that is the quickest to move past this obstacle since the concept of SSCM makes sense to them. They see spatial variability every time they travel over their fields and recognize the potential for SSCM management.

The initial perceived cost of the technology is another obstacle to overcome. Because of the large amounts of data inherent to all SSCM systems and the use of sophisticated equipment and computer technology for others, it is naturally assumed that these systems will be prohibitively expensive. However, when the costs associated with various SSCM strategies are actually determined the real cost is generally less than initially perceived. These costs will almost certainly decline as competition increases.

Other obstacles exist for the various groups previously identified.

Farmer

Even though farmers recognize that variability exists in their fields and it makes sense that portions of the field need to be managed differently, they are still skeptical that the technology will really do what it is supposed to do. After they are convinced that the equipment truly works, they will likely have doubts about being able to manage the required data. Finally, the real cost of the technology can be a stumbling block. Although the real costs may be much less than initially perceived, cash flow and landlord/tenant situations may limit their involvement. Regardless of these obstacles, farmers are generally open to SSCM technology. To them, it just makes sense.

Retail Agricultural Input Supply Dealers

Dealers are currently the most important group to work with on SSCM technologies since they generally have the largest capital investment to make and are reluctant to recommend practices to their clients that are not profitable. Additionally, the complexities associated with managing data and the challenges associated with actually making the applications will fall most heavily on them. Equally important is the higher level of management needed to pull the whole system together and the human resources needed to "make it all happen." This is probably the group that will need the greatest amount of attention and "hand holding" to successfully transfer SSCM technology to production agriculture. By the same token, the obstacles associated with this group are the hardest to address. Some dealers are enthused by this technology and will make it happen while others will let it fail.

University/Public Agency Personnel

While many in this group are obviously enthused about SSCM technology and the promises that it offers, others feel that this technology has not been properly and fully evaluated with research. As a result, questions about the benefits to the farmer will be raised. Also, there is often a feeling of skepticism about what they view as a very complex system.

Agricultural Industry Personnel

Obstacles to overcome with this group will depend on whether they are in the input side of the business or are associated with technology introduction. For both groups, there will be questions not only about farmer profitability but business profitability as well. Additionally, those involved with the technology side of SSCM systems will be sensitive to balancing the need for maintaining a proprietary position against overall market expansion. Those associated with the input supply products may be concerned about a loss of market (lower sales).

All of these, and other obstacles, can and will be overcome. The only thing that is in doubt is the time frame. If cooperation among all of the involved groups can be established, the obstacles of mindset, tradition, skepticism, and perceived cost can be easily handled. Making the software and electronic interfaces as transparent to the user as possible will help minimize obstacles relating to complexity of SSCM technologies. And finally, support and educational efforts involving all of these groups will help overcome the obstacles relating to investments in management, money, and human resources.

METHODS FOR ADOPTION OF SSCM TECHNOLOGIES

Successful adoption of SSCM technology requires another look at the technology transfer model described earlier in the chapter. Soil specific crop management technologies are currently practiced by few of the identified target users. Therefore, an inventory, assessment, and evaluation of this target user group may not be necessary.

A first step in adoption of SSCM technologies will likely come through an awareness phase. An example of creating awareness is by conducting selected meetings targeted to the users. In this case, informational meetings at state agricultural industry association conventions along with written information can most effectively reach that target user group. Information through farm and ag industry trade magazines has also been effective in creating an awareness throughout the agricultural community. Making SSCM technologies an integral part of college classroom curricula and extension education will further build an awareness and understanding of the SSCM goals. However, with the complexities of most SSCM strategies, a certain level of acquired knowledge will be necessary to implement change. One can expect to make an impact on potential target users' awareness. However, impact must be turned into a teachable moment to create the desired change and adoption of SSCM.

Soil specific crop management technology adoption will occur most successfully through workshops that demonstrate the needed information, databases, and equipment with the innovative crop advisors and dealers. These workshops can most effectively be carried out through team cooperation. The complexity will dictate the need for industry and academic community involvement in the planning and conducting of these workshops. Only through cooperation can it be expected that appropriate databases, interpretations, and strategies develop. Interwoven into those developments will be the appropriate engineered equipment capable of simplifying that SSCM strategy.

It is our opinion that a considerable amount of "hand holding" needs to take place for successful adoption of this emerging SSCM technology. The important factor may well be in the appropriate selection of the retail dealer and crop consultant as outlined in a previous section. Workshops should be designed to allow the potential adopter to become comfortable with the technology. As stated in the technology transfer model, the potential to acquire knowledge through practical experience is essential.

The ultimate advantage to successful selection of a potential adopter is in the later networking process that will occur. This individual or company will be sought for experience and information. Also, sharing of experiences must occur among the select users and developers of this emerging technology. Presently, a patchy network of SSCM technology adopters is beginning to form.

SUMMARY

Change in agriculture has generally been promoted by documenting the advantages of one or more technologies in terms of increased productivity, reduced labor, and profitability. This model has been successful in transferring technologies that are straight-forward, do not necessarily require enhanced management skills, or require change in multiple practices simultaneously. Some technologies requiring enhanced management skills or change in a series of practices have experienced a slower rate of adoption, and in some instances resistance to adoption. Soil specific crop management technology fits directly into the category of requiring new or enhanced management skills for adoption.

REFERENCES

McLaughlin, Gerald W., and Josetta S. McLaughlin. 1989. Barriers to information use: The organizational context. p. 21-33. In P. T. Ewell (ed.) Enhancing information use in decision making. New Directions for Institutional Research 64. San Francisco.

Padgitt, Steven C. 1989. Farm practices and attitudes toward groundwater policies: A statewide survey. Iowa State Univ. Ext., IFM 3.

Padgitt, Steven C., Sylvia Watkins, and Dean Grundman. 1990. Assessing extension opportunities for integrated pest management. Iowa State Univ. Ext., IFM 13.

SECTION VII

POSTER SUMMARIES

29 Measuring Yield On-The-Go: The Minnesota Experience

K. Ault
R. H. Dowdy
USDA Agricultural Research Service
University of Minnesota
St. Paul, MN

J. A. Lamb
J. L. Anderson
Department of Soil Science
University of Minnesota
St. Paul, MN

Applying agricultural inputs by soil condition has become an important concept to improve economic and reduce environmental effects. To evaluate this practice researchers need to be able to measure crop yields on-the-go. The objective of this study was to adapt a conical weighing device to a combine to measure corn grain yields continuously in a large research plot (1.8 ha) area. Three transects 200 m long were measured. In one transect, corn was hand harvested from a 1 by 6 m area every 15 m. These grain yields were compared to combined yield from the adjacent transect. In the third transect, corn grain yield was measured with the combine and then data was collected in a second pass down the transect with the combine's thresher engaged but with no grain harvest by the combine. This last measurement was done to estimate the noise in the yield measuring system. Preliminary results are: (i) There was agreement with hand-harvested grain yields and combine yields; (ii) There was considerable noise in the measuring system. Some of this noise was eliminated with a data smoothing function; and (iii) There was lag time between the recording of the grain yield and when it was actually harvested. Also the grain spread out in the combine so the best resolution was 12 m. The lag time and spreading of the grain flow causes difficulty in identifying the location of the grain yield in the plot. More information on lag times and how to reduce noise is needed before this system can be reliable for evaluation grain yields.

Copyright © 1993 ASA-CSSA-SSSA, 677 South Segoe Road, Madison, WI 53711, USA. *Soil Specific Crop Management.*

30 Multi-ISFET Sensors for Soil Nitrate Analysis

Stuart J. Birrell
Agricultural Engineering Department
University of Illinois
Urbana, IL

John W. Hummel
USDA - Agricultural Research Service
University of Illinois
Urbana, IL

The management of a field as a heterogeneous area is now being investigated by many researchers in universities, companies, and research institutions. However, for site-specific application of inputs, it is necessary to develop systems to sense the variation of important properties in real time. Several studies have shown that ion selective electrodes can be used to measure soil nitrates. This work involves an investigation in the use of Ion Selective Field Effect Transistors (ISFETs), to measure soil nitrate levels.

ISFETs have several advantages over ion selective electrodes such as small dimensions, low output impedance, high signal/noise ratio, fast response and the ability to integrate several sensors on a single chip. However, ISFETs have the disadvantage of greater long-term drift and hysteresis than ion selective electrodes. Although these are a potential problem in static measurements, the use of a dynamic measurement system such as Flow Injection Analysis minimizes the effects of drift and hysteresis, and exploits the specific properties of ISFETs. The ability to use small sample volumes and sense multiple species simultaneously makes the ISFET an attractive sensor for the development of a real-time soil nutrient sensing system.

An ISFET with four integrated sensors was tested in a Flow Injection System using four different flow rates (0.04, 0.09, 0.14, 0.19 mL/s), five sample injection times (0.25, 0.5, 0.75, 1, 2 s) and three washout times (0.75, 1, 2 s). The correlation coefficients of a linear regression of the peak height against the logarithm (base 10) of the nitrate concentration of standard solutions were within the range 0.89 to 0.99, except for the lowest flow rate that was in the range 0.72 to 0.99. A cycle period of 1.5 s (0.5 injection, 1.0 washout time) appears to be possible. The major problem encountered was inconsistent opening and closing of the injection valve, and an improvement should increase the precision of the system.

Copyright © 1993 ASA-CSSA-SSSA, 677 South Segoe Road, Madison, WI 53711, USA. *Soil Specific Crop Management.*

31 Yield Variability in Central Iowa

T. S. Colvin
USDA - Agricultural Research Service
Ames, IA

In 1989, workers at the National Soil Tilth Laboratory began to keep records on the yields of small plots within a 160-acre area in Boone County near Ames, IA.

The objective was to determine if there is sufficient variability to warrant the management of small areas within fields differentially. This would be in contrast to managing the field (6 to 500 acre) as a single unit. In theory, this is like managing single animals within herds rather than the herd as a uniform average.

Yields were measured only for corn (*Zea mays* L.) in 1989 on the two south fields. Throughout the 3 years, corn, soybean [*Glycine max* (L.) Merr.], oat (*Avena sativa* L.), and hay yields have been measured depending on the rotation. The two west fields are in a corn-soybean rotation with the two east fields in a corn-soybean-corn-oat-hay rotation.

The yields for corn were obtained with a small commercial combine modified for plot work that had a scale and moisture meter above the normal grain tank. The corn plots were two to three rows wide by 40 ft long. The plots were end to end down eight transects on each 40 acre field. The direction of the transects was determined by the row direction.

RESULTS

For all crops on all four fields the range of yield from highest to lowest plot was at least 2-1 and often 4-1 for each crop or cutting (hay). Grouping the plots by soil type has not been worthwhile because there is as much variability between soil types as shown on the county soil survey as there is within soil types. This is a similar result to one obtained in South Carolina using a much more detailed soil map.

In the Southwest Field, the yield of line 5 has been compared between 1989 and 1991. In 1989, the yields varied from about 115 to 160 bu/acre. In

---

Copyright © 1993 ASA-CSSA-SSSA, 677 South Segoe Road, Madison, WI 53711, USA. *Soil Specific Crop Management.*

1991, the yields varied from about 10 to 175. The lowest yields in the 2 years were not in the same position and neither were the high yields. The primary difference between the years was caused by the excessive wetness in 1991 that caused local flooding and loss of stand which resulted in the low yield for that year.

A similar comparison was made between line 5 in the SW Field and line 5 in the NE Field for 1991. There was not nearly as much variation in the NE Field as there was in the SW Field. Several factors seem to have caused this. Different hybrids, and later planting in the NE Field may have been key.

32 Managing Variability of Climate and Soil Characteristics

Characteristics in Conservation Tillage Systems: Effects on Field Behavior of Herbicides

Thanh H. Dao
USDA - Agricultural Research Service
El Reno, OK

The general public distress over groundwater contamination has grown in the past decade because there is increasing evidence of chemical intrusion in bodies of water bordering and underlying areas of agricultural activity. The concept of Best Management Practices (BMP) evolved in the earnest effort to cope with nonpoint source pollution of the environment from agriculture. Conservation practices have been developed to meet the challenges of erosion of natural resources since the mid-1930s. Conservation production systems are becoming the preferred water conservation and nonpoint source pollution control practices. At El Reno, Oklahoma, we have been examining the characteristics of these conservation practices since 1983. One of our goals is to assess the influence these systems have on the fate of applied pesticides, in particular herbicides because they comprise about 75% of all agricultural pesticides in use. We have observed that methods developed and promoted to maintain soil productivity in major agroecosystems can also meet today's agenda for nonpoint source pollution abatement.

The climate of the Southern Great Plain is characterized by extreme variability in precipitation and temperatures. We found that the impact of climatic variability can be attenuated with respect to soil water storage and thermal regime with no-tillage practices. The crop residue layer has a mulching effect and reduces water evaporation as well as attenuating temperature fluctuations in soil. These alterations in hydraulic and thermal properties of the soil under a crop residue mulch can drastically affect pesticide fate in soil.

At the field scale, there exists a great deal of temporal and spatial variability in soil density, a property that is related to water status in soil, and in many ways influences water use by crops. Intrinsic field variability exists as demonstrated by a grid-sampling method. Spatial variability was evident in both

---

Copyright © 1993 ASA-CSSA-SSSA, 677 South Segoe Road, Madison, WI 53711, USA. *Soil Specific Crop Management.*

conventional and conservation tillage methods. A temporal variation of soil density is expected in plowed soils, with low soil density values occurring soon after mechanical tillage is performed. However, the soil settles, and deforms from wetting and drying, and on impact of raindrops. We observed a lack of temporal dependence and lower density of untilled soils as a result of partial incorporation of crop residues and undisturbed organic debris of previous crops' root systems. There is a build up of soil organic matter content and the development of an elaborate channel system that contribute to improved soil porosity and aeration.

We are finding that conservation production practices also alter soil chemical and biological properties. These practices modify, in a major way, soil organic carbon distribution, and alter the stratification of nutrients and microbial substrates, and microorganisms' numbers, kinds, and activity. Soil retention of organic pesticides and degradation kinetics are much enhanced by management systems that leave a plant residue mulch on the soil surface as it offers a carbon-rich microenvironment that is favorable for microbial-mediated processes. Computer simulations of the fate of the herbicide metribuzin were made to assess the effect of changes induced by residue and tillage management practices. The results affirmed the importance of weather variability on field behavior of herbicides, as well as the influence of soil alterations induced by management practices for a benchmark soil series of central Oklahoma.

Many constraints still affect widespread implementation of BMP's on the farm due to the lack of understanding of the ecology and management of the complexity of alternative agricultural methods. These constraints also indicate continued need for research, and underscore the necessity for new farming systems to increase efficiency of inputs.

33 Soil Landscape Relations and Their Influence on Yield Variability in Kent County, Ontario

K. A. Denholm
J. L. B. Culley
Agriculture Canada
Land Resource Division
Guelph and Ottawa, Ontario

J. D. Aspinall
E. A. Wilson
Ontario Ministry of Agriculture and Food
Guelph, Ontario

Soil productivity is affected by many factors including climate, management, soil properties, and site characteristics. Grain corn (*Zea mays* L.) and soybean [*Glycine max* (L.) Merr.] are the dominant field crops in Kent County. Soils range from fine sands to heavy clay with large areas of very fine sandy loam and silt loam. The specific study objectives were: (i) to quantify the yield differences observed between landscape positions; and (ii) to relate yield differences to soil properties and slope position.

Twenty fields were selected for study based on the major soil types and landscape positions, crop rotation, and tillage practices. A transect was located within each field representing the greatest variation in soil and slope properties. Benchmark locations were selected along each transect to capture the range in soil and landscape variability.

Yields were hand harvested at each benchmark each year. Soil cores were taken at each benchmark and analyzed for particle size distribution, pH, $CaCO_3$, organic carbon, and percent gravel. Bulk density and available water were calculated for each horizon. Residual N was sampled at planting in 1991. Soil moisture was measured using Time Domain Reflectometry (TDR) at selected benchmark locations throughout the 1991 growing season.

Average corn yields for 1989, 1990, and 1991 were 8.68, 9.43, and 8.10 Mg/ha, respectively. Mean yield for 1990 was significantly greater ($P<0.01$) than either 1989 or 1991, which were not significantly different. Mean soybean yields for 1990 and 1991 were 2.99 ($^{+/-}0.072$) and 2.65 ($^{+/-}0.08$) Mg/ha, respectively and were significantly different ($P<0.005$).

Copyright © 1993 ASA-CSSA-SSSA, 677 South Segoe Road, Madison, WI 53711, USA. *Soil Specific Crop Management.*

One field was selected for presentation representing 3 years of continuous corn on loamy fine sand and fine sandy soils. Greatest vertical difference between lowest and highest points along the transect is approximately 1 m. Yield trends were consistent over 3 years at landscape position. Yield is negatively correlated with slope position and elevation and bulk density but positively correlated with percent clay in the Ap horizon. Yields averaged 7.74, 6.67, and 5.01 Mg/ha, respectively for 1989, 1990, and 1991. Mean yield for 1989 was significantly different from 1990 ($P<0.05$) and from 1991 ($P<0.01$).

34 A Field Information System for Spatially-Prescriptive Farming

Shufeng Han
Department of Agricultural Engineering
University of Illinois
Urbana, IL

Carroll E. Goering
Department of Agricultural Engineering
University of Illinois
Urbana, IL

Spatially-prescriptive farming (SPF) permits the automatic control of seeding and chemical application rates to match productivity of the soil in each part of a field. The essential elements of SPF include sensors to measure soil parameters, a field navigation system (FNS) to continuously measure the position of the tractor in the field, and a field information system (FIS) to handle the data required in SPF. In our approach to SPF, each field is subdivided into square cells that are small enough to have uniform soil properties.

Our poster session focuses on a FIS that we are building. The FIS will accept point-by-point data on soil parameters with accompanying coordinates provided by the FNS. The FIS includes provision for using geostatistics to block point-by-point data into cell values. Other features of the FIS include cell editing, preparation of queries to answer questions concerning soil parameters, and generation of maps for visual inspection. Application rate maps can also be created for controlling applicators in the field. Although the FIS is still under development, enough of its features have been completed to support an interesting demonstration.

35 Machine Vision Swath Guidance

John W. Hummel
USDA - Agricultural Research Service
Urbana, IL

Kenneth E. Von Qualen
Agricultural Engineering Department
University of Illinois
Urbana, IL

The industry-wide accepted method for field sprayer swath marking is foam dropped from the end of the spray boom. The pesticide is often applied up to 1.2 m past the end of the boom. On successive passes through the field, the operator aligns the end of the spray boom according to the foam markers. Positioning errors may result in overlapping or skipping regions of chemical application.

A machine vision guidance system has been developed, which incorporates a foam marking system, to increase the accuracy of spray application. A camera, in the field implementation of the system, would be mounted on the end of the spray boom. Tests of the concept were conducted with a camera, equipped with an image splitter to aid in the calculation of offset and heading, mounted on a mast on the front of a large garden tractor. The mounting bracket provided the camera an offset of 1.0 m from the centerline of the tractor at a height of 1.7 m above the ground. The image splitter, mounted in front of the camera, allowed the field of view ahead of the sprayer boom and field of view directly below the boom to be acquired and windowed simultaneously in one image using one camera. The lower window collected an image directly below the camera to calculate the offset. The upper image looked near the horizon to allow the heading of the vehicle with respect to the markers to be calculated. Run length encoding was used to find the location of two markers. A linear regression of the marker locations was then performed to calculate the heading and offset values of the sprayer.

The system accurately measured the heading and offset of the vehicle during stationary data collection. The average heading error for the normal operating conditions of $<5°$ heading and <2 m offset was - $0.160°$ with a 99% confidence interval of $\pm 0.276°$. The average offset error for these conditions was 0.062 m with a 99% confidence interval of 0.074 m.

Copyright © 1993 ASA-CSSA-SSSA, 677 South Segoe Road, Madison, WI 53711, USA. *Soil Specific Crop Management.*

36 MAPS Mailbox - A Land and Climate Information System

J. S. Jacobsen
A. E. Plantenberg
G. A. Nielsen
J. M. Caprio
Montana State University
Bozeman, MT

Soil specific crop management requires geographically referenced soil, climate, and land attributes. MAPS Mailbox is a PC program that provides land resource specialists, farmers, educators, and businesses with information on 150 land and climatic attributes for any location in Montana. MAPS Mailbox has a raster (grid) format, with the state divided into 17,993 cells or "mailboxes." Each cell represents approximately 8 square miles (2+ by 3+ miles). The user is asked to identify a location either by township, range, and section or by latitude and longitude. From 1 to 150 attributes can be selected and displayed for the mailbox cell and the surrounding eight cells. MAPS Mailbox has been used to (i) determine the suitability of areas for specialty crops, (ii) improve agricultural system management, and (iii) evaluate the suitability for agrichemical application to a site.

Copyright © 1993 ASA-CSSA-SSSA, 677 South Segoe Road, Madison, WI 53711, USA. *Soil Specific Crop Management.*

37 Leaching and Runoff of Pesticides Under Conventional and Soil Specific Management

B. R. Khakural
P. C. Robert
Department of Soil Science
University of Minnesota
St. Paul, MN

D. J. Fuchs
Southwest Experiment Station
University of Minnesota
Lamberton, MN

The influence of conventional and soil specific management on leaching and runoff loss of soil applied alachlor [2-chloro-2', 6'-diethyl-N-(methoxymethyl) acetanilide] was studied across a soil catena with varied slope and drainage characteristics. Three soil types: a well-drained Ves (fine-loamy, mixed, mesic Udic Haplustoll) on the backslope (1-4%), a Ves on the sideslope (6-12%), and a poorly drained Webster (fine-loamy, mixed, mesic Typic Haplaquoll) on the toeslope (0-4%) were studied. Soil, surface runoff and soil water samples were collected and analyzed for alachlor concentration. No alachlor was detected in the soil samples from below 30-cm depth. Detectable amount of alachlor (0.6, 1 µg/kg) was found in only 2 out of 60 soil water samples collected at six different dates in 1990. In 1991, an alachlor concentration of ≥ 1 µg/kg was detected in 33% of the earliest batch of samples (collected after one week of application) and 17% of the later samples at a depth of 60 cm, respectively. At 120-cm depth, a concentration of ≥ 1 µg/kg was detected in 33% of the earliest batch of samples only. The highest alachlor concentration in runoff water was observed at the first runoff event and the concentration decreased with time in later events during both years of study. Average alachlor concentration for the Ves (6-12%) soil was 1.5 and 2 times higher than the Ves (1-4%) and Webster (0-3%) soils, respectively. Ves (6-12%) soil also produced 4-10 times higher runoff volume than other two soil types. No significant difference in alachlor concentration of surface (0-15 cm) soil or runoff water was detected between constant and soil specific alachlor rates. Pesticide loss through surface runoff seems to be a serious problem in these soils, especially on steeper slopes. Conservation practices designed to reduce runoff volume should be implemented to reduce pesticide loss through runoff.

Copyright © 1993 ASA-CSSA-SSSA, 677 South Segoe Road, Madison, WI 53711, USA. *Soil Specific Crop Management.*

38 Spatial Regression Analysis of Crop and Soil Variability Within An Experimental Research Field

D. S. Long
Department of Soil Science
University of Minnesota
St. Paul, MN

S. D. DeGloria
Soil, Crop and Atm. Sciences
Cornell University
Ithaca, NY

D. A. Griffith
Department of Geography
Syracuse University
Syracuse, NY

G. R. Carlson
G. A. Nielsen
Plant and Soil Sciences
Montana State University
Bozeman, MT

Spatial autocorrelation lends organization to the spatial variability of crop and soil variables such that observations measured at one location are not independent of observations at nearby locations. The classical design principles of randomization, replication and blocking do not entirely take this spatial autocorrelation into account. Failure to account for spatial autocorrelation violates the independence assumption in statistics and hence, invalidates probability statements in inferential tests for analysis of (co)variance (Glass et al., 1972) and regression (Griffith, 1988).

Spatial regression is a potentially useful technique to allow valid tests of significance in the presence of spatial autocorrelation. Despite advances in literature and methodology, use of spatial regression by the applied community has been nonexistent (Anselin and Griffith, 1988). The objective of this study is to demonstrate inconsistencies between spatial regression versus classical regression analysis of spatially variable crop and soil data. Crop yield and soil

---

Copyright © 1993 ASA-CSSA-SSSA, 677 South Segoe Road, Madison, WI 53711, USA. *Soil Specific Crop Management.*

physical and chemical properties were measured from a 0.52 ha experimental research field in Montana. An autoregressive response model was used to accommodate spatial autocorrelation in the crop yield data. Regressions were carried out in classical and spatial modes.

Autoregressive response modeling captured moderate levels of spatial autocorrelation as indicated by autocorrelation parameter estimates of about 0.6. The effect of spatial autocorrelation on classical regression was to inflate coefficients of determination (R2), deflate standard errors for slope parameters and overestimate the t values for inferential tests. For example, the coefficients of determination for the relationship between yield and selected soil physical properties were .39 versus .12 for classical and spatial models. The underestimated standard errors in classical models produced t values for parameters that were two to four times larger than t values obtained in spatial models. This study concluded that parameter estimates, standard errors and inferential tests are prone to error if based on a false assumption of independence.

REFERENCES

Anselin, L. and D. A. Griffith. 1988. Do spatial effects really matter in regression analysis? Papers of the Regional Science Association 65:11-34.

Glass, G. V., P. D. Peckham, and J. R. Sanders. 1972. Consequences of failure to meet assumptions underlying the fixed effects analysis of variance and covariance. Rev. Educ. Res. 42(3):237-288.

Griffith, D. A. 1988. Advanced spatial statistics. Kluwer Acad. Pub. Norwell, MA. pp. 273.

39 Precision Farm Management of Variable Crop Land in the Pacific Northwest

Baird Miller
Roger Veseth
Department of Crop and Soil Sciences
Washington State University
Pullman, WA

The Palouse region of the Pacific Northwest is characterized by a highly variable, dune-like topography. Farming in this region takes place across slopes ranging from 0 to 45%, with an average of 13%. The soil properties, such as topsoil depth, soil organic matter, moisture availability, texture, bulk density, residual fertility and pH, vary tremendously across the landscape positions. These changing soil properties result in significant differences in crop growth, grain yield, and quality. Miller et al. (1992) reported that soft white winter wheat (*Triticum aestivum* L.) yields varied by more than 50% across four landscape positions. The reduced yields on the ridgetops and north-facing backslopes were associated with reduced head density and lighter kernel weights. The test weights ranged from 53.8 to 61.7 lb/bu and grain protein ranged from 8.6 to 13.1% across the landscape positions. Typically, the eroded ridgetop positions had much lower test weights and higher grain protein percentages. Pan and Hopkins (1991) found that winter barley (*Hordeum vulgare* L.) yields were reduced by 46% between the toeslope and an eroded ridgetop position. Mahler et al. (1979) reported that dry pea (*Pisum sativum* L.) yield declined 77% between the toeslope and ridgetop position.

In spite of these tremendous differences in grain yield and quality differences among landscape positions, the fields are uniformly managed to meet the average crop requirements. The application of "precision farming" technology strives to meet the specific needs of the crop, based on the yield potential and soil conditions within smaller management units.

Precision farming starts with the design of unique management units. Factors and technology to consider when designing management units include: crop growth patterns, yield maps, soil color pattern, soil tests results, soil surveys, color or infrared photography of the crop and farmer's knowledge of the field.

There are numerous options to consider in a variable crop management strategy.

Fertility Management

Fertility recommendations could be based on management unit specific yield potential and soil test results.

Residue Management

Reduce tillage and maintain more surface residue on management units with greater erosion potential, low crop residue production or lower soil organic matter (i.e., leave stubble standing on upper slopes and ridgetops). Adjust tillage practices to increase water infiltration where soil moisture limits yields (i.e., chisel plow mid-slopes or subsoil ridgetops). Reduce surface residue on areas with low erosion potential to accelerate soil drying and warming (i.e., plow low-lying areas). Reduce surface residue in management units with high residue production, low erosion potential, and a greater incidence of residue related diseases and insects (i.e., plow low-lying areas).

Disease Management

Use longer crop rotations on management units with high soilborne disease potential (i.e., in low-lying areas, use a 3-year wheat-barley-pea rotation). Scout the fields and then only use seed treatments and foliar fungicides on the management units that require treatment.

Weed Management

Scout management units within a field and use weed control measures only where necessary. Intensify weed management on field borders. Select the rate of soil active herbicides to match the soil texture and organic matter of the management unit to increase effectiveness and reduce crop injury. Alter crop rotations, seeding dates, tillage, and residue management to optimize weed control (i.e., move from a wheat-pea rotation to a wheat-barley-pea).

Crops Rotations and Variety Selection

Establish cover crops on areas with low yield and high erosion potential (i.e., perennial grass on eroded ridgetops). Establish deep tap-rooted crops, such as rapeseed, on units where soil compaction limits productivity and contributes to runoff and erosion. Plant varieties tolerant to soilborne and foliar diseases, and lodging on management units with higher soil moisture levels (i.e., Madsen wheat, resistant to strawbreaker footrot, in low-lying areas).

REFERENCES

Miller, B. C., T. Feiz, and W. L. Pan. 1992. Impact of landscape variability on grain yield and quality. In R. J. Veseth, and B. C. Miller (ed.) Precision farming variable cropland: An introduction to variable management within whole fields, divided slopes and field strips. Proc. 10th Inland Northwest Conservation Farming Conf. Pullman, WA. 18 February 1992.

Mahler, R. L., D. F. Bezdicek, and R. E. Witters. 1979. Influence of slope position on nitrogen fixation and yield of dry peas. Agron. J. 71:348-351.

Pan, W. L., and A. G. Hopkins. 1991. Plant development, and N and P use of winter barley. I. Evidence of water stress-induced P deficiency in an eroded toposequence. Plant Soil 135:9-19.

40 Management Approaches to Fertility and Biological Variation in the Inland Pacific Northwest

William Pan
Baird Miller
Ann Kennedy
Tim Fiez
Munir Mohammad
Department of Crop and Soil Sciences
Washington State University
USDA-ARS
Pullman, WA

Several soil biological and chemical entities and processes vary across the rolling hills of the Palouse region of the Inland Pacific Northwest. To efficiently use agricultural inputs, and optimize yields while addressing soil, water, and crop-quality goals, these factors will need to be better understood before variable landscape management systems can be implemented. Variable yields (often obtained at different landscape positions) are an end result of the variation in soil productivity across Palouse landscapes. Variable available water is most often cited as explanation for these variable yields, yet there are many other soil factors that contribute to soil productivity in this region. Organic matter is an important soil constituent that influences virtually all chemical, biological, and physical processes that occur in these soils. The level of organic matter depends on temperature, moisture, and degree of erosion. All contribute to variations observed across agroclimatic zones and within fields, ranging from less than 1% to more than 4%. Improving soil organic matter levels will require effective erosion control, production, and recycling of crop residues with proportional N inputs, and an input of organic matter from animal manure, sewage sludge, or green manure. Yields from organic amendments on these soils often exceed those obtained from synthetic fertilizers. Nitrogen fertility status of these soils also varies across landscapes. Nitrogen mineralization rates, as estimated from check plots not receiving N fertilizer, vary by position. Low N mineralization was estimated on severely eroded ridgetops with exposed subsoil and low organic matter. Nitrogen losses during a single growing season also vary. High levels of unaccounted N in the crop-soil budget were measured on the northfacing backslope and footslope positions, areas that receive input of water from lateral flow, and are subject to excessive leaching and denitrification.

Copyright © 1993 ASA-CSSA-SSSA, 677 South Segoe Road, Madison, WI 53711, USA. *Soil Specific Crop Management.*

Differences in soil residual N build up over several cropping cycles. High residual N tends to build up in positions of low productivity such as eroded ridgetops, where plant utilization is lower. North-facing backslopes have been found to have lower levels of residual N, likely due to greater N losses. Uptake efficiency varies with slope position and level of N supply, ranging from 70 to 65% at the south facing slope vs. 60 to 35% at the north-facing slope. Spring N applications avoid overwinter losses. Recent experiments demonstrated point injected spring N in winter wheat increased N-use efficiency over fall-applied N. Phosphorus fertility is closely correlated with organic matter status across these landscapes, and therefore, tends to be lower in positions that have sustained topsoil and organic matter losses. Subsoils of eroded soils can be extremely low in available P. In these situations, we can fertilize the surface layer to accommodate early season P uptake, but as that surface layer dries out, the P fertilizer is stranded, and in this example, P uptake during grain-filling was low to negligible in positions lacking adequate subsoil P. We are currently investigating potential benefits from deep P placement with subsoiling equipment. In addition, mycorrhizal inoculation of eroded soils may provide future directions for P management. VAM inoculation elicited substantial increases in growth and nutrient accumulation of growth chamber-grown wheat in soils from eroded ridgetop and backslope, but not in toeslope soil. Soil microorganisms affect nutrient cycling and plant growth. The dehydrogenase assay is a measure of microbial activity. Activity in the ridgetop and north backslope were less than that in the footslope and south backslope. Differences could not be completely explained by the amount of soluble carbon present at each position. Variable management will need to address all of these factors collectively in defining efficient management schemes across the Palouse landscape.

41 Yield Variation Across Coastal Plain Soil Mapping Units

E. J. Sadler
D. E. Evans
W. J. Busscher
USDA-Agricultural Research Service
P O Box 3039
Florence, SC

D. L. Karlen
USDA-Agricultural Research Service
Iowa State University
Ames, IA

Soil specific crop management is commonly proposed as a means of increasing efficiency of production and reducing environmental damage resulting from overapplication of fertilizers or pesticides to subfield areas prone to leaching or runoff. Large-scale variation in soil characteristics is apparent from soil maps, and small-scale variation has been the subject of geostatistical studies for some time. It is commonly hypothesized that custom-tailoring management to specific soils will result in significant improvement in environmental quality and production efficiency. Clearly, the machinery, sensors, and control technologies are available to implement custom culture. However, it is not clear what adjustments to culture are necessary to achieve such improvements. Experience shows that certain soils need more or less water, seeding rate, fertilizer, or pesticide, but implementation of this experience into control algorithms for intelligent machines requires quantitative knowledge that does not exist at the current time. An initial step was undertaken at the Florence, SC, laboratory to quantify variation, both for soil characteristics and yield. Concurrently, modeling studies have attempted to explain the observed variation. Future work will attempt to propose custom management strategies for the different soils.

Soil specific crop management research at the Coastal Plains Research Center was initiated in 1984 with a detailed soil survey by USDA-SCS personnel. The area was surveyed and flagged on a 15-m grid. The survey team mapped each grid marker, then narrowed the distinctions between different soils down to ~ 3 m. This map was digitized into SAS PROC GMAP. Field studies started in 1985 and include corn (3 yr) (*Zea mays* L.), wheat (3 yr) (*Triticum aestivum* L.), soybean (3 yr) [*Glycine max* (L.) Merr.], and sorghum (1 yr) (*Sorghum* sp). Yield plots were located with surveying equipment and

---

Copyright © 1993 ASA-CSSA-SSSA, 677 South Segoe Road, Madison, WI 53711, USA. *Soil Specific Crop Management.*

superimposed on a soil specific map. From the composite map, databases of soil mapping unit, yield, and other information were generated. Analyses of variation across mapping unit indicated significance in most seasons, although variation within mapping units was surprisingly large. For soils with thicker sandy surface layers, the depth to the clayey subsoil sometimes was correlated with yield. Interactions among water relations, pH of the subsoil, depth to clay, and tillage complicated these relationships.

Modeling studies have been conducted using several daily-time-step models for these crops, including CERES-Maize (V 1 & 2), CERES-Wheat (V 1 & 2), SORKAM, CERES-Sorghum, and SOYGRO. Soil input data was derived from the literature for the typical pedon description of the 1984 survey. Preliminary results indicate that rooting functions must be adjusted for the pH of the subsoil, the strength of the E horizon, and the effect of tillage on the latter. Further studies suggest that the rainfall-runoff partitioning may be critical for these soils. Sensitivity analyses indicate that water stress is much more important than would be assumed in an area with 1100 mm annual rainfall. This appears to be a result of the limited water held in sandy, shallow-rooted soils.

Future work will increase the specificity of the soils input parameters for the models, including changing from typical pedon descriptions and literature values to local measurements of representative soils. Later runs will examine within-mapping unit variation as a result of measured soil variation. The database has been converted into the ARC/INFO Geographic Information System, and further work will integrate the conventional modeling to the GIS.

42 Sensing for Variability Management

Kenneth A. Sudduth
USDA-Agricultural Research Service
University of Missouri-Columbia
Columbia, MO

Steven C. Borgelt
Agricultural Engineering Department
University of Missouri-Columbia
Columbia, MO

Automatic sensing of the spatial variability in soil and crop parameters is an important component of soil specific crop management. Many parameters can change significantly over a small area, making manual sample collection or processing prohibitive in terms of cost and timeliness. Our current research efforts include sensing of organic matter and other soil constituents, grain yield monitoring with an instrumented combine, and use of global positioning system (GPS) receivers for location sensing.

Organic Matter Sensing

A prototype sensor[1] that uses near infrared (NIR) reflectance techniques to determine the organic matter content of the surface layer of soil was developed and tested in the laboratory and field. Calibration of the sensor was accomplished in the laboratory with a test set of 30 Illinois mineral soils prepared at soil moisture tensions ranging from 1.5 MPa (wilting point) to 0.033 MPa (field capacity). Laboratory predictions yielded an r^2 of 0.89 and a standard error of prediction of 0.40% organic matter. The sensor was also able to predict soil moisture and cation exchange capacity. Limited in-furrow field operation of the sensor produced a much higher standard error (0.91% organic matter), due to the movement of the soil past the sensor as scanning was accomplished. Additional laboratory tests indicated that the sensor could predict organic matter contents of soils obtained from across the lower Corn Belt (OH, IN, IL, and MO) with a single calibration equation. The optics and electronics

---

[1]Soil organic matter sensor developed in cooperation with John W. Hummel, USDA - Agricultural Research Service, Urbana, IL; and Agmed Inc., Springfield, IL.

Copyright © 1993 ASA-CSSA-SSSA, 677 South Segoe Road, Madison, WI 53711, USA. *Soil Specific Crop Management.*

of the unit have been refined for improved accuracy and reliability, and are currently undergoing testing. Although improvements in sample presentation techniques would be required to use the sensor in real-time field operatin, the current configuration could provide rapid, on-site organic matter determination for mapping and subsequent application rate control using a positioning system.

Grain Yield Monitoring

A full-size combine has been instrumented with a commercially available yield sensor for monitoring within-field yield variability. Laboratory tests showed good accuracy and repeatability of calibrations with wheat and soybean. The system was used to harvest corn and soybean in 1991. Good agreement ($r^2>0.95$) was obtained between batch weights and totalized yield sensor volumes. Significant variations in yield were monitored on a slightly sloping claypan soil. Data are currently being analyzed to evaluate the effect of combine dynamics on the yield sensing process.

Global Positioning System Location Sensing

Two GPS receivers have been obtained and used in differential mode to precisely locate sampling points. A bar code reader has been interfaced to one receiver to allow ready identification of samples. Work currently underway to evaluate the precision and accuracy of the GPS system has been hampered by short working windows caused by poor satellite availability. Plans are to integrate differential mode GPS with the grain yield monitor for the coming harvest season.

43 Nitrogen Specific Management by Soil Condition[1]

J. A. Vetsch
G. L. Malzer
P. C. Robert
Department of Soil Science
University of Minnesota
St. Paul, MN

W. W. Nelson
Southwest Experiment Station
University of Minnesota
Lamberton, MN

Recent advances in technology has brought about considerable interest in developing the methodology for making variable N rate applications within a field. The objectives of this study were to evaluate yield variability within production fields, determine yield response to applied fertilizer N and differential N loss as influenced by soil conditions. Results are being evaluated to determine what measurable soil conditions are best suited for making site specific N recommendations. Experimental locations of 14 and 18 ha were established in 1990 at Lamberton and Becker, respectively. Different N management treatments were applied in replicated strips across the fields. Yields obtained from segmented harvest areas from control strips (zero N) indicate extreme yield variability. Considerable yield variability was measured within soil map units. Preliminary results would suggest that variable N rate application can reduce N inputs, maintain or improve yield, and increase net return for the producer.

---

[1]The assistance and financial support of the Minnesota Agricultural Experiment Station, USDA-CSRS, Dow Elanco and Soil Teq Inc. is greatfully acknowledged.

Copyright © 1993 ASA-CSSA-SSSA, 677 South Segoe Road, Madison, WI 53711, USA. *Soil Specific Crop Management.*

LIST OF PARTICIPANTS

Ahlrichs, John	Cenex/Land O'Lakes, Box 64089, St. Paul, MN 55164-0089
Allen, C. E.	Vice President, University of Minnesota, St. Paul, MN 55108
Allmaras, Ray	USDA-ARS, 152 Borlaug Hall, Univ. of Minnesota, St. Paul, MN 55108
Alms, Eugene	Alms Consulting, Inc., Rt 2 Box 23D, Lake Crystal, MN 56055
Anderson, Dean	Lor-Al Products, Inc., Box 289, Benson, MN 56215
Anderson, James	Dep. of Soil Science, University of Minnesota, St. Paul, MN 55108-6028
Anthony, C. T.	Land Analysis Laboratory, The Pennsylvania State University, University Park, PA 16802
Ascheman, Robert	Ascheman Associates, 2921 Beverley Drive, Des Moines, IA 50322
Aspinall, J. D.	P O Box 1030, 52 Royal Road, Ontario Ministry of Agriculture, Guelph, Ontario N1H 6N1 CANADA
Ault, Kurtis I.	USDA-ARS, MWA, 439 Borlaug Hall, Univ. of Minnesota, St. Paul, MN 55108
Ball, Rodnie	Dickey-john, 15 S. Country Club Rd., Auburn, IL 62615
Barnes, Robert	American Society of Agronomy, 677 S. Segoe Rd., Madison, WI 53711
Bauer, Norm	Ag-Chem, 5720 Smetana Dr., Minnetonka, MN 55343
Bauer, Ronald	USDA-SCS, 100 Centennial Mall N, Lincoln, NE 68508

Beck, Robert H.	CENEX/Land O'Lakes, Box 64089 St. Paul, MN 55164-0089
Birrell, Stuart	Dep. of Agric. Eng., Univ. of Illinois, Urbana, IL 61801
Bock, Bert	TVA, P O Box 1010, Muscle Shoals, AL 35660
Borgelt, Steve	Agricultural Eng. Dep., University of Missouri, Columbia, MO 65211
Bouma, Johan	Department of Soil Science and Geology, Agriculture University, P.O. Box 37, 6700AA Wageningen, The Netherlands
Brady, Randy	Trimble Navigation, 645 N. Mary Ave., Sunnyvale, CA 94086
Breitbach, David	Farm Credit Bldg., Ste. 600, 375 Jackson St., St. Paul, MN 55101
Buchholz, Daryl	214 Waters Hall, University of Missouri, Columbia, MO 65211
Busscher, Warren J.	USDA-ARS, SSA, Coastal Plains Water Cons. Res. Ctr., Florence, SC 29502
Caprio, J. M.	807 Leon Johnson Hall, Montana State Univ., Bozeman, MT 59717
Carlson, G. R.	Plant and Soil Sciences, Montana State University, Bozeman, MT 59717
Chaplin, Jonathan	Dep. of Agric. Eng., University of Minnesota, St. Paul, MN 55108
Cheng, H. H.	Dep. of Soil Science, University of Minnesota, St. Paul, MN 55108
Clay, Dave	Dep. of Plant Science, South Dakota State University, Brookings, SD 57007
Colvin, Thomas	USDA-ARS, National Tilth Lab, 2150 Pammel Drive, Ames, IA 50011

PARTICIPANTS

Cowgur, Bruce	Mid Tech, 2733 E. Ash St., Springfield, IL 62703
Culley, John	Land Resource Res. Inst., Central Exp. Station, Ottawa, On OC6 CANADA
Dao, Thanh	Ag Research Service, P O Box 1199, El Reno, OK 73036
Day, R. L.	Dep. of Agronomy, The Pennsylvania State Univ., University Park, PA 16802
DeGloria, S. D.	Soil, Crop and Atm. Sciences, Cornell University, Ithaca, NY 14853
Denholm, K. A.	Land Resource Division, Agric. Canada, Ottawa, Ontario K1A 0C6 CANADA
Dickey, Gylan L.	USDA-SCS, TISD, 2625 Redwing Rd., Ste 110, Ft. Collins, CO 80526
Dowdy, Robert	USDA-ARS, 458 Borlaug, University of Minnesota, St. Paul, MN 55108
Duffy, Michael D.	Dep. of Agricultural Economics, Iowa State University, Ames, IA 50011
Enger, Steve	Ferguson Group, Rt 2 Box 68A, Hatton, ND 58240
Evans, Dean E.	USDA-ARS, SSA Coastal Plains Water Cons. Res. Ctr., Florence, SC 29502
Evans, Eric	Dep. of Agriculture, Univ. of Newcastle upon Tyne, United Kingdom NE1 7RV
Fairchild, Dean	Soilteq, Box 25, Waconia, MN 55387
Fiez, Tim	Crop and Soil Sciences, Washington State Univ., Pullman, WA 99164-6420
Finke, P. A.	Dep. of Soil Science & Geology, Agricultural University, Wageningen, THE NETHERLANDS

Fixen, Paul	P O Box 682, Brookings, SD 57006
Forcella, Frank	North Central Soil Conservation Research Lab, Morris, MN 56267
Fuchs, Dennis J.	University of Minnesota, Box 475, Lamberton, MN 56152
Gandrud, Dale	517 East School St., Owatana, MN 55060
Gaultney, Larry	Dep. of Agricultural Engineering, Purdue University, West Lafayette, IN 47907
Gebhardt, Maurice	Agricultural Management Services, 204 E. McKenzie, Suite F, Punta Gorda, FL 33950
Gessler, P. E.	Division of Soils, CSIRO, Canberra, Australia
Goering, Carroll	Dep. Agric. Eng., University of Illinois, Urbana, IL 61801
Griffith, Duane	Extension Ag. Economics, Montana State University, Bozeman, MT 59717
Gustafson, David I.	Rhone Poulenc, 2 Alexander Dr., Triangle Park, NC 27709
Hall, Thomas	American Society of Agronomy, 677 S. Segoe Rd., Madison, WI 53711
Hammond, Max	CENEX/Land O'Lakes, Rt. 1, 398-6 NW Ephrata, WA 98823
Han, Shufeng	Dep. of Agric. Eng., University of Illinois, Urbana, IL 61801
Haneklaus, S.	ILLIT Gmb H, Ludemannstrasse 72, Ste. Nr. 4, D-2300 Kiel, GERMANY
Hanson, Lowell D.	Field Control Systems, P.O. Box 10851, White Bear Lake, MN 55110

PARTICIPANTS

Harrison, W. Douglas	USDA-SCS, Federal Bldg., Room 443, 10 E. Babcock, Bozeman, MT 59715
Havermale, Neil	Farmers Software Association, P O Box 660, Ft. Collins, CO 80522
Holmes, William	R.R. 2, Box 187, Oran, MO 63711
Holzhey, C. Steven	USDA-SCS, 100 Centennial Mall N, Rm 152, Lincoln, NE 68508-3866
Houtsma, James	Farmer Magazine, 7900 International Drive, Bloomington, MN 55425
Hudson, Berman D.	USDA-SCS, Federal Bldg., Rm 152, Lincoln, NE 68508-3866
Huffman, Jerry	Dow Elanco, 103 Tomaras Ave., Suite 1, Savoy (Champaign), IL 61874
Huggins, David R.	University of Minnesota, Box 475 Lamberton, MN 56152
Hummel, John	USDA-ARS, 376 Ag. Engr. Sci. Bldg., Urbana, IL 61801
Hunt, Patrick G.	Coastal Plains Soil and Water Research Center, Florence, SC 29503
Hurtis, Paul	DMI, Rt. 150 East, Box 65, Goodfield, IL 61742-0065
Hutson, John	Cornell University, Bradfield Hall, Ithaca, NY 14853
Jacobsen, Jeff	Dep. of Plant & Soil Science, Montana State Univ., Bozeman, MT 59717-0312
Joerger, Robert	CENEX/Land-O-Lakes, 327 E. Main, Mayville, ND 58257
Johnson, Gregg	Graduate Research Assistant, University of Nebraska, Lincoln, NE 68583
Johnson, Richard	Deere & Co. Technical Center, 3300 River Dr., Moline, IL 61265

Karlen, Douglas L.	National Soil Tilth Lab, 2150 Pammel Drive, Ames, IA 50011-4420
Kennedy, Ann C.	USDA-ARS, 215 Johnson Hall, Washington State Univ., Pullman, WA 99164-6421
Khakural, B. R.	Dep. of Soil Science, University of Minnesota, St. Paul, MN 55108
Klemme, Kent	Dep. Agric. Eng., South Dakota State University, Brookings, SD 57007
Knox, Ellis	USDA-SCS, 100 Centennial Mall N., Lincoln, NE 68508-3866
Knutson, Roger	Soil Teq, Box 25, Waconia, MN 55387
Lamb, John	Dep. of Soil Science, University of Minnesota, St. Paul, MN 55108
Lamp, Juergen	Institute of Soil Science, Christian Albrechts Univ. of Kiel, Olshausen Str. 40, 2300 Kiel, GERMANY
Larson, William	Dep. of Soil Science, University of Minnesota, St. Paul, MN 55108
Leikam, Dale	Farmland Industries, Box 7305, Kansas City, MO 64116
Lindstrom, Michael J.	North Central Soil Conservation Research Lab, Morris, MN 56267
Long, Dan	Dep. of Soil Science, University of Minnesota, St. Paul, MN 55108
Luhn, Dennis	Dickey-john Corp., Box 10, Auburn, IL 62615
Lytle, Dennis J.	USDA-SCS-NSSC, 100 Centennial Mall N., Lincoln, NE 68508-3866
Macy, Ted	Top-Soil Testing, 27 Ash Street, Frankfort, IL 60423

Mailander, Michael P.	Dep. of Agric. Eng., Louisiana State University, Baton Rouge, LA 70803
Malzer, Gary	Dep. of Soil Science, University of Minnesota, St. Paul, MN 55108
Mangold, Grant	Soybean Digest, 540 Maryville Center Drive, St. Louis, MO 51033-9703
Mann, John	Illini FS, 1509 East University, Urbana, IL 61801
Marks, Robbin S.	Nat. Res. Defense Council, 1350 New York Ave. N.W., Washington, DC 20005
Mausbach, Maurice	USDA Soil Conservation Service, Box 2890, Washington, DC 20013
McQuinn, Al	Ag Chem, 5720 Smetana Dr., Minnetonka, MN 55343
McQuinn, Chuck	Ag Chem, 5720 Smetana Dr., Minnetonka, MN 55343
Memory, Russell	Flexi-Coil Ltd., 1000 71st St. E, Saskatoon, Sask. CANADA
Miller, Baird	Washington State University, Crop and Soil Sciences, Pullman, WA 99164-6420
Miller, Gerald	Dep. of Agronomy, Iowa State University Ames, IA 50011
Mohammad, Muir	Crop and Soil Sciences, Washington State University, Pullman, WA 99164
Monson, Robert J.	Ag Chem, 5720 Smetana Dr. Minnetonka, MN 55343
Montgomery, Bruce	Minn. Dep. of Agriculture, 90 W. Plato Blvd., St. Paul, MN 55107
Moore, Ian	Centre for Resource & Environmental Studies, Australian National University, G.P.O. Box 4, Canberra ACT 2601, AUSTRALIA

Mortenson, David	Dep. of Agronomy, University of Nebraska, Lincoln, NE 68583
Motz, Darin S.	Agric. Eng. Dep., Texas A&M Univ., College Station, TX 77843-2117
Mulkey, Lee	Environmental Research Lab/EPA, College Stn Rd., Athens, GA 30613
Mulla, Dave J.	Washington State University, Crop and Soil Sciences, Pullman, WA 99164
Munson, Robert	NFSA, 339 Consort Drive, Manchester, MO 63011
Munter, Robert	Dep. of Soil Science, University of Minnesota, St. Paul, MN 55108
Murphy, Larry S.	Potash & Phosphate Inst., 2805 Claflin Rd., Ste. 200, Manhattan, KS 66502
Nelson, Wallace	Southern Experiment Station, Lamberton, MN 56152
Nielsen, G. A.	Dep. of Plant & Soil Science, Montana State Univ., Bozeman, MT 59717-0312
Nowak, Peter	University of Wisconsin, 212A Ag Hall, Madison, WI 53706
Oberle, Prof. S. L.	Soils Department, University of Wisconsin, Stevens Point, WI 54481
Olson, Glenn	John Deere & Co., John Deere Rd, Moline, IL 61265
Olson, Ron	TopSoil, 27 Ash Street, Frankfort, IL 60423-0340
Onstad, Charles A.	USDA-ARS, North Iowa Avenue, Morris, MN 56267
Pan, Bill	Dep. of Crop and Soils, Washington State University, Pullman, WA 99164

Pavelski, Richard	Pavelski Farms, Inc., Box 68, Amherst Junction, WI 54407
Petersen, Gary	Dep. of Agronomy, Pennsylvania State University, University Park, PA 16802
Peterson, G. A.	Dep. of Agronomy, Colorado State University, Ft. Collins, CO 80523
Pierce, Francis J.	Dep. of Crop and Soils, Michigan State University, East Lansing, MI 48824
Plantenberg, A. E.	718 Leon Johnson Hall, Montana State University, Bozeman, MT 59717
Pollack, J.	Land Analysis Lab, The Pennsylvania State Univ., University Park, PA 16802
Randall, Gyles	Southern Experiment Station, University of Minnesota, Waseca, MN 56093
Rawlins, Steve	USDA/ARS, NPS, Bldg. 005, BARC-West, Beltsville, MD 20705
Reichenberger, Larry	Farm Journal Magazine, RR1 Box 199A, Mt. Hope, KS 67108
Riedell, Walter	USDA-ARS, Insect Lab, RR3, Brookings, SD 57006
Robert, Pierre	Dep. of Soil Science, University of Minnesota, St. Paul, MN 55108
Russell, Mark	DuPont Agricultural Products, Experiment Station, Wilmington, DE 19880-0402
Russo, J. M.	ZedX, Inc., Bealsburg, PA 16803
Rust, Richard	Dep. of Soil Science, University of Minnesota, St. Paul, MN 55108
Sadler, E. J.	Coastal Plains Research Center, USDA-ARS, P.O. Box 3039, Florence, SC 29502
Sawyer, John	GROMARK, 1701 Towanda Avenue, Bloomington, IL 61702

Schafer, Robert L.	USDA-ARS, Box 57, Loachapoka, AL 36865-0057
Schmidt, Berlie	USDA-CSRS, Aerospace Center, Rm 303 901 D St. S.W., Washington, DC 20024
Schmidt, Walter	The Ohio State Univesity, 952 Lime Ave., Findley, OH 45840
Schnug, Ewald	Kiel University, Olshausen Str. 40 2300 Kiel GERMANY
Schrock, Mark	Dep. of Agricultural Engineering, Kansas State University, Manhattan, KS 66506
Schueller, John K.	Dep. of Mech. Eng., University of Florida Gainesville, FL 32611-2050
Schulz, Joe	Ag-Chem, 5720 Smetana Dr., Minnetonka, MN 55343-9688
Schulze, Darrell	Dep. of Agronomy, Purdue University, W. Lafayette, IN 47097
Schumacher, Joe	Dep. of Mechanical Eng., South Dakota State University, Brookings, SD 57006
Searcy, Steve	Agric. Eng. Dep., Texas A&M University College Station, TX 77843
Senjem, Norman	U.S. Agriculture, 7900 International Dr., Bloomington, MN 55425
Sether, Irv	Box 107, Jackson, MN 56143
Sinclair, H. Raymond, Jr.	USDA-SCS, 100 Centennial Mall North, Lincoln, NE 68508
Smith, Scott	Dep. of Soil Science, University of Minnesota, St. Paul, MN 55108
Smith, Wayne	Deere & Co. Technical Center, 3300 River Drive, Moline, IL 61265
Spetzman, Jerry	Minn. Dep. of Agriculture, 90 W. Plato Blvd., St. Paul, MN 55107

Spivey, Lawson D.	USDA-SCS, Box 2890, Washington, DC 20013-2890
Sudduth, Kenneth	USDA-ARS-CSWQRU, University of Missouri, Columbia, MO 65211
Thompson, Wayne	Dep. of Soil Science, University of Minnesota, St. Paul, MN 55108
Tofte	Micro-Trak Systems, Box 3699, Mankato, MN 56002
Tyler, Dave	Dep. of Survey Engineering, University of Maine, Orono, ME 04469
Urevig, Edgar	Tilney Farms, Lewisville, MN 56060
Veseth, R. J.	Crop and Soil Sciences, Washington State Univ., Pullman, WA 99164-6420
Vetsch, Jeff	Dep. of Soil Science, University of Minnesota, St. Paul, MN 55108
Von Qualen, K. E.	Dep. Agric. Eng., University of Illinois, Urbana, IL 61801
Voorhees, Ward B.	USDA-ARS, Morris, MN 56267
Waits, David A.	Space Remote Sensing Ctr., Bldg. 103, Stennis Space Center, MS 39529
Ward, Justin	Natural Resources Defense Council, 1350 New York Avenue NW - Ste 300, Washington, DC 20005
Wesley, D. E.	Key Agri-Services, Inc., 114 Shady Lane Macomb, IL 61455
Wilbur, Robert D., Dir.	American Cyanamid Company, P O Box 400, Princeton, NJ 08540
Wilson, E. A.	Ontario Ministry of Agriculture, P O Box 1030, 52 Royal Road, Guelph, Ontario N1H 6N1 CANADA

Wilson, John	Dep. of Earth Science, Montana State University, Bozeman, MT 59717
Wolkowski, Richard	Dep. of Soil Science, University of Wisconsin-Madison, Madison, WI 53706
Wollenhaupt, Nyle	Dep. of Soil Science, University of Wisconsin-Madison, Madison, WI 53706
Young, Linda J.	Dep. of Agronomy, University of Nebraska, Lincoln, NE 68583-0915
Ziegler, Duane	John Deer Harvester Works, 1800 158th St., E. Moline, IL 61244
Ziegler, Larry	Tyler Manufacturing Co., E. Hwy 12, Box 249, Benson, MN 56215

CONVERSION FACTORS FOR SI AND NON-SI UNITS

Conversion Factors for SI and non-SI Units

To convert Column 1 into Column 2, multiply by	Column 1 SI Unit	Column 2 non-SI Unit	To convert Column 2 into Column 1, multiply by
Length			
0.621	kilometer, km (10^3 m)	mile, mi	1.609
1.094	meter, m	yard, yd	0.914
3.28	meter, m	foot, ft	0.304
1.0	micrometer, μm (10^{-6} m)	micron, μ	1.0
3.94×10^{-2}	millimeter, mm (10^{-3} m)	inch, in	25.4
10	nanometer, nm (10^{-9} m)	Angstrom, Å	0.1
Area			
2.47	hectare, ha	acre	0.405
247	square kilometer, km^2 (10^3 m)^2	acre	4.05×10^{-3}
0.386	square kilometer, km^2 (10^3 m)^2	square mile, mi^2	2.590
2.47×10^{-4}	square meter, m^2	acre	4.05×10^3
10.76	square meter, m^2	square foot, ft^2	9.29×10^{-2}
1.55×10^{-3}	square millimeter, mm^2 (10^{-3} m)^2	square inch, in^2	645
Volume			
9.73×10^{-3}	cubic meter, m^3	acre-inch	102.8
35.3	cubic meter, m^3	cubic foot, ft^3	2.83×10^{-2}
6.10×10^4	cubic meter, m^3	cubic inch, in^3	1.64×10^{-5}
2.84×10^{-2}	liter, L (10^{-3} m^3)	bushel, bu	35.24
1.057	liter, L (10^{-3} m^3)	quart (liquid), qt	0.946
3.53×10^{-2}	liter, L (10^{-3} m^3)	cubic foot, ft^3	28.3
0.265	liter, L (10^{-3} m^3)	gallon	3.78
33.78	liter, L (10^{-3} m^3)	ounce (fluid), oz	2.96×10^{-2}
2.11	liter, L (10^{-3} m^3)	pint (fluid), pt	0.473

CONVERSION FACTORS FOR SI AND NON-SI UNITS

Mass

To convert Column 1 into Column 2, multiply by	Column 1 SI Unit	Column 2 non-SI Unit	To convert Column 2 into Column 1, multiply by
2.20×10^{-3}	gram, g (10^{-3} kg)	pound, lb	454
3.52×10^{-2}	gram, g (10^{-3} kg)	ounce (avdp), oz	28.4
2.205	kilogram, kg	pound, lb	0.454
0.01	kilogram, kg	quintal (metric), q	100
1.10×10^{-3}	kilogram, kg	ton (2000 lb), ton	907
1.102	megagram, Mg (tonne)	ton (U.S.), ton	0.907
1.102	tonne, t	ton (U.S.), ton	0.907

Yield and Rate

0.893	kilogram per hectare, kg ha^{-1}	pound per acre, lb acre^{-1}	1.12
7.77×10^{-2}	kilogram per cubic meter, kg m^{-3}	pound per bushel, lb bu^{-1}	12.87
1.49×10^{-2}	kilogram per hectare, kg ha^{-1}	bushel per acre, 60 lb	67.19
1.59×10^{-2}	kilogram per hectare, kg ha^{-1}	bushel per acre, 56 lb	62.71
1.86×10^{-2}	kilogram per hectare, kg ha^{-1}	bushel per acre, 48 lb	53.75
0.107	liter per hectare, L ha^{-1}	gallon per acre	9.35
893	tonnes per hectare, t ha^{-1}	pound per acre, lb acre^{-1}	1.12×10^{-3}
893	megagram per hectare, Mg ha^{-1}	pound per acre, lb acre^{-1}	1.12×10^{-3}
0.446	megagram per hectare, Mg ha^{-1}	ton (2000 lb) per acre, ton acre^{-1}	2.24
2.24	meter per second, m s^{-1}	mile per hour	0.447

Specific Surface

10	square meter per kilogram, m^2 kg^{-1}	square centimeter per gram, cm^2 g^{-1}	0.1
1000	square meter per kilogram, m^2 kg^{-1}	square millimeter per gram, mm^2 g^{-1}	0.001

Pressure

9.90	megapascal, MPa (10^6 Pa)	atmosphere	0.101
10	megapascal, MPa (10^6 Pa)	bar	0.1
1.00	megagram per cubic meter, Mg m^{-3}	gram per cubic centimeter, g cm^{-3}	1.00
2.09×10^{-2}	pascal, Pa	pound per square foot, lb ft^{-2}	47.9
1.45×10^{-4}	pascal, Pa	pound per square inch, lb in^{-2}	6.90×10^3

(continued on next page)

Conversion Factors for SI and non-SI Units

To convert Column 1 into Column 2, multiply by	Column 1 SI Unit	Column 2 non-SI Unit	To convert Column 2 into Column 1, multiply by
Temperature			
1.00 (K − 273)	Kelvin, K	Celsius, °C	1.00 (°C + 273)
(9/5 °C) + 32	Celsius, °C	Fahrenheit, °F	5/9 (°F − 32)
Energy, Work, Quantity of Heat			
9.52×10^{-4}	joule, J	British thermal unit, Btu	1.05×10^3
0.239	joule, J	calorie, cal	4.19
10^7	joule, J	erg	10^{-7}
0.735	joule, J	foot-pound	1.36
2.387×10^{-5}	joule per square meter, $J\ m^{-2}$	calorie per square centimeter (langley)	4.19×10^4
10^5	newton, N	dyne	10^{-5}
1.43×10^{-3}	watt per square meter, $W\ m^{-2}$	calorie per square centimeter minute (irradiance), $cal\ cm^{-2}\ min^{-1}$	698
Transpiration and Photosynthesis			
3.60×10^{-2}	milligram per square meter second, $mg\ m^{-2}\ s^{-1}$	gram per square decimeter hour, $g\ dm^{-2}\ h^{-1}$	27.8
5.56×10^{-3}	milligram (H_2O) per square meter second, $mg\ m^{-2}\ s^{-1}$	micromole (H_2O) per square centimeter second, $\mu mol\ cm^{-2}\ s^{-1}$	180
10^{-4}	milligram per square meter second, $mg\ m^{-2}\ s^{-1}$	milligram per square centimeter second, $mg\ cm^{-2}\ s^{-1}$	10^4
35.97	milligram per square meter second, $mg\ m^{-2}\ s^{-1}$	milligram per square decimeter hour, $mg\ dm^{-2}\ h^{-1}$	2.78×10^{-2}
Plane Angle			
57.3	radian, rad	degrees (angle), °	1.75×10^{-2}

CONVERSION FACTORS FOR SI AND NON-SI UNITS

Electrical Conductivity, Electricity, and Magnetism

10	siemen per meter, S m^{-1}	millimho per centimeter, mmho cm^{-1}	0.1
10^4	tesla, T	gauss, G	10^{-4}

Water Measurement

9.73 × 10^{-3}	cubic meter, m^3	acre-inches, acre-in	102.8
9.81 × 10^{-3}	cubic meter per hour, m^3 h^{-1}	cubic feet per second, ft^3 s^{-1}	101.9
4.40	cubic meter per hour, m^3 h^{-1}	U.S. gallons per minute, gal min^{-1}	0.227
8.11	hectare-meters, ha-m	acre-feet, acre-ft	0.123
97.28	hectare-meters, ha-m	acre-inches, acre-in	1.03 × 10^{-2}
8.1 × 10^{-2}	hectare-centimeters, ha-cm	acre-feet, acre-ft	12.33

Concentrations

1	centimole per kilogram, cmol kg^{-1} (ion exchange capacity)	milliequivalents per 100 grams, meq 100 g^{-1}	1
0.1	gram per kilogram, g kg^{-1}	percent, %	10
1	milligram per kilogram, mg kg^{-1}	parts per million, ppm	1

Radioactivity

2.7 × 10^{-11}	becquerel, Bq	curie, Ci	3.7 × 10^{10}
2.7 × 10^{-2}	becquerel per kilogram, Bq kg^{-1}	picocurie per gram, pCi g^{-1}	37
100	gray, Gy (absorbed dose)	rad, rd	0.01
100	sievert, Sv (equivalent dose)	rem (roentgen equivalent man)	0.01

Plant Nutrient Conversion

	Elemental	Oxide	
2.29	P	P$_2$O$_5$	0.437
1.20	K	K$_2$O	0.830
1.39	Ca	CaO	0.715
1.66	Mg	MgO	0.602